D1162110

Archimedes

Volume 4

Archimedes

NEW STUDIES IN THE HISTORY AND PHILOSOPHY OF SCIENCE AND TECHNOLOGY

VOLUME 4

Archimedes has three fundamental goals; to further the integration of the histories of science and technology with one another: to investigate the technical, social and practical histories of specific developments in science and technology; and finally, where possible and desirable, to bring the histories of science and technology into closer contact with the philosophy of science. To these ends, each volume will have its own theme and title and will be planned by one or more members of the Advisory Board in consultation with the editor. Although the volumes have specific themes, the series itself will not be limited to one or even to a few particular areas. Its subjects include any of the sciences, ranging from biology through physics, all aspects of technology, broadly construed, as well as historically-engaged philosophy of science or technology. Taken as a whole, *Archimedes* will be of interest to historians, philosophers, and scientists, as well as to those in business and industry who seek to understand how science and industry have come to be so strongly linked.

Archimedes

Volume 4
New Studies in the History and Philosophy of Science and Technology

---

Observations and Predictions of Eclipse Times by Early Astronomers

by

JOHN M. STEELE
University of Durham, United Kingdom

KLUWER ACADEMIC PUBLISHERS
DORDRECHT / BOSTON / LONDON

A C.I.P. Catalogue record for this book is available from the Library of Congress.

ISBN 0-7923-6298-5

Published by Kluwer Academic Publishers,
P.O. Box 17, 3300 AA Dordrecht, The Netherlands.

Sold and distributed in North, Central and South America
by Kluwer Academic Publishers,
101 Philip Drive, Norwell, MA 02061, U.S.A.

In all other countries, sold and distributed
by Kluwer Academic Publishers,
P.O. Box 322, 3300 AH Dordrecht, The Netherlands.

Printed on acid-free paper

Printed in the Netherlands.

For my Parents

Contents

Preface

Most of the research presented in this book was undertaken during my time as a PhD student at the Department of Physics of the University of Durham. I am pleased to record here my sincerest gratitude to my degree supervisor, F. Richard Stephenson, with whom I had the honour to work. I also wish to thank the Department of Physics for its generous provision of a Departmental Studentship, without which this study would not have been possible, and for its continued support of research into the history of astronomy. Also in Durham I wish to thank the members of the Departments of History and Archaeology for the interest they have shown in this work. The manuscript for this book was completed during my time as a Postdoctoral Fellow at the Dibner Institute for the History of Science and Technology. I thank the Institute both for this award, and for providing a stimulating atmosphere in which to complete this work.

There are many friends and colleagues who have helped with this work in different ways. Hermann Hunger generously sent me his editions of the Babylonian eclipse texts in advance of publication. Christopher Walker provided transliterations of several unpublished tablets and answered many questions concerning Babylonian texts. Norbert Roughton provided many references to eclipses in the Babylonian Almanacs and Normal Star Almanacs. Peter Huber and Nathan Sivin kindly made available to me copies of their unpublished manuscripts. For various discussions during the course of this work I thank John Britton, Alexander Jones, Leslie Morrison, Noel Swerdlow, and Benno van Dalen. I also thank Ben Weiss of the Burndy Library for help selecting illustrations.

Some preliminary results from this work have appeared in a series of papers published in the *Journal for the History of Astronomy*,[1] the *Bulletin of the School of Oriental and African Studies*,[2] and *Centaurus*.[3] I wish to thank my co-authors, and the editors and publishers of these journals for permission to revise those works here.

[1]Steele & Stephenson (1997), Steele (1997), Steele, Stephenson, & Morrison (1997), Steele & Stephenson (1998a), Steele (1998b), Steele & Stephenson (1998b).

[2]Steele (1998a).

[3]Steele (1999b).

Part I

Introductory Orientations

Chapter 1

Introduction

1.1 Introductory Remarks

One of the most awe-inspiring of celestial events seen by early man must surely have been the occurrence of an eclipse of the sun or moon. With no apparent warning, one or two times a year, a darkness encroaches upon the bright light of the sun or moon: sometimes to completely cover the heavenly body, sometimes to retreat before the light is fully extinguished. To make the event more ominous, the eclipsed moon may become a dim, blood red colour, or, in the rare event of a total solar eclipse, the day may literally turn into night during which stars become visible and the air turns cold. It is therefore not surprising that eclipses were viewed as important astrological events in many early civilizations.[1]

All over the world, myths and legends have been contrived to explain eclipses. In ancient China, a celestial dragon was believed to eat the sun or moon during the eclipse,[2] and a Babylonian incantation series contains an account whereby the moon-god Sin is encircled by seven demons sent by the god Anu and is then rescued by the god Marduk.[3] Eclipses were often seen as portents, generally, although not exclusively, bad portents. Rituals have been performed, probably since long before the development of writing, to assuage the evil associated with these events. And eclipses have often been used by writers to underline dramatic events.[4]

The main purpose of this book is to collect together all known accounts of timed eclipse observations and predictions made by early astronomers, and to give a detailed description of the sources in which they are found.[5] But what do I mean by "early

[1]For a survey of ethnohistorical data regarding eclipses, see Closs (1989). This survey is confined to the Central American region. To my knowledge no such surveys have been made for other areas.

[2]Thus the Chinese term for an eclipse, *shih*, literally meaning "eaten." See Needham (1959: 409).

[3]*Utukkū lemnūti*, 16. See Kilmer (1978).

[4]Witness the surprising number of eclipses — too many for mere coincidence — said to have been seen during, or on the eve of, major battles in classical antiquity.

[5]Untimed observations and predictions will be mentioned, but not discussed in detail. For a recent

astronomers?" My answer to this question must be divided into two parts. First of all, I take "early" to refer roughly to the period before the invention of the telescope, that is, before circa AD 1600. Other points could equally well have been chosen. For instance, the proliferation of the use of mechanical clocks to time astronomical observations in the late seventeenth and early eighteenth centuries marked a significant change in the way eclipses were observed (although here it is hard to point to a specific year to take as a cut off date). My reasons for choosing AD 1600 are twofold. The use of a telescope to determine the moment of an eclipse contact had a significant effect on the accuracy with which eclipses, particularly lunar eclipses, could be timed.[6] In addition, from about AD 1600, there is a vast increase in the number of timed eclipse observations that are extant, with, in many cases, several records of observations of the same eclipse by different astronomers preserved. Indeed, there are probably more eclipse records from the seventeenth century alone than from the previous twenty centuries discussed in this study.[7] It would therefore have been unmanageable to treat them all adequately here.

I use the term "astronomers" in this book to refer to any individuals who recorded detailed observations or predictions of eclipses, regardless of their motivation for doing so. Thus I make no distinction between, say, Tycho Brahe and the Chinese astrological and calendar officials, or between Ibn Yūnus and the scribes of Mesopotamia. This is not in an attempt to "legitimize" these individuals by hiding their astrological interests and highlighting their scientific roles. It matters little whether they were "doing science" in the modern sense or not. Most of the individuals discussed in this book were observing and predicting eclipses for various reasons, both religious/divinatory and "scientific," but to them there was often no distinction between the various aspects of their work — we should not make one either. My use of the astronomer is simply a matter of convenience — it would be tiresome to write out a literal description of each individual's interests or occupation every time I mentioned them — but I ask the reader to constantly bear in mind the wide range of roles that these astronomers may have had.

As stated above, this book is concerned with extant records of eclipse observations and predictions. Thus, my primarily interest is with what we may call non-mathematical astronomy: records of observations and predictions of specific celestial events, rather than generalised mathematical methods of astronomical calculation. Mathematical astronomy — the various astronomical theories, tables, etc., that are known from early cultures — will only be discussed with regard to their bearing on the extant records (in particular the extant predictions).[8] A study of the development of the methods of eclipse prediction in mathematical astronomy would itself be very interesting, but quite different to that undertaken here.

---

discussion of many untimed observations, see Stephenson (1997b).

[6]There is a considerable improvement in the accuracy of eclipse timings in the latter half of the seventeenth century when telescopes began to be used more widely. See Stephenson & Morrison (1984).

[7]See the collection by Morrison, Lukac, & Stephenson (1981).

[8]It turns out that the eclipse predictions found in extant records were often not made using contemporary mathematical astronomical methods. See, for example, Section 2.8 below.

In addition to collecting together the various accounts of eclipses, I will also compare the eclipse timings with modern computations to give an indication of their accuracy. The purpose of this is not to judge the various groups of observations, or to show that any particular culture reached a particular level of accuracy so many centuries before another culture. Rather, these comparisons will be made to discover what information they may provide regarding the techniques that were used by early astronomers to observe and predict eclipses. In doing so, it will be necessary to discuss several other related topics, such as the techniques used in dating damaged eclipse reports, the units of time used by the early astronomers, and the methods by which eclipses were predicted.

Although this study is primarily historical, it does make use of computations of the circumstances of historical astronomical events based upon modern theories of lunar, solar, planetary, and stellar motion. This will, I hope, act as an example of one of the ways that the techniques of one discipline can be used to help research into another. Furthermore, as modern studies of the long-term variations in the Earth's rate of rotation have made extensive use of historical eclipse observations, the present investigation may have implications for future research in this area.[9] Thus there exists a circle in which science can be used to help in a historical study, and this in turn may be of some benefit to further scientific investigations.

1.2 Early Eclipse Records

According to Aveni (1980: 3), "all developing civilizations exhibit a reverence for the sky and its contents." Whilst this may at first seem to be a somewhat bold statement, it is nevertheless borne out by the available evidence. Virtually all known cultures, both literate and pre-literate, are thought to have gazed into the heavens and to have recorded what they have seen — whether it be in written accounts, drawings, or even architectural representations. I outline below the main cultures from which astronomical records are preserved:

Mesopotamia:

Several thousand cuneiform texts from Mesopotamia contain references to astronomical phenomena. These range from collections of celestial omens, reports of astronomical observations and predictions, and texts of mathematical astronomy. In particular, during the Late Babylonian period (c. 750 BC onwards), detailed astronomical records were being kept by the Babylonian astronomers in the so-called Astronomical Diaries and related texts. These include several hundred records of eclipse observations and predictions.

[9] I will discuss this briefly in Chapter 8.

Ancient Egypt:

We have very little astronomical material from Egypt before the Hellenistic period. Most of what we have relates to time-keeping and the calendar. It has been claimed by Sellers (1992), although without fully convincing evidence, that eclipses played a fundamental role in development of Egyptian religion, but no actual records of eclipses have so far been found from ancient Egypt. Therefore, ancient Egypt will not be considered further in this study.

The Greco-Roman World:

In the latter half of the last millennium BC, Greece and Rome took over as the dominant military and cultural powers in Europe and the Near East. Only a small number of Greek scientific works have been preserved; nevertheless, some of the ones we have are of great interest for the history of science due to the influence they had on later scientists. In the field of astronomy, the most important Greek work is undoubtedly Ptolemy's *Almagest*. Within this work a small number of eclipse observations have been preserved. Several additional references to eclipses are to be found in some of the historical and literary works written by classical authors. Finally, a number of papyri have been recovered from Greek and Roman Egypt over the past hundred years that contain astronomical material. In particular, a few papyri contain records of eclipses.

India:

Indian interest in astronomy seems to have developed in the latter half of the first millennium BC. As with virtually all Indian astronomy, this interest was almost exclusively in the development of astronomical theories; the only role of observation in Indian astronomy appears to have been in checking accepted parameters.[10] As a result, very few records of observations are preserved in Sanskrit sources. Some eclipse records are preserved, for example in the *Mahābhāskarīya-Bhāṣya* of Parameśvara,[11] but they tend to be very brief and do not include detailed timings of the eclipses. Thus, India will not be considered further in this study.

The Islamic Near East:

From shortly after the death of the Prophet Muḥammad in the seventh century AD, the Muslims had established a commonwealth of nations that stretched from India to Morocco and Spain. Islamic society provided a background within which astronomy was able to flourish. Drawing on knowledge preserved from Ancient Greek sources, as well as more recent Persian and Indian developments, Islamic astronomers made significant contributions to astronomy over the next thousand years or so. There are many

[10] Pingree (1996).

[11] *Mahābhāskarīya-Bhāṣya* v, 77. See, for example, Subbarayappa & Sarma (1985: 13–15).

thousands of manuscripts in museums throughout the world that contain the writings of Islamic astronomers. Only a small number have so far been studied in detail. Nevertheless, around fifty detailed records of observations of eclipses by medieval Islamic astronomers are now known.

Late Medieval and Renaissance Europe:

In the period running up to the European Renaissance, elements of Islamic astronomy were transmitted to Europe through Muslim Spain. This was to provide the necessary background in which astronomers such as Copernicus and Kepler could formulate their theories of solar, lunar, and planetary motion. For the first time in Europe, astronomers began to make careful observations of astronomical phenomena. The principal reason for making these observations was to test and formulate improvements to established astronomical tables.

China:

Interest in astronomy in China seems to have developed in the middle of the second millennium BC. By the eighth century BC, systematic astronomical records were beginning to be kept. These continue more or less uninterrupted to the present day, despite the many changes of ruling dynasty. Due to its relative isolation in its formative years, Chinese astronomy differed significantly from that in the western world. In particular, in China, astronomy was primarily the activity of the bureaucratic state, rather than of priests or scholars as in the west. This was a two-edged sword, for it provided a strong motivation for astronomical advancement, but also led to the political manipulation of astronomers and their observations. With the arrival of the Jesuit missionaries in the seventeenth century AD, Chinese astronomy underwent a profound change. The Jesuits brought with them western astronomical theories which quickly became incorporated into Chinese astronomy. Nevertheless, the traditional role of astronomy in Chinese society continued until the establishment of the Republic in AD 1911.

Korea:

The country of Korea was established in about AD 670 when the kingdom of Silla subjugated its two neighbours, Paekche and Koguryo. The earliest history of Korea, the *Samguk Sagi*, describes the period from legendary times down to AD 935. Like most Korean books, it is written in Chinese, for Korean culture at this and later periods was strongly influenced by China. In particular, early Korean astronomy was based wholly upon that practiced in China, even to the extent of the adoption of the Chinese calendar. The *Samguk Sagi*, along with later histories such as the *Koryo-sa*, contains a number of astronomical records, including many eclipses. However, in none of these records is either the observed or predicted time of the eclipse given.[12] Therefore, these

[12] Stephenson (1997b).

records will not be considered further in this study.

Japan:

Japanese civilization developed during a similar period as that in Korea. As with its neighbour, Japanese culture, and in particular its astronomy, was profoundly influenced by that of China. However, from the end of the ninth century AD, Japan entered a state of semi-isolation. It was not until the fifteenth century AD that Japan began again to experience any significant foreign influences on its society. Nevertheless, Chinese astronomical practices continued to dominate in Japan throughout this period of separation. About one hundred Japanese works are known to contain records of astronomical events. Preserved among them are a number of eclipse records, many of which contain the expected time of the eclipse.

Mesoamerica:

It is well known that the Mayan and other Mesoamerican civilizations developed a very advanced astronomy. This included highly accurate calendar schemes, and mechanisms for making predictions of, for example, eclipses and planetary visibilities.[13] Unfortunately, however, our understanding of Mayan astronomy is severely hindered by the relatively small amount of material available for study; only four Mayan books survived the Spanish conquest. One of the surviving Mayan works, now known as the Dresden Codex, contains a table for predicting eclipses, but its exact workings are not yet fully understood.[14] No firmly dated records of either observations or predictions of eclipses are known, and so Mesoamerican astronomy will not be considered further in this study.

It will be clear from my discussion above that there are three basic heritages of astronomy in the world: a "western heritage," founded upon the astronomy of Mesopotamia, that may be traced though Greco-Roman, Indian, and Islamic astronomy up to the astronomers of the European Renaissance; an "eastern heritage" of astronomy developed in China and adopted in Japan and Korea; and a "New World heritage" of astronomy in Mesoamerican cultures.[15] Eclipse records are only preserved from the first two of these groups, and so they will form the basis of the present study. It should be noted, of course, that the division into Eastern and Western heritages is artificial. It implies that these two traditions developed independently which, although partially true, is far

[13] See, for example, Aveni (1980).

[14] See, for example, the differing interpretations of Lounsbury (1976) and Bricker & Bricker (1983).

[15] My use of the word "heritage," rather than the more common "tradition," here is deliberate. "Heritage," it seems to me, better reflects the way in which one culture draws upon material from other cultures, whereas "tradition" seems to imply that material was thrust upon a later culture, this culture having no choice but to accept it.

from being the full story. Cross fertilization of astronomical knowledge did occur — most notably between the Indian, Islamic, and Chinese astronomers. Nevertheless it will be useful to utilize this division in the present study.

1.3 Basic Eclipse Theory

Before it is possible to describe the formation of eclipses, it is necessary to make some remarks regarding the various systems used for keeping track of time. There are three main time systems: terrestrial time (TT), universal time (UT), and local time (LT). Terrestrial time is defined by the motion of the moon and planets. It is an invariant time system. Universal Time is defined by the rotation of the Earth. However, since the Earth's rotation is slowing down due to the effects of tidal friction and other causes, the length of the mean solar day is changing, and so universal time is a non-constant time system. For example, about two thousand years ago the mean solar day was around 0.05 seconds shorter than today. This results in a cumulative clock error of several hours between an ideal clock and one measuring time by the Earth's rotation. This clock error is equal to the difference between terrestrial time and universal time (in the sense TT–UT), and is known as ΔT. Like universal time, local time is also defined by the solar day, with the additional condition that midday is when the sun is at its highest point in the sky. Local times can therefore be converted into universal times by adjusting for the geographical longitude (with reference to the Greenwich Meridian) and the equation of time. Throughout this work all times will be given in hours and decimals.

It is well known that the moon moves round the Earth in an approximately circular orbit. With respect to the fixed background of stars the moon completes its revolution in an average period of 27.3216 days, known as a sidereal month. However, from our position on the Earth, the sun also appears to circle us, returning to the same position with respect to the fixed stars after 365.2564 days, known as a sidereal year.[16] Over the course of a sidereal month, therefore, the sun has moved about 27° ahead of the stars, and so it takes slightly more than another 2 days for the moon and sun to reach conjunction. When the alignment is in the order sun–moon–Earth, the moon is dark. This is known as new moon or conjunction. However, when the alignment is sun–Earth–moon, the full disk of the moon is illuminated. This is known as full moon or opposition. The average time interval between two successive conjunctions or oppositions, known as the synodic month, is equal to 29.5306 days.

If the plane in which the moon orbits the Earth were the same as that in which the sun moves (known as the ecliptic), one luminary would be obscured at every conjunction and opposition. However, the two planes are in fact inclined at an angle of about 5° to one another (see Figure 1.1). The points where the paths of the motions of the two luminaries intersect are known as nodes. The average interval between successive passages of the moon by a given node, known as a dracontic month, is equal to

[16] In the following discussion, it is simpler to think of a stationary Earth orbited by the sun.

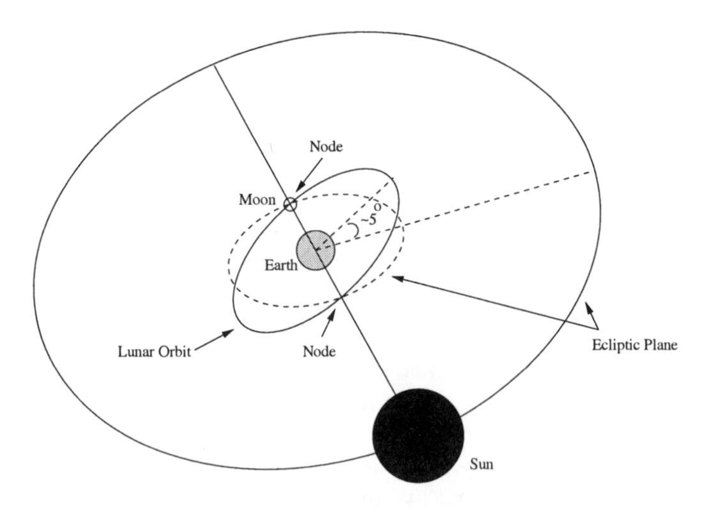

Figure 1.1: The inclination of the moon's orbit to the ecliptic.

27.2122 days. Only when the moon is near to a node will an eclipse be possible. Due to the differing lengths of the synodic and dracontic months, this circumstance does not occur at every conjunction or opposition, but only every six (or occasionally five) syzygies.[17]

If the moon is near to a node at opposition (i.e., if the latitude of the moon is sufficiently small when its longitude is 180° away from the sun), then the moon will enter the shadow of the Earth. This is known as a lunar eclipse. The basic mechanism of a lunar eclipse is shown in Figure 1.2. In this case the moon passes through the light-shaded penumbral shadow until it touches the dark-shaded umbral shadow of the Earth. This is known as the moment of first contact. The moon then continues on its path until it is completely within the umbral shadow. This moment is known as second contact. The moon is now totally eclipsed. When the moon's path brings it to the far side of the umbral shadow, it begins to recover its light. This moment is called third contact. Finally, the moon completely leaves the Earth's umbral shadow, at fourth or last contact, and the eclipse is over. It is sometimes possible to detect a dimming of the moon whilst it is in the penumbral shadow; however, there are no firmly dated observations of penumbral eclipses from the pre-telescopic period.[18]

If the moon's path brings it quite close to a node at opposition, but not sufficiently close to form a total eclipse, it is possible that only part of its surface may enter the Earth's umbral shadow, the remainder passing either completely above or below (see Figure 1.3). This results in a partial eclipse. Only first and last contacts are defined for a partial lunar eclipse. The magnitude of an eclipse is defined as the fraction of

[17]Neglecting penumbral eclipses. In addition, solar eclipses are also possible at one month intervals, but not at the same location.

[18]Contrary to the statement by Stephenson (1997b: 185). See Section 2.6 below.

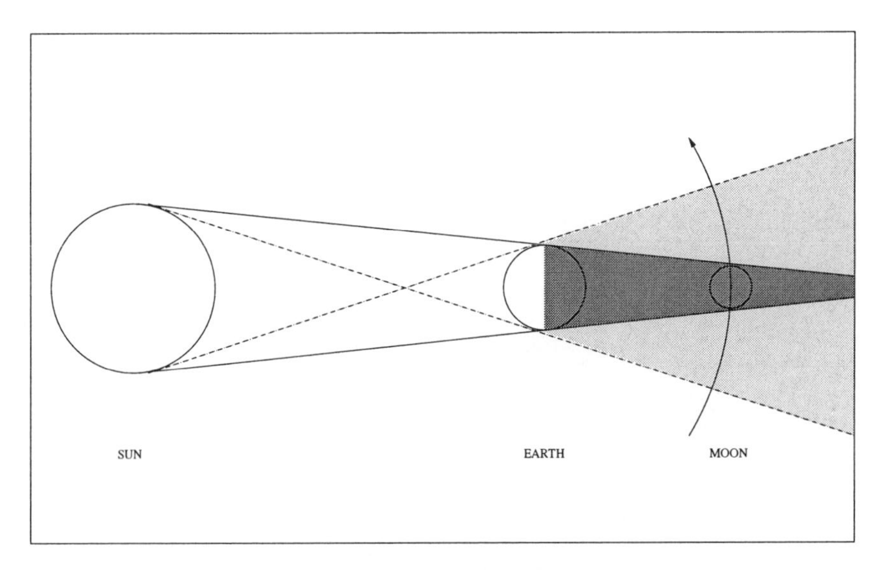

Figure 1.2: The basic mechanism of a lunar eclipse.

the lunar diameter that is eclipsed at maximum phase. Because the moon's orbit is not exactly circular, its distance from the Earth varies. The average interval between successive closest approaches to the Earth (perigee), known as an anomalistic month, is equal to 27.5546 days. Due to this variation, the magnitude of an eclipse is not only dependent upon the moon's path through the Earth's shadow (i.e., upon its latitude), but also upon its distance from the Earth (i.e., upon its anomalistic motion).

Because lunar eclipses are due to the Earth's shadow falling on the moon's surface, they are visible from everywhere on the Earth where the moon is above the horizon (see Figure 1.4). This allows lunar eclipses to be predicted fairly successfully using cycles. Once an eclipse has occurred, another eclipse will take place when (a) the moon is at the same phase again, and (b) the moon is at its same position in its orbit with respect to the node. In other words, the eclipse will occur after there has been both a whole number of synodic months (condition a) and a whole number of dracontic months (condition b). Although there is no relatively small integral common multiple of these two intervals, a number of periods are close. For example, 47 synodic months is only one tenth of a day different from 51 dracontic months, and 135 synodic months is about half a day more than 146 dracontic months. The most successful of these periods is 223 synodic months, which is very close to 242 dracontic months. Furthermore, it is also very close to 239 anomalistic months.[19] This means that after 223 synodic months, lunar eclipses will recur which have similar magnitudes and durations. This period, equal to about $6585\frac{1}{3}$ days or around 11 days over

[19] 223 synodic months = 6585.322 days, 242 dracontic months = 6585.357 days, 239 anomalistic months = 6585.538 days.

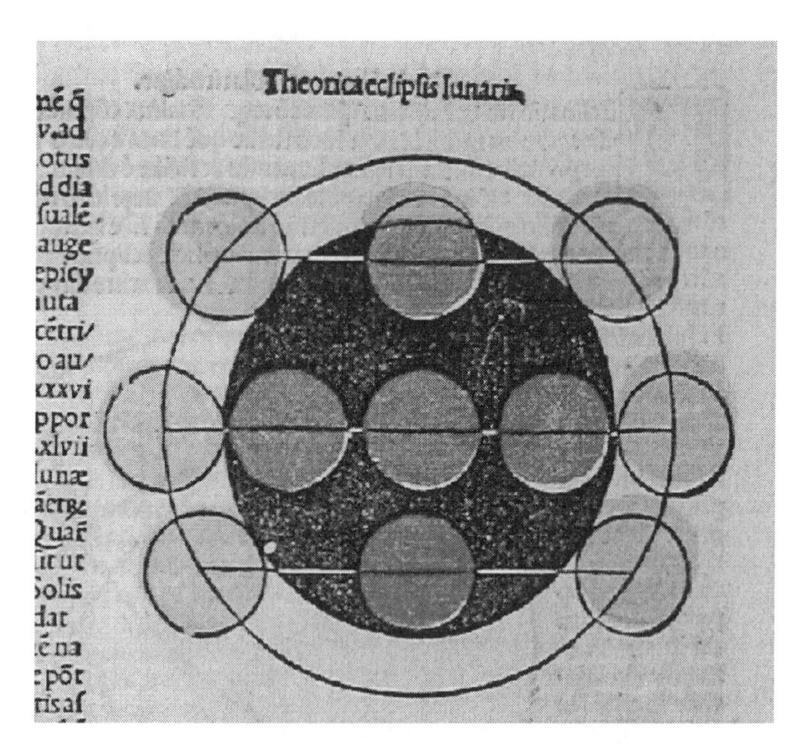

Figure 1.3: Illustration of a range of possible types of lunar eclipses depending on the latitude of the moon, taken from Johannes de Sacro Bosco's *Sphera* (Venetia 1537). [Courtesy: The Burndy Library, Dibner Institute for the History of Science and Technology, MIT]

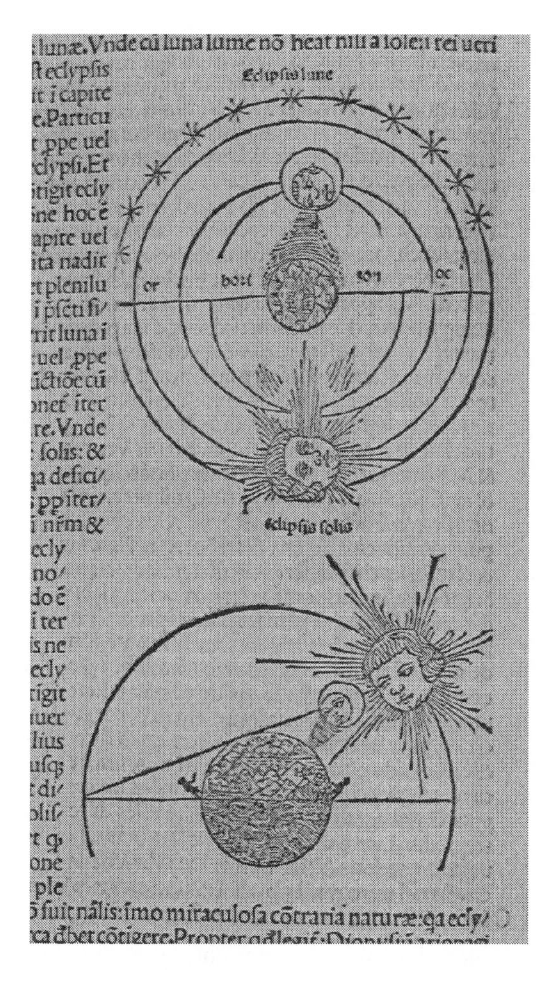

Figure 1.4: Illustration showing that lunar eclipses are seen from all locations on the Earth's surface, but that solar eclipses are only seen within a small region, taken from Johannes de Sacro Bosco's *Sphera* (Venetia 1537). [Courtesy: The Burndy Library, Dibner Institute for the History of Science and Technology, MIT]

18 years, has become known as the "Saros," and was well known in antiquity.[20]

Cycles such as the Saros can not only be used to predict eclipse possibilities at 223 month intervals, but also to determine all of the syzygies at which eclipses are possible within that 223 month period. Given that eclipses can only occur at six or five month intervals, and the observation that six month intervals occur at least five times more often than five month intervals, one can determine the distribution of the eclipse possibilities as follows. If within a 223 month period there are a eclipses at a six month interval and b eclipses at a five month interval, then clearly $6a+5b = 223$. Then, if $a > 5b$, for integer a and b, the only solution to $6a+5b = 223$ is $a = 33$ and $b = 5$, making a total of 38 eclipse possibilities in 223 months. A similar argument can be used for the other eclipse periods. For example, $6a+5b = 135$ yields $a = 20$ and $b = 3$, making 23 eclipse possibilities within 135 months. Distributing the five month intervals as evenly as possible within the eclipse cycle, one finds that the eclipse possibilities fall naturally into groups containing either 7 or 8 eclipses in which the first eclipse comes at a five month interval and the others at six month intervals.[21]

Whereas lunar eclipses occur at opposition, solar eclipses are formed when the moon at conjunction is close to a node. The basic mechanism of a solar eclipse is shown in Figure 1.5. In a solar eclipse, the moon passes between the sun and the Earth, casting a shadow on the Earth's surface. However, as the apparent sizes of the sun and moon are very similar, the apex of the umbral shadow only just reaches the Earth's surface on average. Thus, only a small part of the Earth's surface is covered by the umbral shadow. This is known as the central region (see Figure 1.4). Depending on the relative distances of the sun and moon at the time of an eclipse, the apex of the umbral shadow may or may not touch the Earth's surface. If it does, then a total solar eclipse, in which the whole of the sun is obscured, will be seen within the central region, but if not then an annular eclipse, during which a ring of the sun remains bright surrounding a darkened centre, will be seen. Outside of the central region of a solar eclipse, a partial eclipse may be seen. Maps showing the path of the central regions of solar eclipses have been produced by many authors, including, most famously, von Oppolzer (1887). It should be noted, however, that often these maps do not make sufficient allowance for the changes in the Earth's rate of rotation, and so become increasingly inaccurate at early periods.

During a solar eclipse, the shadow falls on a fraction of the Earth's surface. This means that the appearance of a solar eclipse is not the same to observers in different parts of the world. Furthermore, because of the relative motion of the moon and sun, the eclipse shadow sweeps from west to east across the part of the globe facing the sun, whilst the Earth itself rotates from east to west. The contribution of the Earth's rotation to slowing the speed of the shadow across the Earth's surface depends, however, upon the position of the eclipsed sun in the sky. When the sun is close to the horizon, then the Earth's rotation has a minimal effect on the speed of the eclipse shadow, whereas if the sun is overhead at the equator, it can slow the shadow's passage by about 50%.

[20] As has often been remarked, the name Saros is a modern misnomer. See Neugebauer (1957: 141–143).

[21] These relations can also be obtained theoretically from an analysis of eclipse possibilities equally spaced in longitude. See Aaboe (1972).

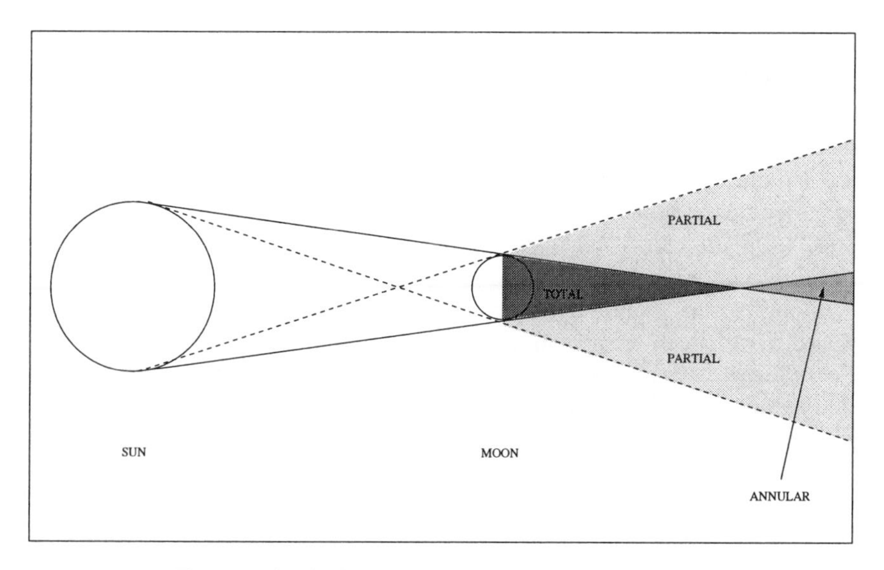

Figure 1.5: The basic mechanism of a solar eclipse.

This makes it much more difficult to predict the circumstance of a solar eclipse than a lunar eclipse as it is necessary to possess a detailed knowledge of the geometry of the Earth, moon, and sun, and the geographical position of the observer on the Earth's surface. Nevertheless, cycles such as the Saros can be used, and certainly were used in antiquity, to predict times when solar eclipses were *possible*, and to exclude times when they were *impossible*.

1.4 Methods of Analysis

As I have stated in the introduction, the main goal of this study is to collect together all known records of observations and predictions of eclipse times made by astronomers in the pre-telescopic period, and to discover what may be learnt from a basic analysis of these times. This will mainly consist of comparing the recorded times with those deduced from modern computations. It is therefore necessary to discuss how these computed times have been derived, and the methods that will be used in the analysis.

The computed times of eclipses have been obtained from original computer programs designed and kindly made available to me by F. R. Stephenson.[22] These programs are based upon the solar ephemeris of Newcomb (1895) and a corrected version of the lunar ephemeris designated *j*=2,[23] incorporating a lunar acceleration of -26" cy^{-2} as determined by Morrison & Ward (1975) from an analysis of the transits

[22] Occasionally minor modifications have been made to these programs by the present author.
[23] IAU (1968).

of Mercury.[24] Values of the Earth's rotational clock error, ΔT, were taken from the spline fit of Stephenson & Morrison (1995). Because this spline fit is obtained from an analysis of both untimed and timed eclipse observations made in various cultures, it is possible to use it to compute the circumstances of eclipses in the past without fear that it will introduce systematic errors into the analysis. According to Stephenson (1997b), these computer programs have been extensively tested against other ephemerides, in particular the JPL ephemeris known as DE102, and found to be sufficiently accurate for historical purposes.

Throughout this study, the term "computed time" has been reserved for those times deduced from modern computations. Times measured by the early astronomers are called "observed" times, and times calculated by the early astronomers from their theories are called "predicted" or "calculated" times. The main tool used in this analysis is to determine the error between two of these times. The error in an observed time is the difference between the computed and the observed time (in the sense computed minus observed). Two errors may be calculated for a predicted time: the "true" error, which is the difference between the computed time and the predicted time (i.e., computed minus predicted time); and the "observed" error, which is the difference between the observed and the predicted time (i.e., observed minus predicted time). I will also use a value which I term the "accuracy" of a time. This I define as the absolute value of an error in a time.

It is necessary at this point to distinguish between the "accuracy" of an observation and the "precision" of an observation. Accuracy is a measure of the quality of an observation when compared with an assumed "correct" value. Precision is a measure of the size of the unit used in the observation. For example, an observation of 13 minutes is precise to 1 minute, but if, say, the correct time is 10 minutes, then it is only accurate to 3 minutes.

When there is more than one time determined by a particular astronomer or group of astronomers, I will frequently calculate the mean error and accuracy. The mean error will give an indication of any systematic error in the timings, whereas the mean accuracy will give a rough measure of their overall quality. More advanced statistics will not be used in this study since it has been repeatedly shown, for example by Swerdlow (1979), Hamilton & Swerdlow (1981), and Bielenstein & Sivin (1977), that they are often inappropriate (and occasionally dangerously misleading) in studying historical data.[25]

A number of stellar and planetary positions and visibilities are computed throughout this work. Planetary positions are determined using the ephemerides of Bretagnon and Simon,[26] and star positions are taken from the catalogue of Hirshfeld & Sinnott

[24]This value of the lunar acceleration is very close to that obtained from lunar laser ranging by Dickey et al. (1994).

[25]Swerdlow (1979: 529) writes: "Probability and statistics have become increasingly fashionable for dressing up history with an appearance of scientific rigor, perhaps because they can be used to prove, indifferently and in merciless detail, either the obvious or the preposterous, perhaps because numbers seem at once too impressive and too objective to question."

[26]See, for example, their discussion in Bretagnon, Simon, & Laskar (1985).

(1982), with suitable corrections for precession and proper motion.

Finally, all of the computed altitudes given in this work have been corrected for refraction to give the apparent rather than the true value. Similarly, when deducing local times from altitudes, all of the observed values have been corrected for refraction. These corrections have been made using the formulas given by Schaefer (1993). At the level of precision to which early astronomers made their altitude measurements, however, these corrections are virtually negligible for altitudes of more than a couple of degrees above the horizon.

1.5 A Note on the Conventions used in this Study

Throughout this work many eclipse records have been translated and dated. Unless specifically credited to another author, all of the translations published here are my own. Text contained in square brackets has been restored from damaged readings by the translator. Parentheses are used to enclose glosses added by the translator (or, occasionally, by myself if I am quoting another author's translation) to aid reading but which have no counterpart in the original text.

Transliterations of cuneiform texts follow the style described by Sachs & Hunger (1988: 36-38), with Sumerian readings of logograms given in small capitals, and Akkadian in lower case italics. Romanization of Chinese characters follows the style of Wade-Giles, with the exception of the names of modern authors who have used the Pinyin rendering in their publications. Chinese text is given in italics except for names of people and places. Latin transcriptions are also given in italics except for proper names.

Throughout this work, sexagesimal numbers will often be quoted. Following earlier authors, sexagesimal numbers are transcribed using commas to separate places, and a semicolon to separate integers from fractions.[27]

Finally, many calendar systems are used in the eclipse records discussed in this work. However, I have converted all of these dates into the western calendar when discussing them. Dates up to 4 October 1582 AD are given in the Julian calendar. Those after are given in the Gregorian calendar.[28] Although all dates given in the text are in the familiar BC/AD style, for convenience dates in tables and figures often use the astronomical dating system. This is a continuous version of the Julian calendar in which year 0 corresponds to 1 BC, year -1 to 2 BC, and so forth.[29]

---

[27] Generally therefore: a,b,c;d,e = $(a \times 60^2) + (b \times 60) + c + (d \times 60^{-1}) + (e \times 60^{-2})$.

[28] The Gregorian calendar differs from the Julian calendar in that leap years are defined to be years which are divisible by 4, except those that are divisible by 100 unless they are also divisible by 400. In the Julian calendar, every year divisible by 4 is a leap year. Thus, in a 400 year period, there are 3 more leap years in the Julian calendar than in the Gregorian calendar. The Gregorian calendar therefore more accurately approximates the solar year.

[29] Mathematically, a year -n = n + 1 BC.

Part II

The Western Heritage

Chapter 2

Mesopotamia

2.1 Introduction

Until the end of the last century the little that was known of Mesopotamian astronomy came from scattered references to the "Chaldaeans," portrayed as mystical astrologers in the works of Greek authors such as Strabo and Diodorus Siculus and the Old Testament,[1] and from the small number of observations contained in Ptolemy's *Almagest* which he describes as having been observed in Babylon. However, after the archaeological exploration of Assyria and Babylonia by Austin Henry Layard and others which began in the 1840s, original cuneiform records have come to light that have transformed this image.[2] Not only were celestial events used for divination in Mesopotamia, they were also systematically observed and recorded in Babylon from around the middle of the eighth century BC,[3] and by the fourth century BC mathematical schemes had been developed that allowed various astronomical phenomena such as the appearance and disappearance of the planets and the time-intervals between successive oppositions and conjunctions of the sun and moon to be calculated.[4] Babylonian astronomy was to have a profound and long lasting legacy: Greek, Indian, Islamic, and even Medieval European astronomy all drew on its influence.[5]

---

[1]See Neugebauer (1975: 607–614).

[2]For a general history of the excavations and the recovery of the cuneiform texts, see Larsen (1996) and Budge (1925). Details of the recovery of the astronomical and astrological tablets are given by Neugebauer (1957: 53–70).

[3]There is no firm evidence for similar long-term systematic observational programmes in the other Mesopotamian cities. However, during the eighth and seventh centuries BC, reports of astronomical observations and their astrological interpretations were regularly sent to the Assyrian kings in Nineveh.

[4]Aaboe (1974) has described this final type of astronomy as "scientific," that is "a mathematical description of celestial phenomena capable of yielding numerical predictions that can be tested against observations."

[5]For a general overview of the legacy of Babylonian astronomy in other cultures, see Pingree (1998) and Neugebauer (1963). For the records of Babylonian astronomical observations in Ptolemy's *Almagest*, see Chapter 3. On the transmission of Babylonian mathematical astronomy to Greece, see Toomer (1988) and Jones (1991, 1993), and to India, see Pingree (1973, 1987).

Mesopotamia, the "land between the rivers," refers to the area of south-west Asia near the Tigris and Euphrates rivers. During the last four millennia BC it was one of the most important powers in the ancient world. By the third millennium BC, a number of city states had come to power, most importantly Sumeria in the south, and Babylonia in the thin strip of land between the two rivers just to the south of present day Baghdad. Around the turn of the second millennium BC the Amorite dynasty succeeded in unifying the whole of Mesopotamia under Babylonian rule. This was to be the zenith of Babylonian power in the region, for in the sixteenth century BC Babylon was sacked. There then followed a period of rule by the foreign Kassites. During this period Assyria, based in the north of the region, began to grow, and by the beginning of the first millennium BCthe Kassites had been overcome, and the Assyrian and Babylonian Empires strove for dominance in the region. By the beginning of the first millennium BC, the Assyrian Empire had become the most important power in the area, extending throughout almost all of Mesopotamia. Nevertheless, the city of Babylon continued to have its own ruler, and existed in a state of semi-independence from its Assyrian overlords.

By the beginning of the Late Babylonian period (c. 750 BC), Assyrian domination began to wane and, in 626 BC, Nabopolassar seized the Babylonian throne. In 612 BC the Assyrian capital Nineveh was sacked and the Assyrian Empire was more or less extinguished. Nabopolassar's Chaldaean dynasty lasted until 539 BC. In that year the Persian Cyrus marched on Babylon, only to be welcomed into the city as a liberator of the people from the unpopular king, Nabonidus. The Persian Achaemenid dynasty lasted until Alexander the Great defeated Darius in 331 BC. He too was to receive a warm welcome from the Babylonian people. When Seleucus gained control of Babylonia, he made a decision to build a new city, Seleucia, about 60 miles to the north of Babylon. The beginning of Seleucus' reign also marked the beginning of a new era in near-eastern chronology for, from after his time, years would continue to be numbered from the Seleucid era, rather than from the reign of each king. Seleucus' successor Antiochus I made the decision to make Seleucia the royal city and began the forced movement of the civilian population of Babylon to this new city. This was a decision from which Babylon was never to recover, for by the end of the first century AD, the city was deserted. By 125 BC, the Greeks had been driven out of Mesopotamia by the Arsacid Parthians; however, this did not halt the decline of the city of Babylon. For reference, the main divisions of the Late Babylonian period are listed in Table 2.1

The primary motivation for the development of astronomy in Babylon, and more generally throughout Mesopotamia, was astrological. For, according to a Babylonian Diviner's manual:

> "The signs on earth just as those in the sky give us signals. Sky and earth both produce portents; though appearing separately, they are not separate (because) sky and earth are related. A sign that portends evil in the sky is (also) evil on earth, one that portends evil on earth is evil in the sky."
>
> [K.2847 Obv. 48–42; trans. Oppenheim (1974)]

Indeed, such was the importance of the diviner's work that in line 71 of the same text

Year	Division
c. 750–626 BC	Assyrian Domination
626–539 BC	Chaldaean Dynasty
539–331 BC	Achaemenid Rule
330–311 BC	Macedonian Rule
311–125 BC	Seleucid Period
125 BC –	Partian Period

Table 2.1: Important periods in Late Babylonian history.

he is warned that he must "pay attention and be not careless!" Astrological interpretations were largely drawn from the great omen series *Enūma Anu Enlil*, which reached its canonical form sometime around the end of the second millennium BC.

I shall begin this chapter by giving an overview of the various sources of astronomical records in Mesopotamian history. This will mainly concentrate on the Late Babylonian period (c. 750 BC – AD 75), and will include a discussion of the classification and structure of the texts, and comments on the techniques used in their dating. I will then proceed to discuss the timed observations and predictions of both solar and lunar eclipses made by the Babylonian astronomers. Finally, I will discuss the possible methods used in the prediction of eclipses by the Babylonians.

2.2 Sources of Astronomical Records

Writing in Mesopotamia developed in Sumeria at some time before the end of the fourth millennium BC. Using either a reed or a wooden stylus, pictographs were impressed into a damp piece of clay which, when allowed to dry, retained the image indefinitely, unless the tablet suffered damage. Because of this enduring quality, the tablets provide one of the few sources of "original" texts preserved from the ancient world.[6] The pictographs used by the early Sumerians developed into a form of wedge-shaped writing known as cuneiform. Initially, each cuneiform sign was a logogram indicating a particular word, but later a single sign came to act not only as a logogram, but also had one or more syllabic readings. Because of this, the cuneiform script was very versatile, and it was possible to use it to write not only Sumerian, but also the Akkadian language. In the Late Babylonian period, texts were usually written in Akkadian, although Aramaic was fast becoming the main spoken language.

The earliest evidence of extensive astronomical activity in Mesopotamia is found in the great omen series *Enūma Anu Enlil*. This series comprises omens for both astronomical and meteorological events such as lunar and solar eclipses, the appearances

[6]Or, at least, near-original texts, as a number were obviously copied in antiquity by the Babylonian scribes. However, they are still only, say, second or third generation copies. By contrast, the writings from ancient China were written on perishable materials, and so we only possess late copies of the works that have gone through the copying process, with all of its inherent problems of corruption to the text, many times.

of planets, and thunder and lightning.[7] It is generally believed that, whilst the omens had long been studied, the series was not compiled in its final form until the beginning of the first millennium BC.

Tablet 63 of the *Enūma Anu Enlil* series, often referred to as the "Venus Tablet of Ammiṣaduqa," contains a sequence of omens derived from the first appearance and the first disappearance of the planet Venus. Unlike other omens in this series, however, they are associated with specific dates covering 21 years in the reign of King Ammiṣaduqa of the Babylonian First Dynasty.[8] Because of this association, it has been suggested that these omens relate to real *observations* of phenomena of Venus. Various attempts have been made to determine their date, which would provide an important link in attempts to establish an absolute chronology for the second millennium BC,[9] but there is still no consensus on this issue. It has also been argued by Huber (1987) that some of the omens relating to lunar eclipses must have been based upon earlier observed events, and he has proposed dates in the latter part of the third millennium BC for them. This approach has been followed by Gasche et al. (1998) who refer to these events as "observations," even though they are written in the usual language of the omen texts (if x then y). However, it seems very unlikely that these eclipse omens refer to specific events in the past. Omen statements are not reports of observations, but refer only to potentially observable events. This is clear from the large number of omens that relate to events which are in fact impossible.[10]

A small number of other works from this early period are concerned with celestial phenomena. For example, the so called "Astrolabes" associate various stars (including some of the planets) with the months of the year and the area of the horizon over which they rise. Literary works, such as the "Epic of Creation," *Enūma eliš*, outline, albeit rather obliquely, a theological cosmology.[11] The most important text, however, is that known by its incipit as MUL.APIN.[12]

MUL.APIN is preserved in a number of copies from the Neo-Assyrian and Late Babylonian periods. In its canonical form it was written on two tablets and comprised a miscellany of astronomical (mainly calendrical) and astrological material, apparently compiled sometime around the end of the second millennium BC, but containing elements that probably go back to the Old Babylonian period (c. 2000–1500 BC). It contains items such as a star catalogue, intercalation schemes, planetary periods, a table listing weights of water associated with a clepsydra, a discussion of a shadow table, and a collection of stellar omens.

In contrast to the relatively small number of astronomical works written before

---

[7] For a survey of the *Enūma Anu Enlil* series, see Weidner (1944a, 1944b, 1956, 1969) and Koch-Westenholz (1995). So far, the tablets concerned with lunar eclipses have been edited by Rochberg-Halton (1987, 1988), parts of those dealing with the planetary omens by Reiner & Pingree (1975, 1981, 1998), and those containing solar omens by van Soldt (1995).

[8] The Venus Tablet has been published by Reiner & Pingree (1975).

[9] For example, Langdon & Fotheringham (1928), Weir (1972), Huber et al. (1982), and Gasche et al. (1998).

[10] For a discussion of empiricism in the omen texts, see Rochberg (1999a).

[11] Lambert (1975), Horowitz (1998).

[12] MUL.APIN has been edited by Hunger & Pingree (1989).

the middle of the eighth century BC, from this period until the first century AD a vast number of astronomical tablets have been preserved.[13] These may conveniently be split into two groups: Letters and Reports sent by the astrological scholars to the Assyrian kings of the eighth and seventh century BC, which have been found in Nineveh; and a large number of records of astronomical observations and predictions and texts of mathematical astronomy dating from the eighth century BC to the first century AD uncovered in Babylon. A small number of astronomical texts have also been found in other cities, most notably around 50 texts of mathematical astronomy from Uruk. However, these generally reflect methods also used in Babylonian astronomical texts and so will be considered along with the Babylonian material. About 90% of the astronomical tablets that have been recovered are now held by the British Museum in London.

The Assyrian texts almost all date from the reigns of the Kings Esarhaddon (681–669 BC) and Assurbanipal (669–648 BC). They may be divided into two classes: Letters and Reports. The Letters, which have been published by Parpola (1993),[14] were written by scholars on various matters, usually in response to a question from the king. Many of the Letters concern celestial observations and divination, for example:

> "To the king, my lord: your servant Nabû-ahhe-eriba. Good health to the king, my lord! May Nabû and Marduk bless the king, my lord! Concerning what the king, my lord, wrote to me: 'Is it favourable for the crown prince to come into the presence of the king?' — it is very favourable. The crown prince may come into the presence of the king, my lord, this (very) day. May the gods Bel and Nabû lengthen his days, and may the king, my lord, see him prosper! The month is good, this day is good: the planet Mercury (signifies) the crown prince, (and) it is vis[ib]le in the constellation [Ari]es; Venus [is] visible in [Bab]ylon, in the home of [his] dynasty (lit. father); and the moon will complete the day in the month Nisannu (I). We count this together: it is propitious."
>
> [LABS 70; trans. Parpola (1993: 55)]

The second group of Assyrian texts are the astrological Reports of the specialists in divination employed by the king to report both their celestial observations, and to give an interpretation. Unlike Letters, Reports were often sent unsolicited on the basis of the observations made by the diviners. The Reports, which have been published by Hunger (1992),[15] often simply contain a quotation from the omen series *Enūma Anu Enlil*. This is sufficient for it to be inferred that the observation was made, for the protasis of an omen always implies an observation in the Reports.[16] In a number of

[13]It should be noted that although works such as *Enūma Anu Enlil* and MUL.APIN were written well before this time, they are mainly preserved in copies made during this later period.

[14]This is a revised version of his earlier edition, Parpola (1970). See also his commentary, Parpola (1983). In the following discussions, the Letters are denoted by their LABS number in Parpola (1993). For a table of concordances with museum numbers, see Parpola (1993: 408–412).

[15]This replaces the earlier edition by Thompson (1900). In the following discussions, the Reports are denoted by their ARAK number in Hunger (1992). For a table of concordances with museum numbers, see Hunger (1992: 374–379).

[16]Hunger (1992: xvi).

cases, however, the observation is also described, and sometimes explanatory remarks about the omen are given. For example:

> "[If] on the 14th day the moon and sun [are seen together]: reliable speech; the land will become happy; [the gods] will remember [Akkad] favorably; joy [among the troops]; the king will reach the highest rank; the cattle of [Akkad wi]ll lie in the steppe undisturbed. If the moon and sun are in opposition: the king of the land will widen his understanding. I.e., on the 14th day each [month one god] will [be seen with the other]. (The moon) came into opposition with [the sun] on the appropriate day, and [its] position is equal (to the sun's) portending a reign of long days, (and) well-being of the king of the world and [his] people. [From] Rašil the elder, servant of the king."
>
> [ARAK 395; trans. Hunger (1992: 226)]

A number of eclipses, both observed and predicted, are mentioned in the Letters and Reports. Unfortunately, the texts generally do not include the date, and the whole corpus of Letters and Reports will have to be systematically redated on the basis of the astronomical and historical data they contain before their astronomical content can be studied in detail.[17] Furthermore, in a number of cases the location of the site from which the observations were made is not known.

The large number of astronomical tablets recovered from Babylon reflect a much greater diversity of date and content than the Assyrian texts. Unfortunately there is no archaeological record for the discovery of these texts; most were in fact bought by the British Museum from Baghdad antique dealers, and even those that were recovered from excavations made on behalf of the Museum by H. Rassam around the turn of the present century did not have their provenance systematically recorded.[18] Nevertheless, it is clear from other evidence — the fact that other tablets contained in the same museum collections do have their provenance expressly mentioned, the character of the personal names which appear in the colophons, and the particular deities mentioned in the introductory invocations — that they all originated from either Babylon or the neighbouring city of Borsippa.[19] Furthermore, there is no difference in style between the fifty or so astronomical texts excavated by Rassam at Babylon, and those of uncertain provenance.

The first attempt to classify the Babylonian astronomical tablets was by Sachs (1948). Working from only a handful of examples he classified the tablets into "Astronomical Tables," later to become known as ephemeris ACT texts after their publication under the title *Astronomical Cuneiform Texts* by Neugebauer (1955), "Almanacs," "Normal Star (NS) Almanacs," "Goal-Year Texts," and "Astronomical Diaries." In *Late Babylonian Astronomical and Related Texts* (LBAT) Sachs (1955) expanded his survey to include over 1500 tablets, and published copies of many of them drawn by T. G. Pinches and J. N. Strassmaier.[20] Sachs' earlier classification remained valid for

[17] The dates proposed by Parpola (1983) for many of the texts can not be justified. See, for example, de Meis & Hunger (1998) and Brown (1999b).

[18] Reade (1986).

[19] Sachs (1948).

[20] In the following discussions, I will denote texts published in Neugebauer (1955) by their ACT number,

most of the new material, with the addition of two new classes of texts which he described as "Planetary and Lunar Observations," and "Horoscopes." A few other texts exist which do not fall into any of these categories. These I will simply refer to as "Miscellaneous Texts."

It is possible to group the various categories of astronomical texts into two main divisions: the ACT texts of mathematical astronomy, and those texts that contain actual observations and predictions made by the Babylonian astronomers, known, for want of a better name, as Non-Mathematical Astronomical Texts (NMAT).[21] Unsurprisingly, the ACT texts developed much later than the NMAT texts; the earliest ACT texts come from shortly before the beginning of the Seleucid era (c. 350 BC), whereas the NMAT texts contain records dating from as early as the eighth century BC.

Before I proceed to discuss the various types of Late Babylonian Astronomical Text in detail, it is necessary to consider when systematic astronomical records began to be kept in Babylon. In his *Almagest* (III, 3), the second century Greek astronomer Ptolemy stated that he had access to Babylonian astronomical records from the beginning of the reign of Nabonassar in 747 BC.[22] There is also an Hellenistic tradition reported by Georgius Syncellus that:

> "From the time of Nabonassar, the Chaldaeans accurately recorded the times of the motions of the stars. The polymaths among the Greeks learned from the Chaldaeans that — as Alexander (Polyhistor) and Berossus, men versed in Chaldaean antiquities, say — Nabonassar gathered together (the accounts of) the deeds of the kings before him and did away with them so that the reckoning of the Chaldaean kings would begin with him."
>
> [*Chronographia*, 207; trans. Brinkman (1968: 227)]

Whilst the suggestion that Nabonassar deliberately destroyed all earlier records is probably no more than an attempt by later historians to explain the fact that the ages before his time were so poorly documented,[23] the statement that Nabonassar initiated the Babylonian tradition of systematic astronomical records may be borne out by other evidence.[24] The Babylonian Chronicle Series, which details important historical events, in particular those involving the king, begins with Nabonassar. The Astronomical Diaries, the earliest of which has been dated to 652 BC, also contain, at the end of the astronomical observations for each month, a brief account of the important historical events that have occurred. Grayson (1975) has suggested that these entries in the Astronomical Diaries may have been the source for the chronicles, or, at the very

---

and texts listed in Sachs (1955) by their LBAT number. LBAT numbers preceded by one star refer to tablets listed in Sachs (1955), but for which copies have not been published. Double starred LBAT numbers refer to texts listed in Sachs (1955) which have been published elsewhere — predominantly in Epping (1889), Kugler (1900, 1907, 1909, 1912, 1913, 1914, 1924), and Kugler & Schaumberger (1935). Texts not catalogued in either LBAT or ACT will be denoted by their museum numbers. Where joins have been made to previously catalogued fragmentary tablets, these are denoted by plusses. For lists of concordances between ACT, LBAT and museum numbers, see Neugebauer (1955: 453–459) and Sachs (1955: xi–lvi).

[21] The term NMAT was coined by Aaboe (1980).

[22] See, for example, the translation of Toomer (1984: 116).

[23] Brinkman (1968: 227). .

[24] Hallo (1988).

least, that the chronicles and the Diaries had a common source. However, Brinkman (1990) has argued strongly against this interpretation. The tablet LBAT *1414, which lists successive lunar eclipse observations and predictions, contains a record that can be firmly dated to 9 April 731 BC, just after the end of Nabonassar's reign. Another tablet, LBAT 1413, also contains successive lunar eclipses, and was tentatively dated by Sachs (1955) to 747 BC, Nabonassar's ascension year. However, as Huber (1973) has noted, there are problems with establishing the date of this tablet, and so I shall postpone further discussion of it until Section 2.3 below.

As I have mentioned above, the Late Babylonian Astronomical Texts may be divided into a number of categories within two main groups: the ACT texts of mathematical astronomy, and the NMAT texts of non-mathematical astronomy. As this study will primarily be concerned with the NMAT texts, I will begin by discussing these texts. The following descriptions are largely based upon those given by Sachs (1948), Sachs (1974), Sachs & Hunger (1988), and Hunger (1999).

The fundamental observational text of the Babylonian astronomers was the Astronomical Diary. As discussed above, it is believed that these began to be kept from the time of the reign of Nabonassar. However, as also noted, the earliest extant Diary has been dated to 652 BC, and there is only one more example (568 BC) preserved from before the fifth century BC. Survivals occur with gradually increasing frequency during the next three centuries, until in the second century BC, examples from about three-quarters of the years are preserved. During the first century BC the frequency of survivals drops off again, and no Diaries are extant which are dated later than 61 BC, although it seems likely that the Diary tradition continued into the first century AD since other astronomical texts are known from this date.[25] In many of these cases, however, the preserved Diaries are only fragments of the original texts, and contain information for only part of the period which they originally covered. Furthermore, not all of the known Diary fragments have proved dateable, although the others are generally too small to be of great interest in any case. All of the dateable Diaries have been published in transliteration and translation by Sachs & Hunger (1988, 1989, 1996).

Typically, an Astronomical Diary contains a day by day account of the observations made by the Babylonian astronomers, for a period of six or seven months. Each month begins with a statement about the number of days in the previous month, and then a measurement of the time between sunset and moonset on the evening of that first day. The time interval between the moon and the sun crossing the horizon is also recorded a further five times during the month, four at full moon, and the final measurement at the end of the month. These intervals, which were first understood by Epping (1889) and dubbed by Sachs (1948) the "Lunar Six," are listed below:[26]

---

[25]The distribution of years with at least one dated Diary fragment is shown graphically in Figure 2 of Sachs (1974). Although a small number of additional Diary fragments have subsequently been dated, these would not significantly alter this figure.

[26]These definitions are based upon those of Sachs & Hunger (1988: 20).

na	The time between sunset and moonset when the moon was visible for the first time after conjunction
ŠÚ	The time between moonset and sunrise when the moon set for the last time before sunrise
na	The time between sunrise and moonset when the moon set for the first time after sunrise
ME	The time between moonrise and sunset when the moon rose for the last time after sunset
GE_6	The time between sunset and moonrise when the moon rose for the first time after sunset
KUR	The time between moonrise and sunrise when the moon was visible for the last time before conjunction

Throughout the month, other lunar observations, such as the passing by of the "Normal Stars"[27] and eclipses of the sun or moon; planetary phenomena such as appearances, disappearances and stations — called "Greek-Letter" phenomena by Neugebauer (1955: 180) — and positions relative to the Normal Stars; solstices and equinoxes; Sirius phenomena; and occasionally events such as comets, meteors and bad weather, were also recorded in the Diaries. Finally, at the end of every month, a brief summary of the level of the river Euphrates, the value of six essential commodities, and any important events in the life of the city, would be reported.

Whilst most of the contents of the Diaries represent observations, it is important to note that the Diaries do also contain a number of predictions. Generally, these predictions are in place of events which were watched for but, often on account of bad weather, could not be seen. For example, a number of lunar six measurements are followed by the words NU PAP, meaning "not seen" and indicating that the time was calculated, whereas some others end with *muš*, "measured." Similarly, a number of eclipses are predicted that could not be seen because the luminary was below the horizon at the time of the eclipse (i.e., a lunar eclipse during the hours of daylight, or a solar eclipse during the night). At least in the Seleucid period, solstices and equinoxes were always computed on the basis of the so called "Uruk scheme."[28] Similarly, the dates of rising and setting of Sirius were not observed but calculated using a simple scheme.[29]

From the Astronomical Diaries, the Babylonians abstracted records to compile texts that have become known as "Goal-Year Texts." These texts contain information that was to be used in making predictions for a specific "Goal" year.[30] This informa-

[27]The Normal Stars are a group of bright stars located close to the ecliptic that were used as reference points in the sky. For a list of the Babylonian Normal Stars, see Sachs & Hunger (1988: 17–19).

[28]Neugebauer (1947b, 1948).

[29]Sachs (1952b).

[30]Although Hunger (1999) has noted that there are a small number of cases where the observations recorded in the Goal-Year texts do not correspond exactly with the descriptions given in the Diaries, he has plausibly suggested that these Goal-Year texts drew on other Diaries than the ones that happen to be preserved, as there are cases where parallel Diaries give slightly different accounts of the same event.

tion comprises lunar and planetary observations, or a prediction if no observation was made, taken from a number of years ago corresponding to one cycle in the following periods of the moon and planets:[31]

Jupiter	71 years for Greek-Letter phenomena
	83 years for conjunctions with Normal Stars
Venus	8 years
Mercury	46 years
Saturn	59 years
Mars	79 years for Greek-Letter phenomena
	47 years for conjunctions with Normal Stars
Moon	18 years
Lunar Six	18 years
	Sums ŠÚ + *na* and ME + GE_6 for last 6 months of 19 years

Although all of the extant Goal-Year texts date from the Seleucid Era, most, if not all, of the planetary periods they use were known in Babylon from well before this time. For example, BM 45728, which probably dates from before the fourth century BC, mentions the 8 year Venus, the 47 year Mars and the 59 year Saturn periods; BM 41004, probably from the fourth or fifth century BC, mentions them all (although the 79-year Mars period is given only implicitly).[32] Also, parts of an unusual astrological omen text (whose protases are given exclusively in numbers) dating from the time of the Assyrian King Ashurbanipal (7th century BC), probably relate to the 71 year Jupiter, the 59 year Saturn, and the 8 year Jupiter period.[33]

The earliest extant Goal-Year Text contains entries for a Goal-Year of the 76th year of the Seleucid Era, which is equivalent to 236 BC, whilst the latest is for the 288th year of the Seleucid Era, or 24 BC. Although copies of many of the Goal-Year Texts were included in Sachs (1955), only a few of these tablets have as yet been published in transliteration or translation.[34]

The Astronomical Diaries were also used to compile lists of individual planetary and lunar phenomena.[35] Most important among these lists are those containing successive eclipse observations and predictions, which I shall call "Eclipse Texts." As I have mentioned, the earliest reliably dated list contains eclipse records stretching back to 731 BC. The latest continues down to 160 BC. Complementary to these lists are a number of tablets which are dedicated to describing an individual eclipse. These range in date from 284 BC to 10 BC. Transliterations and translations of these texts are currently being published by Sachs & Hunger (1999).

[31]See Sachs (1948: 283, 1955: xxv) and van der Waerden (1974: 108–110).

[32]For BM 45728, see Kugler (1907: 45–48) and P. Huber in van der Waerden (1974: 107). For BM 41004, see Neugebauer & Sachs (1967: 200-208).

[33]For this text, see Gadd (1967) and Hunger (1969).

[34]However, Huber (1973) has edited most of the eclipse records found in the Goal-Year texts, together with many from the Diaries and the Eclipse Texts.

[35]It should be noted that in many cases the records have obviously been reworded to give entries within a compilation uniform style.

Two further categories of NMAT are the Almanacs and the Normal Star Almanacs. Observations are never found in these texts. Instead, they detail predictions of many of the elements of an Astronomical Diary for a forthcoming year. Until recently, it had generally been assumed that these predictions were made using the information collected in the Goal-Year texts, and that they in turn provided the source of the predictions in the Diaries. Thus, there was taken to be a closed circle in which records were abstracted from the Diaries into the Goal-Year Texts, which in turn were used to produce the Almanacs and Normal Star Almanacs, and these would finally be fed back into the Diaries as the predictions of unobserved events. However, Hunger (1999) has shown that it may not have been as simple as this. If the Almanacs and Normal Star Almanacs were produced from the Goal-Year texts, then the data in the Goal-Year texts were not simply put into the Almanacs and Normal Star Almanacs, but were modified in some fashion. What this modification was and how it was applied are not yet clear, but it is possible that it involved using the ACT type texts, which otherwise appear to have been of no practical application. Dated Normal Star Almanacs range from the 31st year of the Seleucid Era to the 212th year (281–100 BC), whereas the Almanacs range from the 92nd year of the Seleucid Era (220 BC) to as late as the 385th year (AD 75).[36] Once again, whilst copies of many of the Almanacs and the Normal Star Almanacs were included in Sachs (1955), very few have been published in transliteration or translation.

Finally, we have a small number of Horoscopes which record the position in the zodiac of the five planets, the moon, and the sun on the date of birth of a child. Additionally, the length of the lunar month, the date that the moon set for the first time after sunrise, and the date of the last visibility of the moon (known together as the lunar three), solstices and equinoxes, and eclipses may be included in a horoscope; occasionally a prediction for the life of the child is also given.[37] Rochberg-Halton (1989a) has proposed that the astronomical data in the horoscopes was largely taken from the Almanacs, with some additional material abstracted from the Diaries. The Horoscopes range in date from 410 BC to 69 BC. The complete corpus — 28 Horoscopes plus 4 "Birth Notes" which give only the date of the child's birth — has recently been edited with commentary by Rochberg (1998).

Let me now move on to the ACT type texts, most of which have been published by Neugebauer (1955).[38] These may be split into three classes named by Neugebauer "Ephemerides," "Auxiliary Functions," and "Procedure Texts." The Ephemerides list, for successive phenomena (such as moon-sun conjunctions, oppositions, or Greek-Letter phenomena of a specific planet), various functions needed to calculate the position, time, and visibility of those phenomena. In order to construct the Ephemerides, texts of Auxiliary Functions, such as the velocity of the sun or moon, were written. Finally, the Procedure Texts outline the rules for computing the Ephemerides. The ACT texts range in date from 319 BC, shortly before the beginning of the Seleucid

[36] Sachs (1976).

[37] Sachs (1952a).

[38] For references to additional texts, most of which have been published by Aaboe, Neugebauer, and Sachs, see Neugebauer (1975).

Era,[39] to AD 43.[40]

The relationship between the ACT and the NMAT texts is at present far from clear. We might assume that the observations contained in the Diaries would provide the raw data from which the ACT schemes were developed, and indeed Aaboe (1980) and Swerdlow (1998) in the field of the ACT planetary theories, and Brack-Bernsen (1997) and Britton (1999) for elements of the lunar theory, have proposed methods by which this may have been done; however, these are only partial solutions to this problem and more work remains to be done. Furthermore, there is little evidence that the ACT schemes were used to make the predictions in the NMAT texts.[41]

Finally, let me pose the question: who were the Babylonian astronomers, and from where did they observe? The earliest evidence in answer to this is found in the description of the Babylonian Ziggurat, the famed "Tower of Babel," by Diodorus Siculus:

> "... in the centre of the city (is) a temple of Zeus whom, as we have said, the Babylonians call Belus. Now since with regard to this temple the historians are at variance, and since time has caused the structure to fall into ruins, it is impossible to give the exact facts concerning it. But all agree that it was exceedingly high, and that in it the Chaldaeans made their observations of the stars, whose risings and settings could be accurately observed by reason of the height of the structure."
> [*Bibliotheca Historica*, II, 9; trans. Oldfather (1933: 381)]

The reliability of this account must, however, be questioned. According to Arrian, the Ziggurat was destroyed by Xerxes in the fifth century BC (*Anabasis*, VII, 16), and although Alexander made plans to have it rebuilt in the fourth century BC, these were never carried out. It seems more likely that the astronomers observed from either a building within the city (perhaps the temple), or from the city walls. In the ancient world, these walls were considered to be among the great technological achievements of the Orient. Their fame can probably be attributed to the account of Babylon given by Herodotus:

> "(Babylon) is surrounded by a broad deep moat full of water, and within the moat there is a wall fifty cubits wide and two hundred high. ... On top of the wall they constructed, along each edge, a row of one-roomed buildings facing inwards with enough space between for a four-horse chariot to pass. There are a hundred gates in the circuits of the wall, all of bronze with bronze uprights and lintels."
> [*Historia*, I, 181; trans. de Sélincourt (1979: 113)]

Plainly, Herodotus' account cannot be considered to be the literal truth; a height of 200 cubits, or about 100 meters, for the wall is inconceivable. Archaeological evidence, discussed by Ravn (1942), suggests a height of between 10 and 20 metres.

The identity of the observers at this early period is also unclear. In the Greek and Latin classics, the word "Chaldaean" is often used as a synonym for astrologers and diviners.[42] The writings of Herodotus make it clear that the Chaldaeans are to be

[39] BM 40094, published by Aaboe (1969).

[40] BM 34083 = ACT 53.

[41] See, for example, Sachs & Hunger (1988: 25).

[42] de Kuyper (1993).

identified with the priests of Bel. This suggests that the early Babylonian astronomers were linked to the temples, a suggestion which will reappear again for the astronomers of the Seleucid period. One other possibility is that the astronomers were employed by the king. This certainly happened in the Assyrian capital of Nineveh, where the king employed a number of scholars to make celestial observations and to interpret the omens for him.[43]

By the Seleucid period, the situation had changed. Babylon was no longer a royal city, and so there can be no suggestion that scholars were employed by the king to make observations and perform divination. The development of mathematical astronomy in the ACT texts also leads us to the question: were the ACT texts compiled by the same group of astronomers as made the observations recorded in the Diaries, or were there two independent groups of astronomers? Neugebauer (1989) was firmly of the latter opinion. "Nothing," he wrote, "compels us to assume that these two groups of professional men considered one another with particularly kind feelings." However, Rochberg (1993) has shown, on the basis of the colophons of the astronomical texts, that a group of scholars called *ṭupšar Enūma Anu Enlil*, "Scribes of *Enūma Anu Enlil*," who were responsible for making the observations that were recorded in the Diaries and producing the Almanacs for a forthcoming year, were also the authors of some of the ACT type texts. She also argues that these scribes were based in the Babylonian temples, namely the Esagila in Babylon and the Bit Reš sanctuary in Uruk.

Some information concerning these individuals is given in three tablets describing sessions of the Assembly of the Esagila temple.[44] These documents, which date from the second century BC, record verdicts given by the Assembly whereby the sons of a *ṭupšar Enūma Anu Enlil*[45] are to take over his role, and to have his income shared out between them. Before this verdict, however, the sons had to prove that they could "make the observations, calculations and measurements."[46] But the word used for "observations" here, *nasāru*, is the same as in the rubrics to the Astronomical Diaries. Similarly, the terms for "calculations and measurements" are those used in the rubrics of the ACT ephemerides and the Almanacs.[47] This indeed suggests that the astronomers were the *ṭupšar Enūma Anu Enlil*, and that they were under the jurisdiction of the Esagila temple, at least in this late period. Furthermore, these documents imply that the position of astronomer was, at least partially, hereditary.

For further support of Rochberg's (1993) conclusion that the astronomers were based in the Esagila temple, I may mention the accounts of Strabo and Pliny. Strabo visited the city in 24 BC and wrote that it was:

> "... in great part deserted, so that no-one would hesitate to apply to it what one of the comic writers said of Megalopolitae in Arcadia, 'The great city is a great

[43] Oppenheim (1969).

[44] BOR 4, 132; CT 49, 144; and CT 49, 186. See, most recently, van der Spek (1985) and Rochberg (1999b). An earlier discussion is given by McEwan (1981: 17–21).

[45] Translated rather ambiguously as "astrologer" by van der Spek (1985) and McEwan (1981).

[46] CT 49, 144, Rev. 23; trans. van der Spek (1985: 552).

[47] See Rochberg (1999b).

> desert.' "
>
> [*Geographica*, XVI, 1, 5; trans. Hamilton & Falconer (1906: 145)]

Later Pliny was to write:

> "The temple of Jupiter Belus in Babylon is still standing — Belus was the discoverer of the science of astronomy; but in all other respects the place has gone back to a desert."
>
> [*Historia Naturalis*, VI, 30, 121; trans. Rackham (1942: 431)]

The "temple of Jupiter Belus" referred to by Pliny is the Esagila. From these two accounts it would therefore appear that the Esagila was one of the few buildings still inhabited in the latter half of the first century BC. The fact that we have an account of an observation of a solar eclipse in 10 BC[48] indicates that the astronomers were still active in the city at that time. This provides additional support in favour of the argument that the Esagila temple was the centre of this astronomical activity. Furthermore, the fact that Almanacs have been found from as late as AD 75, suggests that the astronomers were still active as late as the latter half of the first century AD. By AD 116, however, Trajan found that the city was finally deserted.[49]

2.3 Dating Babylonian Astronomical Tablets

As I have mentioned above, the majority of the tablets known to contain astronomical material are now held by the British Museum. They were recovered, either directly by excavations sponsored by the Museum, or indirectly from antique dealers in Baghdad, from the site of Babylon during the late nineteenth and early twentieth century.[50] Sadly, many of these tablets were badly damaged when first dug up, and, despite many successful restorations by joining fragments of the same tablet together, a large number of texts still remain in a woeful condition. In many cases, the line of writing containing the date of the record, which is usually one of the first lines on the tablet, is either broken away or is only partially preserved. Sometimes it is possible to date the tablet on account of the historical remarks contained within it, but this is the exception rather than the norm. Fortunately, however, because of the nature of the astronomical texts — in that they contain observations of the heavens — it is often possible, at least if the tablet contains a reasonable amount of astronomical information, to date it using astronomical means. The basic principle of this technique is to calculate, using modern theories of celestial motion, the positions of the heavenly bodies at times in the past to try to obtain a unique date at which the observations contained in the text could have been made. In his classification of the LBAT, Sachs (1955) derived dates in this fashion for a number of tablets. As a result he was able to publish dates, which were either preserved on the tablet or that he had derived, for about half of his catalogue.

---

[48] See Section 2.3.1 below.

[49] Oates (1986: 143).

[50] Reade (1986).

Month	Name	Abbreviated Name
I	Nisannu	BAR
II	Ajjaru	GU_4
III	Simānu	SIG
IV	Du'ūzu	ŠU
V	Abu	IZI
VI	Ulūlu	KIN
VII	Tešrītu	DU_6
VIII	Araḫsamnu	APIN
IX	Kislīmu	GAN
X	Ṭebētu	AB
XI	Šabāṭu	ZÍZ
XII	Addaru	ŠE

Table 2.2: Babylonian month names (after Sachs & Hunger (1988: 13–14)).

Before I describe the procedure of dating a Babylonian astronomical text by astronomical methods, it is necessary to outline some details of the Babylonian calendar. Essentially, the Babylonians, in common with many ancient civilizations, used a luni-solar calendar, that is a month length based upon the motion of the moon, and a year length based upon the motion of the sun. A Babylonian month was defined as beginning on the evening when the lunar crescent becomes visible for the first time. Each month therefore contained either 29 or 30 days, a day lasting from one sunset to the next. If the lunar crescent was still not visible after 30 days, say because of cloud cover, then the month was begun anyway; thus there are no months lasting 31 days in the Babylonian calendar. However, many early texts, such as MUL.APIN, use an ideal calendar of twelve 30 day months in a year. The names of the 12 Babylonian months are given in Table 2.2. In the astronomical texts the months are usually given by their logograms, rather than in their full Akkadian form. It has become customary in translations to render the months names as a roman numeral (i.e., Nisannu or BAR is usually given as "Month I.")

Because the length of a solar year is about 11 days longer than that of 12 lunar months, the Babylonians inserted an extra "intercalary" month every three years or so. Initially these could be inserted at any time of the year, but by the Late Babylonian period, they were only placed after the sixth and the twelfth months. These intercalary months are denoted in translations as "Month VI_2" and "Month XII_2." Before the sixth century BC there is no evidence for any scheme to calculate when to insert an intercalary month being used, despite the fact that possible schemes are given in MUL.APIN; instead, they were simply added whenever it was considered necessary. Starting around 527 BC, however, Britton (1993) has found evidence that an 8-year intercalation cycle was used, to be replaced in 503 BC by a 19-year cycle. Although there are occasional exceptions when this scheme was not applied consistently,[51] it seems to have been used continually from the fifth century BC, down through the Seleucid Era.

[51] For details, see Figure 2 of Britton (1993).

From an early time, regnal years were used in the Babylonian calendar. The first regnal year of a king's reign began with the first month of the year following his accession to the thrown. The preceding year was therefore both the final year of the previous king and the accession year of the new king. For example, the 21st year of Nabopolassar, his final, lasted from April 605 BC to March 604 BC in the Julian calendar. It was followed by the 1st year of his successor, Nebuchadnezzar II, who had acceded to the thrown in the previous year. Thus the accession year of Nebuchadnezzar II is the same as the 21st year of Nabopolassar. The practice of using regnal years continued until the reign of Seleucus I (311 BC), after which date years were generally given in terms of the number of years from this epoch. This was known as the Seleucid Era. After the Parthians came to Mesopotamia in the middle of the second century BC, dates are often given both in the Seleucid Era, and in the Iranian Arsacid Era.

In converting from the Babylonian calendar to the Julian calendar, it is necessary to know not only the length of each reign, but also the length of each month, and when months were intercalated. Parker & Dubberstein (1956) have made an extensive investigation of these problems, and their tables based upon astronomical computation of month lengths allow dates from the beginning of Nabopolassar's reign (626 BC) to the 386th year of the Seleucid Era (AD 75) to be converted to the Julian calendar with ease. Occasionally, their computed month lengths differ from those actually observed in Babylon by one day, as has been noted by Sachs & Hunger (1988, 1989, 1996).

Let me return now to the method of dating a Babylonian astronomical tablet if the date is either wholly or partially destroyed. The basic principle in this procedure is to compare any preserved observations with the results of retrospective computations. Observations that may be of use in this context include solar and lunar eclipses, planetary visibilities and positions, stellar visibilities, lunar positions, and any measurements of the lunar six. Before commencing this process, however, it is necessary to establish a range of possible dates for the tablet. From the arguments presented in Section 2.2 above, it is safe to assume that the Babylonian texts are unlikely to come from before 750 BC or after AD 100. For some tablets, it may be possible to further restrict this date range on linguistic or historical grounds. Modern theories of the motion of the sun, moon, and planets are then used to accurately determine their positions over dates in this period.

The procedure used to date the tablets depends upon their contents. Almanacs and Normal Star Almanacs, for example, contain a large amount of planetary data which can be used to obtain the approximate longitudes of the various planets over the period covered by the text. Since the outer planets move only slowly — a complete passage of Saturn through the zodiac takes about thirty years, for example — even a rough measure of their longitude provides considerable information for dating. To continue our example with Saturn, knowledge of just the sign of the zodiac in which the planet is located reduces the possible dates of the text to only two or three years out of every thirty. When these are combined with the other planets — which take different amounts of time to return to the same longitude — and with other data such as the month and lunar positions, only a few, or, better still, sometimes only one date will correspond to the information.[52] This must then be the date of the text.

Diaries may also often be dated on the basis of the planetary information they contain. If the Diary also contains a reference to an eclipse observation, this dramatically reduces the possible date for the text. Lunar eclipses are only visible from a given location on average about once every year (although the distribution of visible eclipses within a given time period is not regular); solar eclipses even less frequently. Combining an eclipse observation with the time of the year (obtained either directly from a preserved month name, or derived from the longitude of the moon or an inner planet), generally reduces the possible dates for an eclipse record to under twenty during the Late Babylonian period. The magnitude of the eclipse may further reduce the possibilities — a reference to totality, for example, cuts the number by about a half. Finally, references to the visibility of planets or Sirius may allow us to find a unique date on which the observation could have been made.

It should be noted that in dating astronomical texts, one should not use exact information to decide between possible dates. Times of eclipses, for example, may be subject to errors due to inaccuracies in the clock used to time the observation. Similarly, precise measurements of the distance between, say, the moon and a star may also contain errors. Thus it is best to only use rough information. For example, if an eclipse is reported to begin 20° after sunset,[53] one should not use this information to discard those eclipses which did not begin at this exact time, but only those that did not begin in, say, the first half of the night.

Unfortunately, it is not possible to give a general set of rules by which any text may be dated — there is too great a variety in the content of preserved fragments. It is largely a matter of experience as to what techniques will work best with any particular text. I therefore give below examples of three texts dated astronomically to illustrate the process. Since this book is primarily concerned with eclipse records, I have chosen to use Eclipse Texts.[54] For examples of dating other texts, see Stephenson & Walker (1985).

2.3.1 LBAT 1456

LBAT 1456 is a small tablet containing a record of an observation of a solar eclipse. Sadly, the tablet is badly damaged and both the date and a number of details of the eclipse have been lost. A copy of the tablet by T. G. Pinches, published by Sachs (1955), is reprinted here as Figure 2.1. Due to the unusual terminology used in the text, Sachs described it as possibly being astrological rather that observational in nature. However, subsequent readings have resolved this uncertainty. I reproduce below a preliminary translation of the twelve lines of badly damaged text on this tablet kindly

[52] For similar techniques of dating astronomical papyri, see Jones (1999: v.1 47–55), and horoscopes, see Neugebauer & van Hoesen (1959: 1–2).

[53] The Babylonian unit of time was the UŠ. As there were 360 UŠ in one day, it is customary to render UŠ as "degree" or ° in translations. See Section 2.4 below for further details.

[54] The tablets have been dated at the request of Prof. H. Hunger for their publication in Sachs & Hunger (1999). In all cases, I have worked from his translations of the texts. Further details are given in this volume.

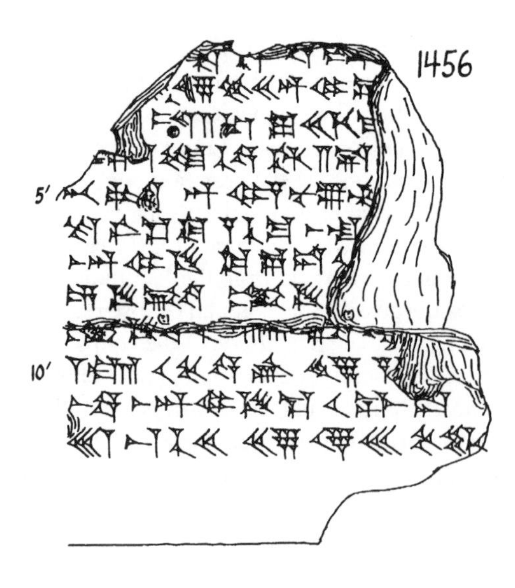

Figure 2.1: Pinches' copy of LBAT 1456. [Courtesy: The Trustees of the British Museum]

supplied by H. Hunger.

1. [...] ... [...]
2. [...] the 28th, solar eclipse; fr[om ...]
3. [...] it began; 23° of day to the inside of the sun ... [...]
4. its ... were clear(?); 2° [...]
5. Venus, Mercury, eclipse ...; the remainder(?) [...]
6. Sirius, which had set, in its non-[...]
7. In its eclipse, ... [...]
8. people broke pots [...]
9. they broke. In 23° of day it cleared from north [and west]
10. to south and east. 48° onset, [maximal phase,]
11. and clearing. In its eclipse, the north and west wind blew.
12. 1,30° (=90°) of day before sunset. The 28th, moonrise to sunrise: 17° 30', measured.

The text describes a solar eclipse during which the planets Venus and Mercury and the star Sirius were visible. The eclipse began at 90°, or 6 hours, before sunset, and lasted for 48°, or 3.2 hours. The date of this tablet can be established primarily because of the observation that Venus, Mercury, and Sirius were visible during the solar eclipse. For this to be the case, over 0.95 of the apparent solar diameter would probably have been obscured at the maximum phase of the eclipse.[55] Large solar eclipses are only infrequently visible at any given location on the Earth's surface, and so these details considerably reduce the range of possible dates for the eclipse.

There were only eight solar eclipses with magnitudes greater than 0.95 visible in Babylon between 750 BC and AD 100. Their dates are 19 May 557 BC, 18 January 402 BC, 15 June 242 BC, 10 October 174 BC, 15 April 136 BC, 30 June 10 BC, 24 November 29 AD, and 30 April 59 AD. Tablets containing observations of the eclipses on two of these dates, 15 June 242 BC, and 15 April 136 BC, have already been securely dated. As the description and the timings of the eclipse found on LBAT 1456 radically differ from the reports found on these other tablets, it is possible to immediately discard these two dates as possibilities for this record. The dates 18 January 402 BC, 10 October 173 BC, and 24 November 29 AD may be discarded as Sirius would have been below the horizon at the time of the eclipse. On 19 May 557 BC Mercury would have set during the eclipse and so would probably not have been visible. Also the timings given in the report are very different to those given by modern computations. On 30 April 59 AD Mercury, Venus and Sirius would have been above the horizon but Mercury would be too faint (mag. = +0.5) to be visible as the computed magnitude of this eclipse was only 0.96. There is also a poor agreement between the recorded and computed time of the eclipse. Thus this date may also be discarded.

The computed circumstance of the eclipse on 30 June 10 BC is in excellent agreement with the observed details of the eclipse. Mercury, Venus and Sirius were all visible during the eclipse. There is also good agreement between the recorded and computed times. By computation, the eclipse began at a local time of 12.80 hours. This corresponds to about 95° before sunset, very close to the 90° quoted in the last line of the tablet. The eclipse is computed to have lasted for about 40°, slightly shorter than the 48° given for the total duration in the tenth line. However, errors of this size in the timing of eclipses are in no way unusual.[56] It is therefore possible to confidently assign this date to the observation on this tablet.

Figure 2.2 shows the position of the observed heavenly bodies at the time of mid-eclipse. By computation, Babylon was just within the path of totality of the eclipse. The track of totality of the eclipse is shown in Figure 2.3.[57] If the eclipse was total, it may explain the dramatic nature of the language used in the report — the act of smashing pots and making noises in response to an eclipse is known worldwide.[58] In a seventh century BC letter to the Assyrian king describing the observation of an eclipse, Mar-Issar wrote that "a bronze kettle drum was set up."[59] A number of descriptions of the ritual playing of a kettledrum during lunar eclipses are preserved from as late as the Hellenistic period,[60] although the present text contains the only known reference

---

[55] This criterion is due to Muller (1975), and is a conservative estimate.

[56] Steele, Stephenson & Morrison (1997).

[57] This map was kindly prepared by Mrs. Pauline Russell and Prof. F. Richard Stephenson of the Department of Physics, University of Durham.

[58] See, for example, the instances quoted by Beaulieu & Britton (1994: 77).

[59] LABS 347; trans. Parpola (1993: 282).

[60] See, for example, Beaulieu & Britton (1994) and Brown & Linsenn (1999).

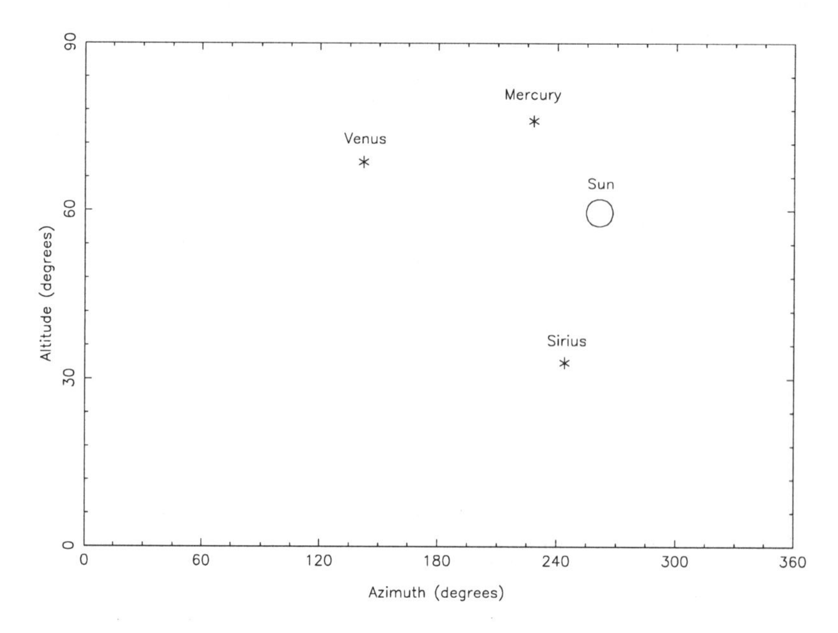

Figure 2.2: The positions of Mercury, Venus and Sirius at the maximum phase of the solar eclipse on 30 June 10 BC.

to a ritual for an eclipse of the sun. In contrast to other parts of the world, eclipse rituals in Mesopotamia were intended to ward off the evil associated with the eclipse, rather than to frighten away, for example, a mythical creature that was extinguishing the moon's light.[61]

The report of the eclipse on LBAT 1456 is the latest yet known from Babylon. Furthermore, it may be one of only two total solar eclipses for which we have a record of its observation preserved in Babylonian history.[62] The question of whether the eclipse was total will be addressed in Section 2.5 below.

2.3.2 LBAT 1452

This is a small, badly damaged tablet containing a report of a lunar eclipse. A copy of the tablet by Pinches, published by Sachs (1955), is reproduced here as Figure 2.4. I

[61] For a detailed discussion of the nature of apotropaic rituals in Mesopotamia, see Reiner (1995: 81–96).

[62] It should be noted that at no stage in establishing the date of this tablet has it been assumed that the eclipse was total, which would require the Earth's rotational clock error, ΔT, to be confined to a small range of values. The only assumption that was made was that the eclipse must have a magnitude of greater than 0.95, which is given for any sensible value of ΔT. Thus, if it can be decided that the eclipse was total, it can be used to refine our present knowledge of ΔT at this period. This will be discussed further in Section 8.2 below.

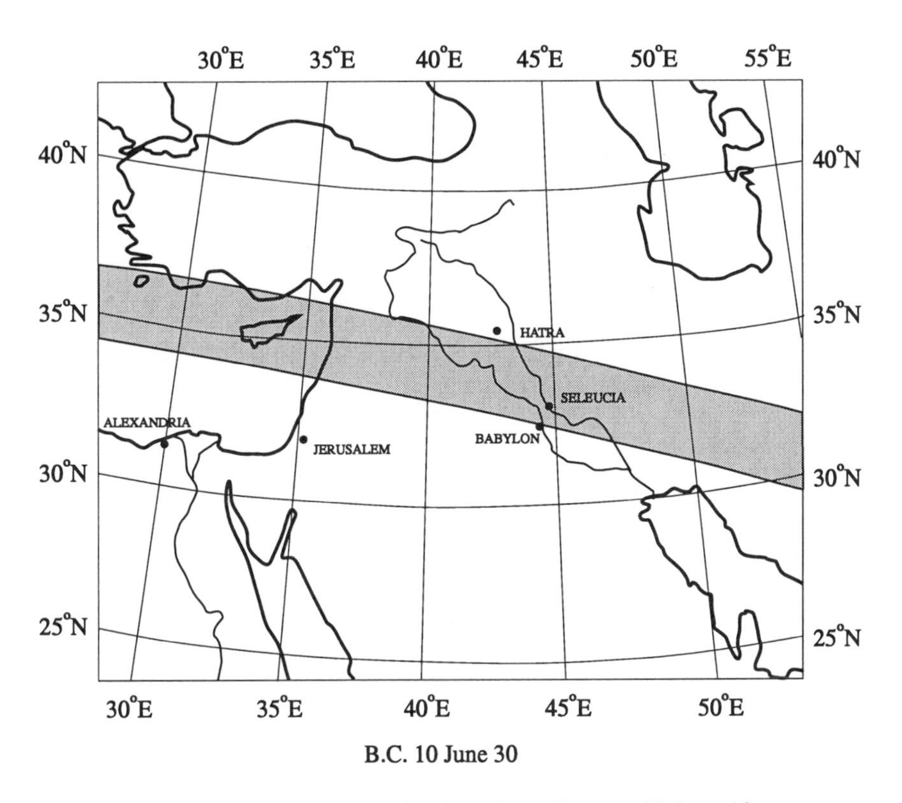

Figure 2.3: Track of totality for the solar eclipse on 30 June 10 BC.

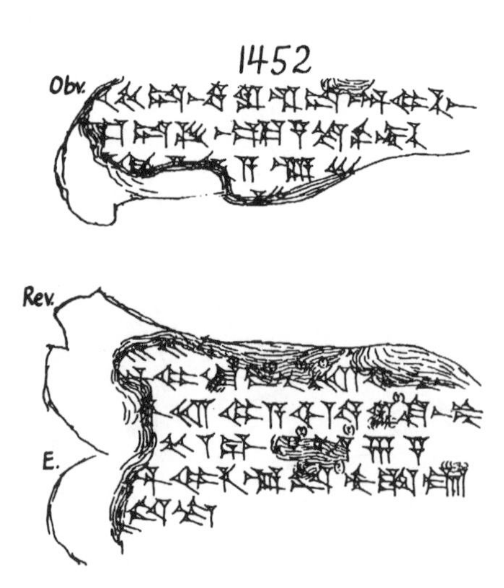

Figure 2.4: Pinches' copy of LBAT 1452 (interchange obverse and reverse). [Courtesy: The Trustees of the British Museum]

quote below H. Hunger's translation of the 10 lines of damaged text:

1'. [...] ... [...]
2'. [... lunar] eclipse; when it began, in 22° of night all
3'. [...] ... was covered. 22° of night maximal phase. When it began to clear,
4'. it cleared [in 21° of night from] east to west. 65° onset,
5'. [maximal phase, and clearing.] Its eclipse was red. Lightning on the south side
6'. [...] flashed (?).
7'. [...] and east wind blew; during clearing, the north wind blew. In its eclipse, ...
8'. [... and Sat]urn stood there; in the beginning of onset, Sirius set.
9'. [...] α Virginis it was eclipsed ... [...]
10'. [...] ... [...]

The text describes a lunar eclipse during which Saturn was visible. Sirius set during the eclipse, which lasted for a total of 65° or 4.33 hours. At the time of the eclipse, the moon was in the region of the star α Virginis. The individual phases of the eclipse are quoted to the nearest UŠ, which suggest a date after about 560 BC for the eclipse. Before this period, times are generally only quoted to the nearest 5 or 10 UŠ.

The critical observations in establishing the date of this tablet are that the moon was near α Virginis, that Saturn was visible during the eclipse, and that Sirius set during the course of the eclipse. At a lunar eclipse, the sun and moon are in opposition. Therefore, if the moon has a longitude of about 173° (i.e., in the neighbourhood of α Virginis), the sun must have a longitude of about 353°, and so the eclipse must have taken place during the months of March or April. Between 560 BC and AD 100 there were only 12 lunar eclipses during these months that were visible from beginning to end in Babylon. They are: 2 April 508 BC, 14 April 425 BC, 6 April 378 BC, 5 April 359 BC, 7 April 294 BC, 17 March 284 BC, 28 March 247 BC, 20 March 135 BC, 1 April 52 BC, 22 March 5 BC, 24 March 61 AD, and 4 April 79 AD. Only during seven of these eclipses was Saturn visible, and so the other dates may be discarded. In all but two of these dates, Sirius was either not visible during the eclipse, or else it was visible during the whole of the eclipse. This only leaves two dates: 17 March 284 BC and 1 April 52 BC. On the former date, the lunar longitude was about 172°, which is within 1° of that of α Virginis. For the latter date, however, the lunar longitude was about 189°, which, while still being in the general region of α Virginis, is much closer to α Libra, and so it would be expected that, in this second case, the moon would have been said to be near α Libra rather than α Virginis. LBAT 1452 can therefore be dated to 17 March 284 BC.

2.3.3 LBAT 1413

This tablet records observations of four successive lunar eclipses observed in the accession, 1st and 2nd years of an unknown reign. This is followed by two predictions for the 2nd and 3rd year. The tablet, Pinches' copy of which was published by Sachs (1955) (reproduced here as Figure 2.5), is very badly damaged. As discussed in Section 2.2 above, Sachs (1955) dated this tablet to the reign of Nabonassar, making it the earliest known observational text from Babylon. However, the date has been questioned by Huber (1973), and so it is appropriate to discuss its dating in detail.

I give below H. Hunger's translation of the tablet:

0. At the command of Bel and Beltija may it go well.
1. 1,40. Accession year [of ...]
2. Month XII, (after) 5 month, the 14th, morning watch, ... [...]
3. 2,10. Year 1. Month VI, [the 1]5th (?), onset (?). It began in the north [...]
4. [...] the south wind blew. It set eclipsed. Month VI was in[tercalary.]
5. [Month XI, the 1]4th, onset (?). 1,40° remained to clearing.
6. [Year 2. Month] V, the 14th, it made a total (eclipse).
7. [Month XI,] omitted.
8. [Year 3. Month V, omitt]ed. Month VI was intercalary.
9. [...] ... [...]

The essential observations in attempting to obtain the date of this tablet are that in month XII of the accession year of an unknown king, there was a lunar eclipse in

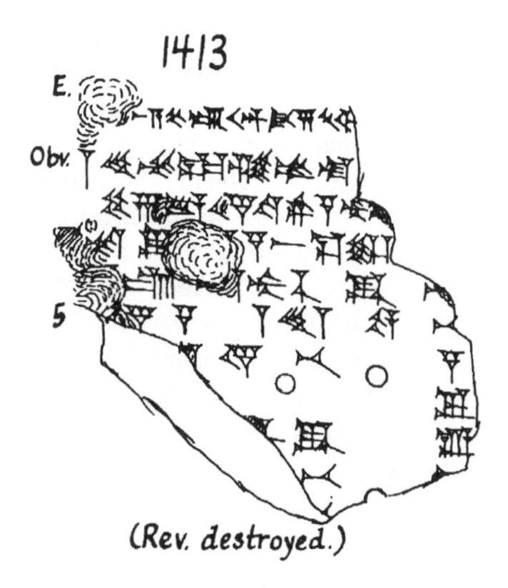

Figure 2.5: Pinches' copy of LBAT 1413. [Courtesy: The Trustees of the British Museum]

the latter part of the night. This was followed by a lunar eclipse in month VI which set whilst it was still eclipsed. In month XI, there was another lunar eclipse, and then in the following month V, there was a total lunar eclipse. It is quite unusual for four lunar eclipses to be visible in a row, and so for this to happen at the beginning of a king's reign is a very rare event, and there should be a good chance that the tablet is dataeble.

Huber (1973) found that there were only three possible dates for the text between 930 BC and 311 BC (after which time accession years were no longer used): 801–800 BC, 747–746 BC, and 693–692 BC. Independently, I have found that between 850 BC and 311 BC only 801–800 BC and 747–746 BC are possible. The third eclipse in the 693–692 BC group could not have set whilst being eclipsed. I will go through both cases in detail below.

On 4 January 801 BC there was a lunar eclipse in the latter part of the night. This was followed six months later by an eclipse on 29 June 801 BC. This eclipse set during totality. There was a third eclipse on 23 December 801 BC, and then a fourth on 18 June 800 BC. This final eclipse rose just after it began and became total.

On 5 February 747 BC the moon was eclipsed. Six months later, on 1 August 747 BC, the moon set eclipsed just before it had cleared. A third eclipse was visible on 25 January 746 BC, and then a fourth on the 21 July 746 BC. This final eclipse was total, but only rose above the ground just as it began to clear. It is therefore questionable whether the total phase of this eclipse could have been seen.

The first set of dates, that of 801–800 BC, gives the best fit to the observation.

However, they would suggest that the first month of the Babylonian year began around 20 January. Typically, the Babylonian year began around the end of March or the beginning of April. The second set of dates would result in a Babylonian year beginning around 20 February, which, whilst still being rather early, is more plausible than 20 of January. It is also known that 747 BC was the accession year of Nabonassar, which would fit the mention of an accession year in line 1. Unfortunately, the Babylonian chronology of the period around 800 BC is insufficiently well known to comment as to whether 801 BC was also an accession year. I am therefore forced to tentatively accept Sachs' (1955) date for LBAT 1413 of 747–746 BC.

2.4 Units of Time

The earliest unit of time attested in Babylonian sources is the EN.NUN, translated as "watch."[63] Each day contained six watches, three for the day and three for the night. Thus the watch was a seasonal time unit; it varied in length both with the time of year and with whether it was during the night or day. At the equinox, the day and night watches were equal in length. This length of one sixth of a day was defined to be equal to 2 *bēru*. The *bēru* is an equinoctial unit of time; it does not vary in length with the seasons. The original definition of the *bēru* was probably as a distance of length, corresponding to about seven miles, but it also came to be a measure of the time it takes to travel this distance.[64]

The *bēru* was divided into thirty smaller units called UŠ. There were 360 UŠ in a day and so one UŠ was precisely equal to 4 minutes of time.[65] The UŠ is therefore equivalent to the "time-degree" used by Ptolemy, and so it has become customary to render the term as "degree" in translations. The UŠ was itself subdivided in sixty parts called NINDA. By the Late Babylonian period, the UŠ had become the fundamental unit of time used in both mathematical and non-mathematical astronomy.

From the Old Babylonian period there exist a number of mathematical problems that have been interpreted as referring to the outflow from a clepsydra.[66] Clepsydras are also mentioned in a diviner's manual in the context of measuring time intervals during lunar observations,[67] and so it seems likely that they were used extensively in Babylon.[68] However, it is not known exactly what form the clepsydras took.[69] The situation is made additionally confusing by a number of tables which record the length of the night throughout the year. These tables vary in the number of intervals into which the year was divided, but all of them fall into one of two categories, that is they

[63] Watches are found, for example, in the omen series *Enūma Anu Enlil*, see Rochberg-Halton (1988: 44–47); and in the letters and reports sent to the Assyrian king, see Parpola (1993) and Hunger (1992).

[64] Neugebauer (1941).

[65] Stephenson & Fatoohi (1994).

[66] Thureau-Dangin (1937).

[67] Oppenheim (1974).

[68] Although there exist tables which have been interpreted as referring to measuring time by use of a gnomon, devices such as this could not have been used during the night.

[69] See Brown, Fermor, & Walker (1999) for a general discussion of clepsydras in Mesopotamia.

are based on the assumption that the ratio between the longest and the shortest night is either 2:1 or 3:2, and generally make use of the "ideal" year of 360 days split into twelve 30-day months. The former ratio is attested in early texts such as MUL.APIN and tablet 14 of *Enūma Anu Enlil*,[70] whilst the latter is implied in an explanatory work known as i.NAM.giš.ḫur.an.ki.a,[71] ACT, and other sources. Neugebauer (1947a) and others have tried to reconcile these two ratios by arguing that the 2:1 ratio was not a ratio of time, but of the weight of water contained in a cylindrical outflowing clepsydra; however, the type of clock proposed by Neugebauer would not work in practice.[72] Furthermore, Pingree & Reiner (1977) have discovered a tablet from the seventh century BC which shows that the 2:1 ratio does indeed refer to time. This ratio is very inaccurate for the latitude of Babylon, and it is hard to see how any clock could have measured it. Instead, it would seem that it is simply an ideal ratio expressing the fact that night-lengths vary throughout the year using small numbers. The later 3:2 is much closer to the true ratio; in fact it is the best ratio that can be obtained using single digit numbers.

Adding further confusion to the matter is the statement by Herodotus that "knowledge of the sundial and the gnomon and the twelve divisions of the day came into Greece from Babylon."[73] This seems to suggest that seasonal hours were used in Babylon, as well as the equinoctial time systems just discussed. However, there is only a small amount of evidence that this was the case. Pingree & Reiner (1977) have interpreted their tablet from the seventh century BC as a table giving the number of UŠ and NINDA in a seasonal hour. Similarly, Fotheringham (1932) claimed that there was a table for converting from *bēru* and UŠ to seasonal hours on a small, badly damaged fragment that has become known as the "Ivory Prism."[74] Finally, Rochberg-Halton (1989b) has claimed that seasonal hours were used in some of the Babylonian horoscopes produced in the Seleucid period. Van der Waerden (1951) has called these seasonal hours "popular units of time" which were used by people in their everyday lives, as opposed to the *bēru* and UŠ which were "astronomical units of time" used by the astronomers. However, there is such limited evidence for the use of seasonal hours in Babylon that I would disagree with his conclusions, and suggest that if there was any "popular" unit of time, which may well be an anachronistic notion in any case, it was the watch which is well attested in the literature.[75]

As I mentioned in Section 2.3 above, the Babylonians began their day at sunset.

[70]MUL.APIN has been published by Hunger & Pingree (1989), and the relevant parts of *Enūma Anu Enlil* by Al-Rawi & George (1992).

[71]i.NAM.giš.ḫur.an.ki.a has been published by Livingstone (1986).

[72]This is mainly due to the surface tension of the water preventing the clock from fully emptying. For the experimental evidence of this statement, see Brown, Fermor, & Walker (1999), and for a theoretical argument, see Høyrup (1998).

[73]*Historia*, II, 109; trans. de Sélincourt (1979: 169).

[74]Fotheringham's reconstruction of the Ivory Prism was published by Langdon (1935), and also discussed by Smith (1969). Fotheringham's interest in the Ivory Prism was in the possibility of its use in the eclipse records reported by Ptolemy in his *Almagest*, and so I will defer further discussion of it until Chapter 3.

[75]Of course, a seasonal hour is simply a quarter of a watch, but I do not think it was ever extensively used as a unit of time in its own right.

However, they did not measure all times from this point. Instead, the day was divided into four parts: sunset to midnight, midnight to sunrise, sunrise to midday, and midday to sunset. Time was measured relative to sunrise or sunset within each division, that is as a number of UŠ after sunset (GE$_6$ GIN), before sunrise (GE$_6$ *ana* ZALAG), after sunrise (ME NIM), or before sunset (*ana* ŠÚ *šámaš*).[76] To my knowledge, nowhere in Babylonian literature is there a definition of what is meant by sunrise or sunset. Britton (1992) has assumed that it was when the middle of the sun was crossing the horizon. However, most cultures have defined sunrise or sunset to be the moment when the upper limb of the sun crosses the horizon (about $1\frac{1}{2}$ minutes earlier or later), and so this has been assumed by many other authors. Fortunately, it is possible to determine with some confidence which definition was adopted by the Babylonians from their lunar six measurements.

Some years ago, the late Prof. A. J. Sachs compiled an extensive list of lunar six measurements and predictions from the Astronomical Diaries and the Goal-Year Texts dating from after 300 BC. This unpublished manuscript has formed the basis of the following investigation.[77] To supplement it I have extracted many further lunar six records from the earlier Diaries, and from two tablets, LBAT 1431[78] and LBAT 1433, which contain collections of lunar six data from parts of the years 323–319 BC and 241–239 BC respectively. Copies of these tablets by T. G. Pinches have been published by Sachs (1955). Due to the scarcity of data before 400 BC, only records from after this date will be considered. Furthermore, only those lunar six timings that appear to have been measured have been considered.[79]

Initially assuming that sunrise and sunset, and moonrise and moonset, were defined as the moment when the upper limb of the luminary crossed the horizon, in Figure 2.6 I show the error in each of the measured lunar six intervals. The mean value of the error in each case is shown as the dotted line on the plot. The lunar six measurements made around new moon turn out to have greater errors than those observed around full moon. This is probably caused by two factors: the difficulty of observing the thin lunar crescent as it nears the horizon, and the fact that the new moon intervals are typically of greater length than their the full moon counterparts, which leads to larger errors if there are clock drifts in the clepsydras used.

As it takes both the sun and the moon about 3 minutes to set at the latitude of Babylon, the length of the full moon intervals will depend upon the definition of the

[76] Neugebauer & Sachs (1967).

[77] Sachs sent this manuscript to F. R. Stephenson, who generously made it available to me.

[78] The lunar six data recorded on LBAT 1431 have previously been investigated by Stephenson (1974) who found evidence for systematic clock drifts in the measurement of each of the individual lunar six. He concluded that the Babylonian astronomers may have used a slightly different clock, possibly labeled for the purpose, to measure each of the six intervals.

[79] The lunar six measurements may be divided into three groups: those said to have been measured using the term *muš*, those said to have been not seen using the term NU PAP, and those that have no comment attached. Unless there is some mention of bad weather in the record, I have assumed that the timings in the third category were measured. This may have caused some predicted material to be included in the analysis, but as there is no significant change in the result if this group is ignored, it would appear that, on the whole, these do indeed represent measured lunar six values.

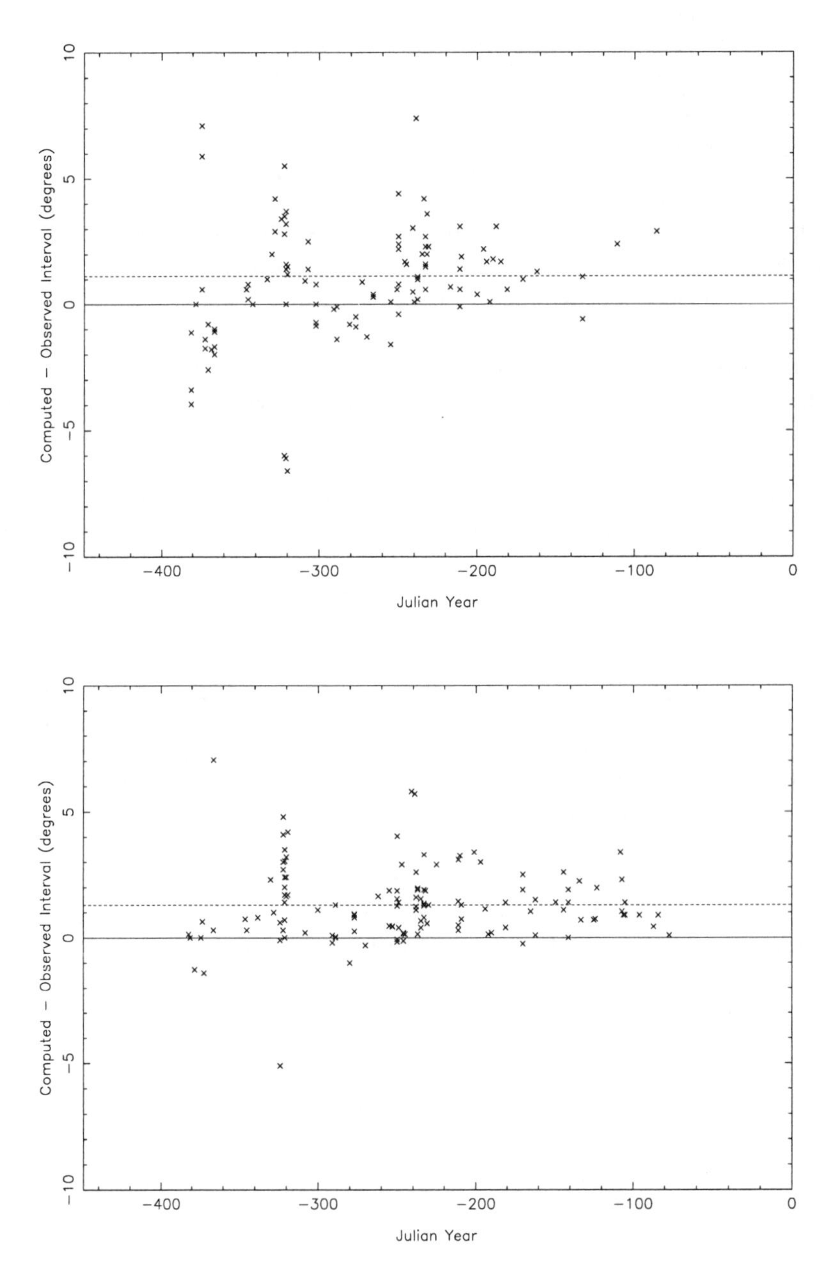

Figure 2.6: The error in the measured lunar six intervals (a, top) *na* and (b, bottom) ŠÚ.

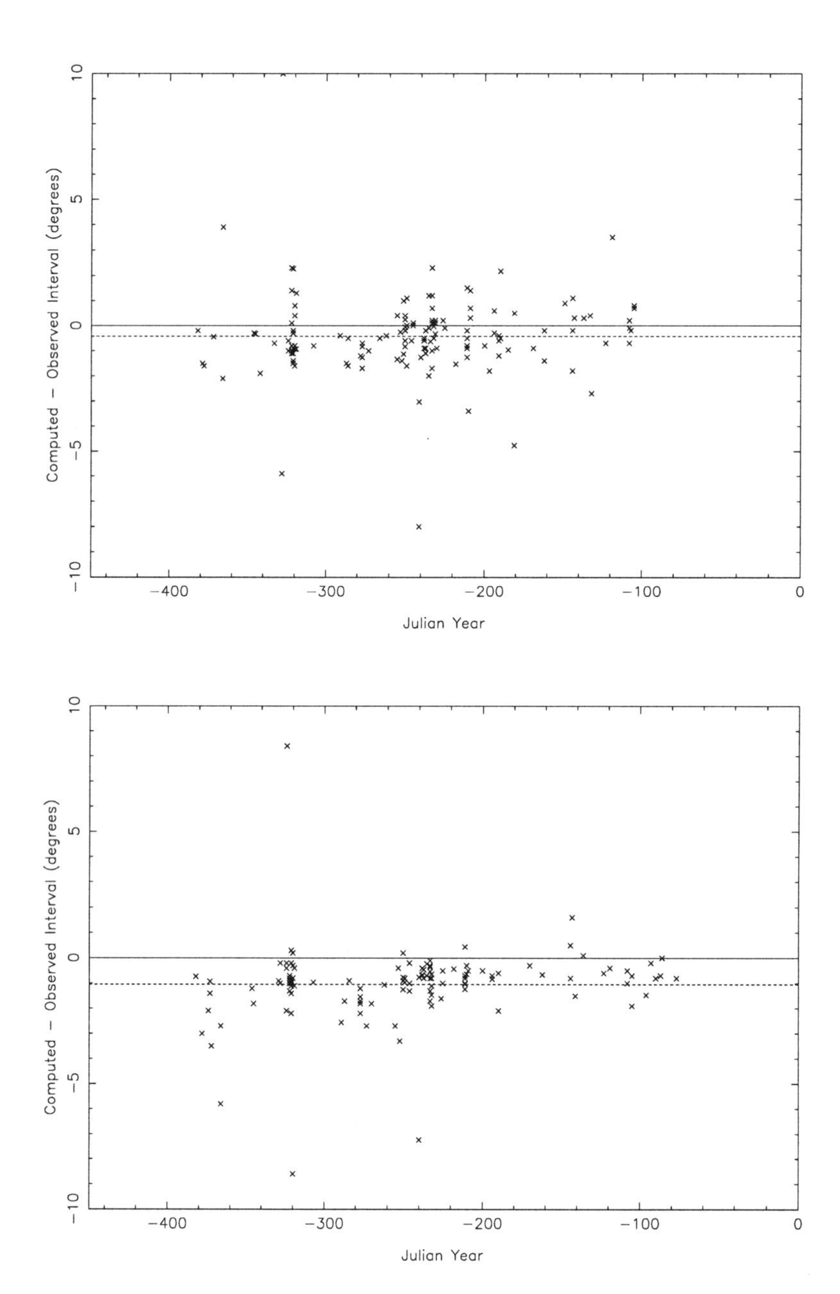

Figure 2.6 (cont.): The error in the measured lunar six intervals (c, top) *na* and (d, bottom) ME.

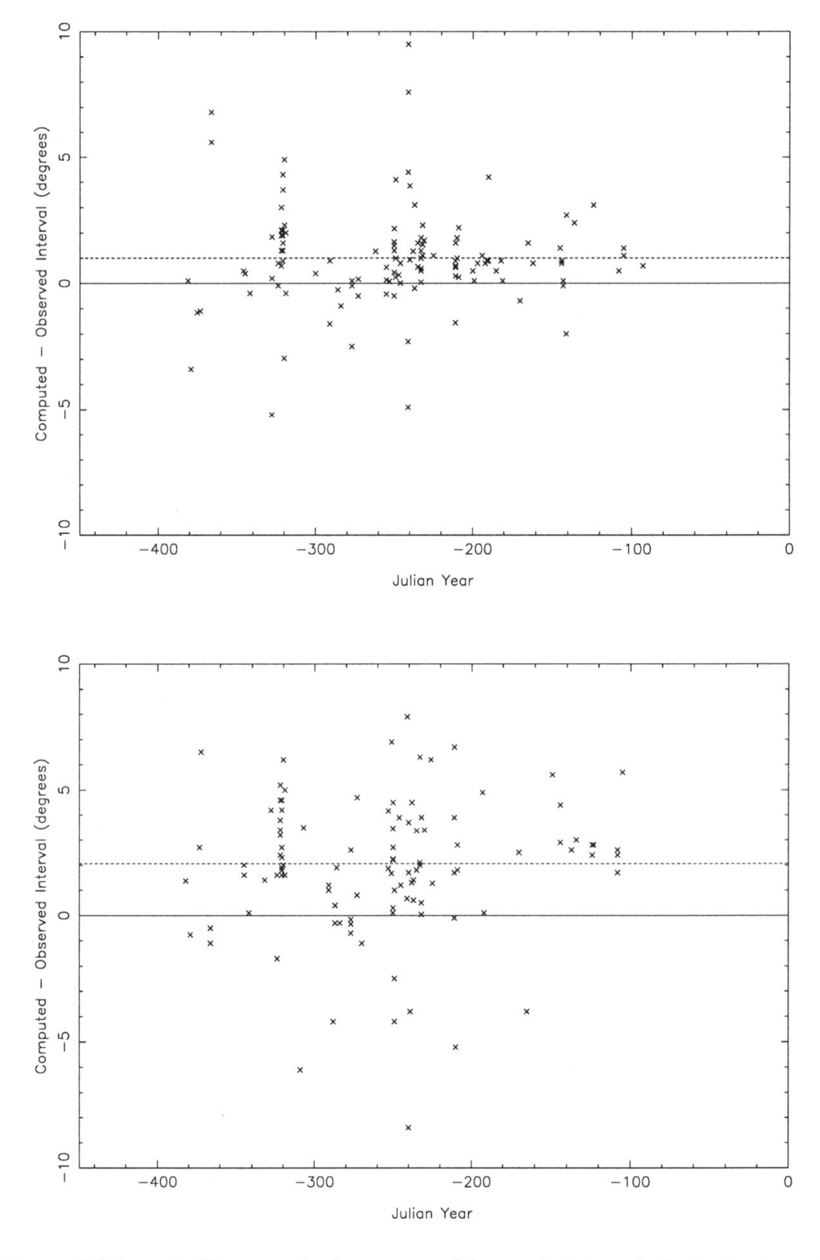

Figure 2.6 (cont.): The error in the measured lunar six intervals (e, top) GE_6 and (f, bottom) KUR.

Lunar Six	Mean Error A (°)	Mean Error B (°)
na	1.1	1.1
ŠÚ	1.3	2.0
na	-0.4	-1.1
ME	-1.1	-1.7
GE6	1.0	1.7
KUR	2.1	2.1

Table 2.3: Mean errors in the lunar six measurments assuming (A) that the time of rising and setting was defined as the moment when the upper limb of the luminary crossed the horizon, and (B) that the time of rising and setting was defined as the moment when the middle of the luminary crossed the horizon.

time of rising or setting of the luminary. The new moon intervals are not affected by changing this definition as they concern the same phenomena for the sun and moon. If the alternate definition of rising or setting as the moment the middle of the luminary crossed the horizon is accepted, two of the full moon intervals, ŠÚ and GE6, would be increased by about 0.7 UŠ, and the other two, *na* and ME, would be decreased by the same amount. The effect of making this assumption to the mean errors of the measurements of each of the lunar six is shown in Table 2.3. Clearly, the mean error is minimized when the first assumption is adopted, strongly suggesting that this is the correct definition.[80] Therefore, throughout the rest of this chapter, the time of sunrise or sunset will be assumed to have been defined by the upper limb of the luminary.

During the night, time can also be measured by means of the fixed stars. In the Islamic Near East, for example, times of astronomical observations were often determined from a measurement of the altitude of a bright star.[81] However, the Mesopotamian astronomers chose to use the culmination (i.e., the crossing of the meridian near the zenith) of a group of bright stars. The Akkadian term for this is *ziqpu*, and so the stars which were used have become known as *ziqpu*-stars.

The earliest list of *ziqpu*-stars is found in MUL.APIN, I, iv, 1–30.[82] Here are detailed the constellations which culminate on different dates of the ideal calendar, and, separately, thirteen dates on which a given star culminates at the same moment that another given star rises. Several later lists are preserved, the most well known being AO 6478, which associate a weight in *mina*s and a value in UŠ with each *ziqpu*-star.[83] These almost certainly relate to the time intervals, measured in both weights of water and UŠ (with a constant ratio of 1 *mina* equal to 6 UŠ), between the culminations of the consecutive *ziqpu*-stars, presumably measured by means of a clepsydra.[84] Also given are values in *bēru* which appear to be some measure of (although not necessarily

[80]The remaining errors are presumably due to the inherent inaccuracies of the clepsydras used for the measurements.

[81]See Section 4.3 below.

[82]See Hunger & Pingree (1989: 57–67 & 141–144).

[83]For these texts, see Schaumberger (1952, 1955) and Horowitz (1994).

[84]For an alternative interpretation of the values given in UŠ, see Brown (1999a).

measured) spatial distance between consecutive stars.

Using lists that give the time intervals between the successive culminations, it would be a fairly straightforward task to obtain time measurements throughout the night. All one would need to do is observe which star was culminating, or close to culminating, at sunset, and then look up the difference in UŠ between this culminating star and the star that culminated at the time of the observation. There are, however, practical difficulties in using the *ziqpu*-stars in this way, not the least of which is establishing a meridian line to determine exactly when stars are culminating.

The *ziqpu*-stars were being used for timing events by at least the Neo-Assyrian period. A letter from the time of Sargon II reports a storm during the night which is said to start at the culmination of the constellation known as "the Circle" and end at the culmination of the constellation "the Triplet."[85] Also, two Assyrian letters describing lunar eclipses, LABS 134 and LABS 149, mention that the eclipse took place as particular stars culminated. In the Late Babylonian NMAT texts, *ziqpu*-stars are sometimes used to note the time of lunar eclipses. The preserved examples all date from between 226 BC and 80 BC.

2.5 Eclipse Records in Late Babylonian History

Late Babylonian eclipse records are found in all types of the NMAT texts: Astronomical Diaries, Goal-Year Texts, Elipse Texts, Almanacs, Normal Star Almanacs, and Horoscopes. All of the references to dated lunar and solar eclipse observations and predictions in these texts are listed in Appendix A. In the present section I shall outline the terminology used in describing the eclipses, and make some general comments on the records contained in each type of source.

In the Late Babylonian Astronomical Texts, observed eclipses are denoted by either *sin* AN-KU$_{10}$, for a lunar obscuration, or *šamáš* AN-KU$_{10}$, for a solar obscuration. If the eclipse was predicted rather than observed, then the order was reversed: AN-KU$_{10}$ *sin*, for a lunar eclipse, or AN-KU$_{10}$ *šamáš*, for a solar eclipse.[86] Furthermore, predicted eclipses are usually described as being *šá* DIB ("omitted") meaning that they were not expected to prove visible as the sun or moon was below the horizon at the predicted time of the eclipse, or *ki* PAP NU IGI ("watched for, but not seen") when the anticipated eclipse failed to appear.

Both observations and predictions of lunar and solar eclipses are recorded in the Diaries, Goal-Year Texts, and the Eclipse Texts, following the general rule that if the eclipse was seen, the observation would be noted, but if not then the advance prediction would be inserted in its place. As mentioned in Section 2.2 above, it appears that eclipse records in the Goal-Year Texts and the Eclipse Texts were compiled from information contained in the Diaries.

[85] Lanfranchi & Parpola (1990: no. 249).

[86] The first discussion of this terminology was, I believe, by Kugler (1900b). More recently, see Sachs & Hunger (1988: 23).

The Eclipse Texts are of particular interest for not only are they the largest archive of eclipse records in Late Babylonian history, they may also offer some information on the schemes used by the Babylonian astronomers to predict future eclipses.[87] The Eclipse Texts may be divided into four categories: individual reports, straight-forward lists of consecutive eclipses and eclipse possibilities, lists arranged in Saros cycles, and "theoretical" texts listing the dates of eclipse possibilities. Among the tablets arranged in Saros cycles are three tablets, LBAT *1414, LBAT 1415 + 1416 + 1417, and LBAT *1419, which Walker (1997) has suggested were part of a set compiled by a single scribe and containing lunar eclipses and eclipse possibilities from twenty-four Saros cycles starting in 747 BC.

These three texts are all set out in rows and columns separated by horizontal and vertical rulings, giving them the appearance of a matrix. Each cell in this matrix contains details of one eclipse possibility. Reading down the columns, the eclipse possibilities are separated by six or occasionally five lunar months. The five month intervals are invariably highlighted by the phrase 5 ITU after the date. Along the rows, eclipse possibilities are separated by one Saros of 223 months. Unlike most tablets, these texts all turn sideways, with the rows continuing from the obverse onto the reverse. From the preserved fragments, we can reconstruct the whole of this large compilation. It would have contained 24 columns on each tablet, and 5 or 6 rows. Over 900 eclipse possibilities dating from 747 BC to 315 BC would have been contained in the compilation which probably stretched over 8 texts.

In the preserved fragments of these texts there are four observations of lunar eclipses that we can compute were only penumbral. In one of these cases, the moon is said to be near β Capricorni, but computations show that the moon was more than 80° away from this star at the time of the eclipse. It would therefore appear that these eclipses were wrongly filed in the compilation. However, if Walker (1997) is correct in his suggestion that the tablets formed part of the same series, which seems extremely likely, then the whole compilation contained over 900 entries, and so it is not surprising that there would be the occasional scribal mistake. Otherwise, though, the Eclipse Texts do contain real observations, where an eclipse was visible, and predictions when they were not.[88]

The observed accounts in the Diaries, Goal-Year Texts, and Eclipse Texts usually give details such as the time of the eclipse measured relative to sunset or sunrise or occasionally with reference to a culminating *ziqpu*-star, an estimate of the magnitude of the eclipse expressed in fingers or twelfths of the lunar or solar diameter, a measurement of the duration of the eclipse and its phases, and sometimes meteorological remarks.[89] The visibilities of stars and planets may also be noted during the eclipse. For example, on 1 August 226 BC:

"Night of the 14th, moonrise to sunset: 4°, measured (despite) mist; at 52° after

[87] I will discuss this aspect of the Eclipse Texts in Section 2.8 below. A fuller description of the structure of the Eclipse Texts is given in my contribution to Sachs & Hunger (1999).

[88] Contrary to the statement by Rochberg-Halton (1989b: 148).

[89] These remarks concern not only factors such as clouds that may have affected the observation, but also wind directions which play an important role in the astrological interpretation of the eclipse.

> sunset, when α Cygni culminated, lunar eclipse; when it began on the east side, in 17° nighttime it covered it completely; 10° nighttime maximum phase; when it began to clear, it cleared in 15° nighttime from south to north; in (its) onset it was slow, in (its) clearing it was fast; 42° onset, maximum phase, and clearing; its eclipse was red; (in) its eclipse, a gusty north wind blew; (in) its eclipse, all the planets did not stand there; 5 cubits behind δ Capricorni it became eclipsed."
>
> [BM 33655, Rev. 4–8; trans. Sachs & Hunger (1988: 141)]

It is not immediately clear in this example whether the time of 52° after sunset refers to the beginning, the middle, or the end of the eclipse. However, it seems most natural to suppose that it is the time of the start of the eclipse, and this is confirmed by, for example, the record of the lunar eclipse on 22 November 353 BC:

> "Month VIII, the 14th. When it began on the south and east side, in 23° all was covered. 18° maximal phase. (After) 16° of night, one-fourth on the east side cleared; it set eclipsed. The eclipse was red. $1\frac{1}{2}$ cubits behind ζ Tauri it was eclipsed. During the eclipse Saturn stood there; the remainder of the planets did not stand there. The north wind which was slanted to the west blew. At 47° before sunrise."
>
> [LBAT *1414, Rev. III', 2'–10'; trans. Sachs & Hunger (1999)]

In this example, the moon was totally eclipsed. It took 23° for the eclipse to become total, and totality lasted for 18°. 6° after the end of totality, the moon set. Therefore, first contact is 23° + 18° + 6° = 47° before the moon set, the same time as stated at the end of the report, as moonset and sunrise were more or less simultaneous on this date. Hence it may be inferred that, in the general case, the time stated in a report of an eclipse observation is the time that the eclipse began.

In the lunar eclipse reports, it is customary for the Babylonian astronomers to give timings of the durations of the following phases: onset, maximal phase, and clearing. For total eclipses it is easy to understand that these relate to the time between the start of the eclipse and totality, the duration of totality, and the end of totality to the end of the eclipse. However, the Babylonian astronomers also give timings of these three phases for partial lunar eclipses. Thus, they defined a "maximal phase" for a partial eclipse when they were unable to detect any change in the extent of the shadow. This is a purely psychological and physiological effect dependent of the acuity of the eye of the observer, and so it will be impossible to compare any timings of such phases with modern computations.

There are only two texts which mention measurements of a maximal phase for a solar eclipse; one is in the report of the total eclipse on 15 April 136 BC, and the other is in the description of the eclipse on 30 June 10 BC. The record of the eclipse on 15 April 136 BC is one of the most remarkable of all historical eclipse observations for it is the only example of a detailed timed observation of a total solar eclipse in the ancient world. It is recorded on two tablets, a Diary (LBAT *429) and a Goal-Year Text (LBAT 1285):

> "The 29th, at 24° after sunrise, solar eclipse; when it began on the south and west side, [... Ven]us, Mercury, and the Normal stars were visible; Jupiter and Mars,

which were in their period of invisibility, were visible in its eclipse [...] it threw off (the shadow) from west and south to north and east; 35° onset, maximal phase, and clearing."

[LBAT *429, Rev. 13'–15'; trans. Sachs & Hunger (1996: 185)]

"Month XII_2 ... The 29th. Solar eclipse, when it began on the south-west side, in 18° of daytime in the morning it became entirely total. At 24° after sunrise."

[LBAT 1285, Rev. 24–28; trans. Hunger (personal communication)]

The fact that the report of the eclipse on 30 June 10 BC, which is quoted above in Section 2.3.1, also includes a timing of a maximal phase strongly suggests that it was a total eclipse. Other solar eclipse reports merely report the durations of two phases of the eclipse: the beginning to the middle (or more likely maximum) of the eclipse, and from the middle to the end. For example, the report of the eclipse on 31 January 254 BC:

"The 28th, 56° before sunset, solar eclipse; when it began, in 12° of daytime [...] when it began [to cl]ear, it cleared from south to north in 11° daytime; 23° onset and clearing; during its eclipse blew the west wind which was slanted to the north."

[LBAT 596 + 258, Rev. 11–12; trans. Sachs & Hunger (1989: 29)]

It is interesting to note that in a number of the examples where durations of onset and clearing are preserved, the two time intervals do not match. This provides strong evidence that the times were indeed measured, for if they were calculated by the Babylonian astronomers, it is unlikely that they would have obtained different durations for these phases.

As I have mentioned, reports of observed eclipses usually include the time and duration of the eclipse and its magnitude. In addition, they generally include a description of the colour of the eclipse, the path of the eclipse shadow during the course of the eclipse, the location of the eclipsed moon, the visibility of stars and planets, and the direction of the wind at the time of observation. With the exception of the stellar and planetary signs, these are the same factors that are commonly mentioned in the eclipse omens of *Enūma Anu Enlil*. Indeed, tablet 4 of the series *Šumma Sin ina tāmartīšu*, which describes how one must observe and interpret an eclipse omen, reads almost like instructions for how to observe an eclipse for a Diary:

"When Sin (i.e., the moon) makes an eclipse, you must also consider the month, the day, the watch, the wind, the path (i.e., area of the sky in which it took place), and the positions of the stars as they stood during the eclipse, and then you can give [... the deci]sion in accordance with its month, its day, its watch, its wind, its path, and its star."

[ACh 2. Supl. Sin 19, 9'–20'; trans. Koch-Westenholz (1995: 106–107)]

In fact, we seem to have a direct link between the factors that affected the interpretation of an eclipse in the *Enūma Anu Enlil* omens and the details of the observations that were recorded in the Diaries. An even closer agreement is found between the eclipse reports in the Diaries and the omens in a late astrological text, BM 36746.[90] In this text, the visibility of stars and planets during the eclipse form an important part of the omen protases. For example:

> "If the moon is eclipsed in Leo and finishes the watch and the north wind blows, Jupiter does not stand (in) the eclipse; Saturn and Mars stand in Aries or in Sagittarius or the Field; variant: in its eclipse [a halo surrounds (the moon) and Regulus stands within it]. For this sign: [the king] of Akkad will experience hardship/*šibbu*-disease; variant: it will seize him, and in a revolt they will oust him from his throne. His people will experience great famine; brother will kill his brother, friend his friend, in battle. For three years [...] will not return [to the throne of Akkad]; the gods will [abandon] the country..."
>
> [BM 36746, Obv. 5'–6'; trans. Rochberg-Halton (1984: 136)]

Returning to the actual accounts of the eclipses in the Diaries and related texts, there is a complete contrast to the detailed descriptions of the observed eclipses in the accounts of predicted eclipses.[91] These are usually very brief stating no more than the time that the eclipse was expected to have occurred, and whether it had been watched for and not seen, or had been omitted. For example, the account of the predicted eclipse on 26 May 194 BC is typically brief:

> "Night of the 29th, at 1,17° after sunset, solar eclipse which was omitted."
>
> [LBAT *322, Rev. 5'; trans. Sachs & Hunger (1989: 277)]

Comparison with the preceding examples of observational records suggests that the eclipse was expected to begin at 1,17° (= 77°) after sunset. There are also a number of examples where the record contains a mixture of an observation and a prediction. For example, on 19 June 67 BC the moon was observed to rise eclipsed. The end of the report gives the time when the eclipse began, which clearly must calculated:

> "Year 180, which is year 244, Arsaces king and Piriwuštanā, his wife, queen. Month X, night of the 15th, moonrise to sunset: 1°, measured (despite) mist. When the moon came out, two thirds of the disk on the north and east side were covered. 6° of night maximal phase. When it began to clear, it cleared in 16° of night from south and east to north and west. 23° maximal phase and clearing. Its eclipse had the 'garment of the sky.' In its eclipse, the north wind blew. In its eclipse, Venus, Saturn and Sirius stood there; the remainder of the planets did not stand there. $1\frac{1}{2}$ cubits in front of α Leonis it was [eclips]ed. At 16° before sunset."
>
> [LBAT 1448; trans. Sachs & Hunger (1999)]

As I have mentioned in Section 2.2 above, the Almanacs and Normal Star Almanacs contain purely predicted material. Thus the eclipse records they contain are advance predictions that are either to be watched for, or which will be omitted. Occasionally, a preserved Diary or other source will contain an observation of an eclipse

[90] BM 36736 has been published by Rochberg-Halton (1984).

[91] It is trivial to note that, of course, not all of the observed accounts will include all of the items I have discussed here, and furthermore that many reports are only partially preserved.

that was predicted in an Almanac or Normal Star Almanac.[92] For example, the lunar eclipse on 3 September 134 BC, was predicted in the Almanac LBAT 1134, and was recorded as observed in the Diary LBAT *435:

> "Night of the 15th, 1 *bēru* after sunset, lunar eclipse."
> [LBAT 1134, Obv. 12']
>
> "[...] when it began [to clear], in 10° night it cleared from east to west ... [...] at 32° after sunset."
> [LBAT *435, Obv. 8'–9'; trans. Sachs & Hunger (1996: 199)]

It is interesting to note that in this case the Babylonian astronomers would have noted only a small difference between their predicted and observed times. 1 *bēru* equals 30 UŠ and so there is only 2 UŠ between the prediction and the observation.

It seems reasonable to suppose that the eclipse predictions made at any particular period would have been made using the same methods. This is supported by the fact that where we have overlapping texts, they contain eclipse predictions on the same dates. For example, predictions of a solar eclipse on 7 June 111 BC are found in both the Normal Star Almanac LBAT **1059 and in the Almanac LBAT 1151:[93]

> "The 29th, 62° before sunset, solar eclipse"
> [LBAT **1059, Obv. 13]
>
> "The 29th, 2 *bēru* before sunset, solar eclipse"
> [LBAT 1151, Obv. 6']

Evidently, the details of these two predictions are almost identical, the only discrepancy being in the expected time of the eclipse. However, again since 1 *bēru* equals 30 UŠ, this discrepancy is small and I would suggest that both predictions were made using the same procedure and that the time has simply been rounded down to the nearest *bēru* in the Almanac.

Finally, a word needs to be said about the eclipse records in the Horoscopes. The eclipses are exclusively denoted by the form AN-KU$_{10}$ *sin/šamás*,[94] which is normally used for predicted eclipses in the NMAT. It seems quite possible that these eclipse predictions were taken, at least in part, from either the Almanacs or the Diaries.

2.6 Accuracy of the Observed Times

Of the many reports of lunar and solar eclipses in Late Babylonian history, more than one-hundred contain a measurement of either the time interval between the beginning

[92]Unfortunately, due to the vagaries of preservation, there is only a small amount of overlap between years contained on dated Diaries, and those on the dated Almanacs or Normal Star Almanacs. See Hunger (1999).

[93]The similarity of the records is even clearer when one considers the texts themselves. LBAT **1059, Obv. 13: 29 1 2 ME *ana šamáš* AN-KU$_{10}$ *šamáš*; LBAT 1151, Obv. 6': 29 2 KASGAL-BU ME *ana šamáš* AN-KU$_{10}$ *ša*[*máš*].

[94]Rochberg (1998: 40).

of the eclipse and sunrise or sunset, or the duration of the phases of the eclipse, or in some cases both. These times, we assume, were measured using some form of clepsydra. By comparing these measurements with modern computations it is possible to evaluate the accuracy with which the Babylonian astronomers were able to time the eclipses, and to note any trends in the errors in the clocks that they used. Previously, Stephenson & Fatoohi (1993) have analysed many of the lunar eclipse timings and found some evidence for clock drifts of about 13%. However, their analysis contains significant errors, in part caused by the use of a preliminary approximation to the Earth's rotational clock error, ΔT, and in part by double counting some of their data. In addition to the times measured using a clepsydra, 16 reports dating between 225 BC and 80 BC mention that the eclipse began as a particular *ziqpu*-star (or occasionally a point so many degrees in front of or behind a star) was culminating.[95]

Tables 2.4 and 2.5 list respectively the lunar and solar eclipses for which at least one timing of either the beginning of the eclipse or a phase duration by a clepsydra is preserved. Also given are the equivalent times as deduced from modern computations, rounded to the nearest 0.25°. As I mentioned in Section 2.5 above, a number of records of partial lunar eclipses contain estimates of the duration of a maximal phase when no apparent change in the extent of the eclipse could be detected. However, as this is a physiological effect dependent on the observer, these phases cannot be compared with computation. Instead, if the onset and clearing phases have also been reported, then the three intervals have been added together to give the total duration of the eclipse. If the total duration and all of the phases of an eclipse are reported, then only the individual phases are given, except as noted above. In all cases, times in these tables are given in UŠ, even if the original record quotes the times in *bēru*.

It is immediately evident from Table 2.4 that before about 570 BC, most of the measured times appear to have been rounded to the nearest 5°. This suggests that the Babylonian astronomers had a growing confidence in their timing methods after this period. However, from Figure 2.7, which shows the error in all of the measured lunar and solar eclipse times, it is clear that there was no corresponding improvement in the accuracy of the times. This implies that any changes in the clocks used by the Babylonian astronomers around 570 BC resulted only in an improvement in precision of quoted measurement and not in real accuracy. In fact, from Figure 2.7 it is clear that over the whole of the Late Babylonian period, there was no improvement in the accuracy of the eclipse timings. Furthermore, there is no significant difference between the accuracy of the solar eclipse timings and the lunar eclipse timings. This suggests that the same devices were used to time both types of event, and that it was these devices, and not factors such as the difficulty in determining lunar eclipse contacts due to the diffuse nature of the Earth's shadow, which was the limiting factor in the accuracy of timing eclipses.

Figure 2.8 shows the error in the measured times plotted against the month of the year in which they were observed. This should reveal any seasonal influences on the accuracy of the timings. Possible causes of this may include changes in the

[95] A few more reports from this period make reference to *ziqpu*-stars, but are too damaged to be analysed.

Date	Observed Details					Computed Details				
	1st Contact	Phase Durations				1st Contact	Phase Durations			
	Interval	1st to 2nd	2nd to 3rd	3rd to 4th	1st to 4th	Interval	1st to 2nd	2nd to 3rd	3rd to 4th	1st to 4th
-685 Apr 22	100° after sunset	-	-	-	-	111.00° after sunset	-	-	-	-
-684 Oct 3	20° after sunset	-	-	-	-	16.75° after sunset	-	-	-	-
-631 May 24	-	-	-	-	20°	-	-	-	-	19.25°
-602 Oct 27	45° after sunset	-	-	-	45°	37.25° after sunset	-	-	-	38.75°
-600 Apr 11	95° after sunset	-	-	-	-	93.25° after sunset	-	-	-	-
-598 Feb 20	105° after sunset	-	-	-	-	88.25° after sunset	-	-	-	-
-593 May 23	10° after sunset	-	-	-	-	24.00° after sunset	-	-	-	-
-591 Apr 2	-	-	-	-	36°	-	-	-	-	36.00°
-587 Jan 19	20° before sunrise	-	-	-	-	30.25° before sunrise	-	-	-	-
-586 Jan 8	35° before sunrise	-	-	-	-	32.25° before sunrise	-	-	-	-
-579 Aug 15	45° after sunset	-	-	-	-	49.50° after sunset	-	-	-	-
-576 Dec 8	105° after sunset	-	-	-	-	110.00° after sunset	-	-	-	-
-575 Jun 3	40° before sunrise	-	-	-	-	31.50° before sunrise	-	-	-	-
-572 Apr 2	90° after sunset	-	-	-	-	90.25° after sunset	-	-	-	-
-561 Mar 3	90° after sunset	-	25°	18°	-	82.50° after sunset	-	26.00°	15.75°	-
-554 Oct 6	55° after sunset	17°	28°	20°	-	55.75° after sunset	16.25°	24.50°	16.25°	-
-536 Oct 17	14° before sunrise	-	-	-	-	20.00° before sunrise	-	-	-	-
-528 Nov 17	24° after sunset	-	-	-	45°	44.00° after sunset	-	-	-	43.25°
-525 Sep 17	60° after sunset	18°	14°	-	-	68.75° after sunset	14.50°	23.75°	-	-
-500 Nov 7	77° after sunset	15°	25°	25°	-	69.25° after sunset	16.75°	23.50°	16.75°	-
-482 Nov 19	10° before sunrise	-	-	-	-	6.75° before sunrise	-	-	-	-
-464 Jun 5	-	-	-	-	40°	-	-	-	-	46.25°
-423 Sep 28	50° after sunset	-	-	-	50°	53.50° after sunset	-	-	-	35.00°
-409 Dec 21	-	-	-	-	60°	-	-	-	-	47.75°
-407 Oct 31	15° after sunset	-	-	-	27°	14.25° after sunset	-	-	-	22.75°
-406 Oct 21	48° before sunrise	-	-	-	56°	49.25° before sunrise	-	-	-	55°

Table 2.4: Lunar eclipse timings.

Date	1st Contact Interval	Phase Durations 1st to 2nd	2nd to 3rd	3rd to 4th	1st to 4th	1st Contact Interval	Phase Durations 1st to 2nd	2nd to 3rd	3rd to 4th	1st to 4th
-405 Apr 15	-	25°	19°	-	-	-	16.25°	18.50°	-	-
-405 Oct 10	14° before sunrise	-	-	-	-	9.25° before sunrise	-	-	-	-
-396 Apr 5	48° after sunset	-	-	-	27°	48.75° after sunset	-	-	-	16.25°
-377 Apr 6	-	15°	21°	19°	-	-	15.75°	19.00°	15.75°	-
-370 May 17	66° after sunset	-	-	-	-	57.00° after sunset	-	-	-	-
-370 Nov 11	30° after sunset	21°	20°	21°	-	38.50° after sunset	17.00°	21.00°	17.00°	-
-363 Jun 29	40° before sunrise	-	-	-	-	33.25° before sunrise	-	-	-	-
-363 Dec 23	14° before sunrise	-	-	-	-	11.75° before sunrise	-	-	-	-
-362 Jun 18	41° before sunrise	-	-	-	-	37.00° before sunrise	-	-	-	-
-352 Nov 22	47° before sunrise	23°	18°	-	-	41.50° before sunrise	17.00°	21.00°	-	-
-345 Jan 14	-	-	-	-	23°	-	-	-	-	23.75°
-326 Jan 14	-	-	7°	16°	-	-	-	21.50°	15.00°	-
-316 Jun 18	10° after sunset	-	-	-	-	15.75° after sunset	-	-	-	-
-316 Dec 13	44° after sunset	19°	5°	16°	-	54.00° after sunset	17.00°	20.75°	17.00°	-
-307 Jul 9	10° before sunrise	-	-	-	-	9.75° before sunrise	-	-	-	-
-283 Mar 17	-	22°	22°	-	65°	-	20.00°	11.00°	-	45.25°
-272 Feb 16	-	-	19°	22°	-	-	-	19.25°	15.75°	-
-239 Nov 3	3° before sunrise	-	-	-	-	1.00° *after* sunrise	-	-	-	-
-238 Apr 28	80° after sunset	-	-	-	40°	61.25° after sunset	-	-	-	35.50°
-225 Aug 1	52° after sunset	17°	10°	15°	-	70.75° after sunset	16.75°	16.00°	16.75°	-
-214 Dec 25	15° after sunset	21°	16°	19°	-	35.00° after sunset	15.75°	20.75°	15.75°	-
-211 Apr 30	20° before sunrise	-	-	-	-	24.50° before sunrise	-	-	-	-
-211 Oct 24	28° after sunrise	-	-	-	-	62.25° after sunrise	-	-	-	-
-193 Nov 5	12° before sunrise	-	-	-	-	8.00° before sunrise	-	-	-	-
-189 Feb 28	30° before sunrise	-	-	-	-	37.75° before sunrise	-	-	-	-
-188 Feb 17	34° before sunrise	16°	-	-	-	41.75° before sunrise	18.00°	-	-	-

Table 2.4 (cont.): Lunar eclipse timings.

Date	1st Contact	Phase Durations				1st Contact	Phase Durations			
	Interval	1st to 2nd	2nd to 3rd	3rd to 4th	1st to 4th	Interval	1st to 2nd	2nd to 3rd	3rd to 4th	1st to 4th
-184 Nov 24	44° after sunset	-	-	-	-	36.75° after sunset	-	-	-	-
-162 Mar 30	85° before sunrise	-	-	-	-	96.25° before sunrise	-	-	-	-
-159 Jan 26	48° after sunset	-	-	-	-	55.25° after sunset	-	-	-	-
-156 Nov 15	-	-	42°	-	-	-	-	24.25°	-	-
-153 Mar 21	4° after sunset	-	-	-	44°	6.50° after sunset	-	-	-	49.50°
-149 Jul 2	-	20°	12°	-	-	-	20.25°	13.00°	-	-
-142 Feb 17	7° after sunset	-	-	-	-	8.75° after sunset	-	-	-	-
-135 Apr 1	30° before sunrise	-	-	-	-	63.75° before sunrise	-	-	-	-
-134 Mar 20	-	-	-	-	60°	-	-	-	-	58.25°
-133 Mar 10	9° before sunrise	-	-	-	-	13.50° before sunrise	-	-	-	-
-133 Sep 3	32° after sunset	-	-	-	-	33.50° after sunset	-	-	-	-
-130 Jul 2	-	-	-	10°	-	-	-	-	16.75°	-
-128 Nov 5	55° before sunrise	-	-	-	40°	55.50° before sunrise	-	-	-	43°
-123 Aug 13	-	19°	24°	19°	-	-	16.75°	24.50°	16.75°	-
-119 Jun 2	66° after sunset	-	-	-	54°	68.75° after sunset	-	-	-	47.75°
-109 Nov 5	25° after sunset	-	20°	-	-	22.00° after sunset	-	25.00°	-	-
-108 May 1	8° after sunset	-	-	-	-	9.25° after sunset	-	-	-	-
-105 Feb 28	66° after sunset	-	-	-	60°	61.00° after sunset	-	-	-	53.00°
-105 Aug 24	50° before sunrise	21°	21°	-	-	44.50° before sunrise	16.50°	25.50°	-	-
-95 Aug 3	57° after sunset	-	-	-	-	63.50° after sunset	-	-	-	-
-86 Feb 28	-	-	-	-	30°	-	-	-	-	33.25°
-80 Apr 21	60° after sunset	-	22°	-	-	49.50° after sunset	-	26.25°	-	-
-79 Apr 10	40° before sunrise	-	-	-	40°	39.75° before sunrise	-	-	-	21.50°
-79 Oct 5	30° after sunset	-	-	-	-	32.50° after sunset	-	-	-	-
-72 Nov 16	37° before sunrise	-	-	-	-	30.50° before sunrise	-	-	-	-
-40 Mar 2	-	21°	-	-	-	-	15.00°	-	-	-

Table 2.4 (cont.): Lunar eclipse timings.

Date	Observed Details				Computed Details			
	1st Contact	Phase Durations			1st Contact	Phase Durations		
	Interval	1st to Max	Max to 4th	1st to 4th	Interval	1st to Max	Max to 4th	1st to 4th
-368 Apr 11	-	6°	-	-	-	15.00°	-	-
-321 Sep 26	3° before sunset	-	-	-	3.50° before sunset	-	-	-
-280 Jan 30	6° after sunrise	-	-	20°	2.75° after sunrise	-	-	23.25°
-255 Sep 6	-	-	-	32°	-	-	-	39.25°
-253 Jan 31	56° before sunset	12°	13°	-	64.00° before sunset	16.00°	15.00°	-
-248 May 5	90° after sunrise	-	-	-	93.25°	-	-	-
-241 Jun 15	-	-	18°	-	-	-	16.25°	-
-240 Nov 28	-	-	-	30°	-	-	-	36.50°
-194 Jan 6	60° after sunrise	-	-	-	38.25° after sunrise	-	-	-
-189 Mar 14	30° after sunrise	15°	15°	-	31.25° after sunrise	17.75°	19.00°	-
-169 Jul 28	20° before sunset	12°	-	-	19.50° before sunset	11.50°	-	-
-165 May 17	-	13°	-	-	-	23.50°	-	-
-135 Apr 15	24° after sunrise	-	-	35°	26.50° after sunrise	-	-	33.50°
-132 Feb 13	51° before sunset	20°	18°	-	50.50° before sunset	21.75°	19.25°	-
-125 Sep 9	-	-	-	35°	-	-	-	41.00°
-111 Jun 18	-	-	8°	-	-	-	14.25°	-
-88 Sep 29	45° after sunrise	-	-	24°	33.50° after sunrise	-	-	31.50°
-9 Jun 30	90° before sunset	-	-	48°	95.25° before sunset	-	-	40.00°

Table 2.5: Solar eclipse timings.

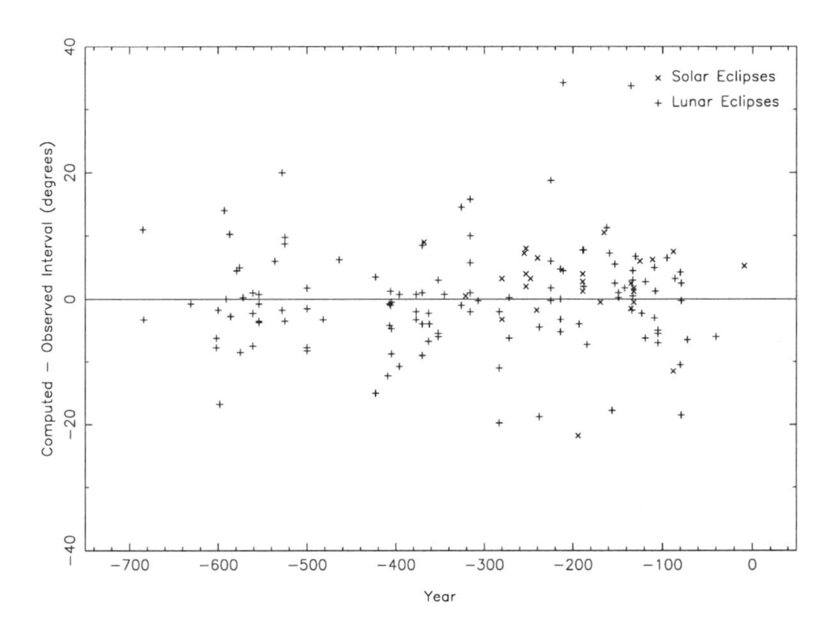

Figure 2.7: The error in the observed solar and lunar eclipse timings over the Late Babylonian period.

viscosity of the water used in a clepsydra with change in temperature, or use of a unit of time that was based upon the inaccurate 2:1 ratio for the longest to the shortest day used by the Babylonians. The annual range of temperature in Babylon today is more than 20°C, and there is also a considerable temperature gradient between sunset and sunrise. The change in water viscosity from the annual temperature range alone is more than 60%. Assuming that the discharge from a waterclock was characterized by laminar flow, this would imply that the discharge rate would also change by more than 60%. From Figure 2.8, however, it is clear that there was no significant seasonal effect on the accuracy of the timings. It is unlikely that the viscosity effect would be cancelled so effectively by a non-uniform time unit, and so we must look for other explanations of this null-effect. One possibility is simply that the clepsydras were insulated from temperature changes in some way — perhaps they were placed inside a windowless room, but this would undoubtedly cause practical problems in using the clock to time observations that would have to be made outside. A more likely answer lies in the design of the clocks themselves. There are certain conditions when laminar flow breaks down, the most important being that laminar flow requires a certain length of spout before it can fully develop. Below this length turbulence develops which may modify the relationship between temperature and rate of flow. In an attempt to clarify this issue, John Fermor of Glasgow Caledonian University has recently performed a

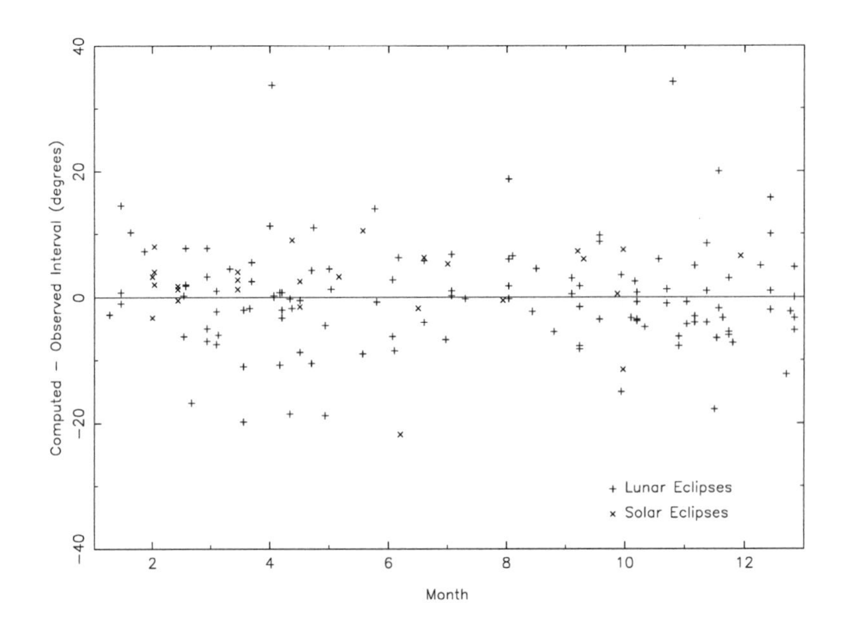

Figure 2.8: The error in the observed solar and lunar eclipse timings over the seasons.

series of experiments with model clepsydras.[96] He found that with sufficiently short and thin spouts, the rate of flow effectively become independent of temperature.[97] It seems quite possible that the Babylonian waterclocks were constructed with just such short spouts.

The lack of any seasonal change in accuracy provides yet further proof, if it were needed, that the UŠ was indeed invariant in length. One may also have expected that there would be a difference in accuracy between measurements made during the summer months, when there is usually little cloud on the horizon at Babylon and sunset or sunset are well defined, and those made during the winter months when, for example, clouds may have obscured these moments. Good weather is never mentioned in the Late Babylonian Astronomical Texts,[98] but there are a number of examples, such as the report of the lunar eclipse on 30 April 212 BC, when cloud partially interfered with the observation:

> "... Lunar eclipse, beginning on the south side. Around maximal phase cloudy, not observed. It set eclipsed. At 20° before sunrise."
>
> [LBAT **1237; Rev. 48–53; trans. Huber (1973: 56)]

[96] The results are reported in Fermor & Steele (1999).

[97] In particular he found that with a spout made of steel tubing of bore about 0.68mm and length 10mm, and a head of water 236mm, the time taken to discharge a fixed weight of water increased by only 6% over the temperature range 30 °C to 10°C. See Fermor & Steele (1999).

[98] Sachs (1974).

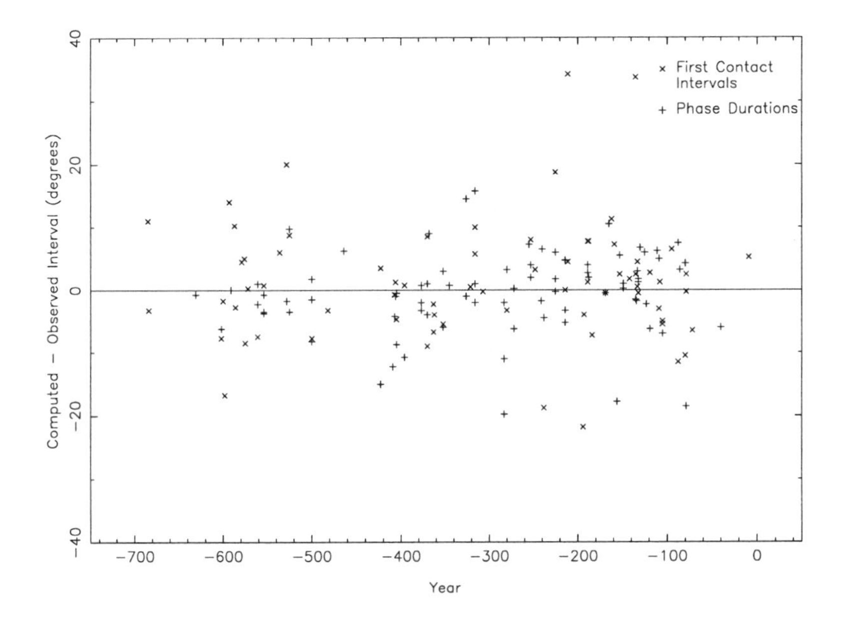

Figure 2.9: The error in the observed first contact intervals and phase durations.

Presumably the sky cleared sufficiently towards the end of this eclipse for moonset to be seen.

Unlike most early astronomers who made timings of eclipses, the Babylonians did not time all of the contacts of the eclipse relative to the same point. Instead, the time of first contact was measured from either sunrise or sunset, and the other phases of the eclipse were measured relative to this. It is therefore interesting to ask whether the same clocks were used for making each set of measurements. In Figure 2.9, the error in the first contact intervals and in the phase durations are plotted separately against the year. There does not appear to be any noticeable difference between the two sets of data. This suggests that the same clocks were used to measure all of the timings associated with the eclipses.

Given that the timings were made using a clepsydra, a major source of their error might be due to some form of clock drift. Clock drifts would imply that the longer the time interval measured, the greater the error in the measurement might be. This is illustrated by Figure 2.10 which shows the error in the observed timings plotted against the computed time intervals. The dispersion of the error appears to increase with the length of the computed interval. This is characteristic of a random error in clock drift. Because this error is random, it is more appropriate to think in terms of the accuracy of the observations (i.e., their error irrespective of whether the times are early or late), rather than of the error in the times. The change in the accuracy with

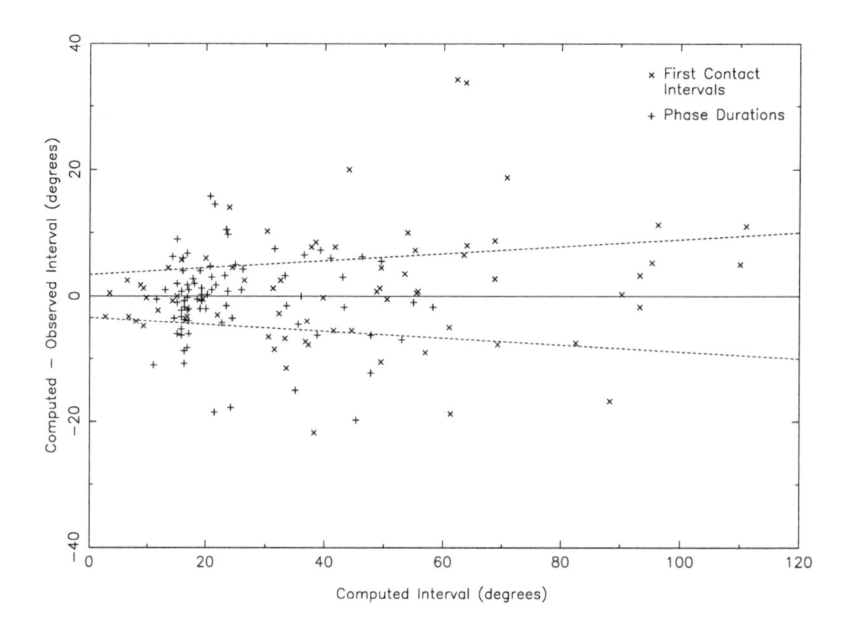

Figure 2.10: The error in the observed first contact intervals and phase durations plotted against the computed interval.

interval can be approximated by fitting a straight line to the absolute values of the errors in Figure 2.7. This line has a slope of about 0.083 degrees per degree of interval and a zero-point of 3.4 degrees. This is illustrated by the two dashed lines in Figure 2.10.

To summarize, it appears that the Babylonian astronomers used the same clepsydras to measure the time of first contact relative to sunrise or sunset as they did for the durations of the various phases of the eclipses, and that the same clocks were used for their observations of both lunar and solar eclipses. There was no noticeable improvement in the accuracy of these clocks over the whole of the Late Babylonian period, despite an improvement in the precision with which the Babylonians quoted their times in the middle of the sixth century BC. These clocks were subject to random errors in drift arising from the fact that they could only be rated to an accuracy of about $8\frac{1}{2}\%$, and could only be read with an accuracy of about 3° (12 minutes). Thus a typical Babylonian eclipse timing of about 40° (2 hours 40 minutes) has an accuracy of just under 7° (24 minutes).

Finally, let me discuss the 16 timings made by means of a *ziqpu*-star.[99] Sometimes,

[99] In his discussion of whether the *ziqpu*-star timings of eclipses can be used to determine ΔT, Stephenson (1997b: 185) incorrectly dates the observation on 12 July 178 BC to 1 August 188 BC. This is based upon an incorrect reading of the date (SE 124 instead of SE 134) in the copy of LBAT 1439 published by Sachs (1955). The eclipse of 12 July 178 BC was only penumbral and so Stephenson remarks that this is the only

Date	*Ziqpu*-Star	*Ziqpu* LT (h)	Clepsydra LT (h)	Computed LT (h)
-225 Aug 2	α Cyg	23.00	22.41	23.94
-214 Dec 25	α Per	19.00	17.99	19.32
-193 Nov 5	$\alpha, \gamma, \eta, \zeta$ Leo	5.67	5.77	6.04
-184 Nov 24	π, σ, ξ, ν Cas	19.05	20.10	19.62
-182 Oct 4	β Aur	3.00	-	1.20
-177 Jul 12	β Cyg	22.95	-	23.17
-162 Mar 30	3° behind ξ Boo	1.01	0.17	23.37
-149 Jul 3	4° in front of 15 Lac	2.62	-	2.46
-142 Feb 17	5° behind β Aur	17.91	17.97	18.11
-135 Apr 1	α Cor	1.60	3.79	1.54
-133 Mar 10	α Lyr	6.28	5.56	5.27
-122 Aug 2	5° behind μ Her	20.18	-	20.65
-119 Jun 1	5° behind α Her	23.63	23.46	23.64
-90 Nov 5	σ And	20.04	-	19.83
-86 Feb 28	δ, θ Leo	23.26	-	22.80
-79 Apr 11	5° in front of μ Her	3.01	2.97	2.97

Table 2.6: Lunar eclipses timed relative to a *ziqpu*-star.

the eclipse is said to begin at the moment when a star culminates, but in other cases a point so many degrees either in front of or behind a star is said to culminate. Presumably, these points were determined by measuring the number of UŠ of time relative to the moment of the culminating star since these time degrees translate directly into degrees of right ascension. The *ziqpu*-star timings are listed in Table 2.6.[100] In this Table I also give the local time associated with the culmination of the *ziqpu*-star (deduced from the right ascensions of the star and the sun), the same time as was measured by a clepsydra (where preserved), and the computed local time. I should first remark that it is very unlikely that the times in UŠ discussed above, which I assumed were measured with a clepsydra, were calculated from the *ziqpu*-star observations. A time relative to sunrise or sunset could in theory be obtained from the *ziqpu*-star observations by using the texts which give the time intervals between consecutive culminations.[101] But this would require that one knew which star was culminating at sunrise or sunset, observations which are never preserved in the Diaries (and are in any case difficult to make during twilight).

The error in the *ziqpu*-star timings is shown in Figure 2.11. The mean accuracy of these timings is about 0.53 hours. This is noticeably poorer than the timings deduced from measurements of the altitude of stars made by, for example, the astronomers of the Islamic Near East.[102] The main source of error in the *ziqpu*-star timings is probably the simple practical difficulty in establishing when a star is actually culminating.

known ancient or medieval report of a penumbral eclipse. However, this remark should now be discounted based upon the improved reading of LBAT 1439 by Sachs & Hunger (1999).

[100]The identification of stars is a notoriously difficult procedure, and can often not be achieved with full confidence. See, for example, the different identifications proposed by Reiner & Pingree (1981) and Koch (1989). I have accepted the identification of the *ziqpu*-stars by Sachs and Hunger (1988, 1989, 1996, 1999).

[101]See Section 2.4 above.

[102]See Section 4.4 below.

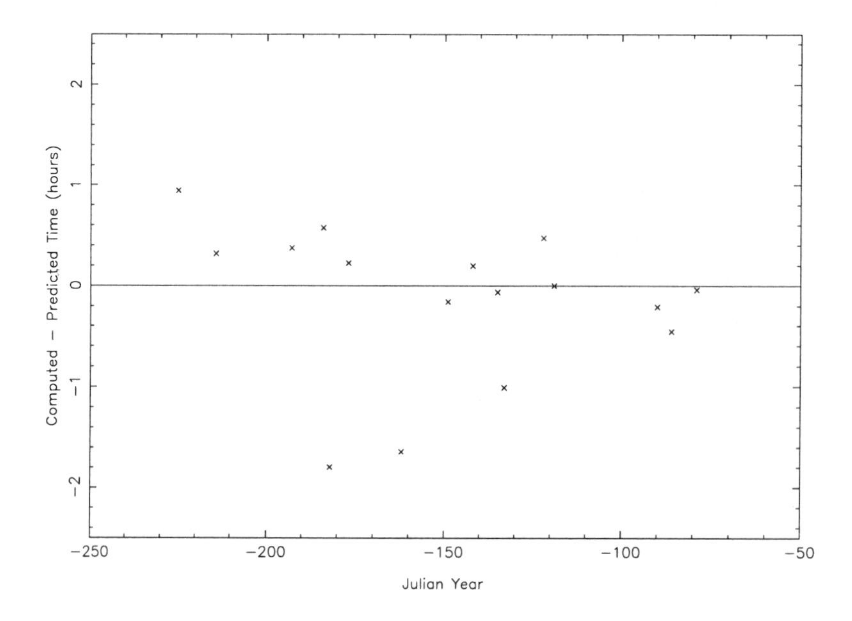

Figure 2.11: The error in the *ziqpu*-star timings of lunar eclipses.

Near the meridian, the path of a star is close to flat. It is not a simple matter, therefore, to determine when the star reaches its highest point in motion. Instead, one would probably need to construct some form of meridian line. This is also no easy task. Cardinal directions did not exist as such in Mesopotamia. The four directions, SI, KUR, ULÙ, and MAR, are conventionally translated as north, east, south, and west. However, they actually relate to the prevailing wind-directions, and these have azimuths somewhat lower than the corresponding cardinal points.[103] They could not, then, have been used to establish a meridian line. Thus, it must have been obtained from either a series of culminating stars themselves, or from bisecting the angle between the rising and setting position of a celestial body (probably the sun).

2.7 Accuracy of the Predicted Times

In addition to the hundred or so timed eclipse observations recorded in Late Babylonian history, there are about as many timed predictions. The majority of these relate to events that could not be observed in Babylon, for if the eclipse was seen then the observation rather than the prediction was recorded in Astronomical Diaries; the other predictions in the Diaries refer to eclipses that were not seen. Only in the Almanacs

[103] See, for example, Horowitz (1998: 196–198). I thank C. B. F. Walker for bringing this point to my attention.

and Normal Star Almanacs do we find predictions that refer to eclipses that may have been seen at a later time by the Babylonian astronomers. In analysing these records, therefore, I am really trying to evaluate the Babylonian methods of making eclipse predictions largely from their failures. However, there is no reason to suppose that the Babylonian schemes would be less reliable when they predicted an eclipse which was not visible than when it was seen. I shall begin by considering the records of timed lunar eclipse predictions.

Table 2.7 lists all of the lunar eclipse predictions contained in the dated Late Babylonian Astronomical Texts for which a time is fully preserved. The records may be divided into three categories: umbral lunar eclipses that were visible somewhere on the Earth's surface but not necessarily at the longitude of Babylon (A), penumbral lunar eclipses (B), and failed predictions (F). Only the first category can be considered as being "successful" in the context of Babylonian astronomy. There are no firmly dated records of Babylonian observations of penumbral eclipses so the category B predictions can perhaps be thought of as "near-misses." Of the 56 eclipses listed in Table 2.7, 37 are in category A, 17 in B, and 2 in F. Thus, in this sample, the Babylonian predictions of lunar eclipses were successful about 66% of the time, with a further 30% being near-misses. Only 4% of the time did their predictions completely fail.

For the category A predictions, Table 2.7 gives both the local time of the eclipse as predicted by the Babylonian astronomers, and that deduced from modern computations. In making these computations, I have assumed that the predicted times relate to the expected time of first contact. The errors in the predicted times are shown in Figure 2.12. The mean error in the predicted times, about -0.40 hours, is shown by the dashed line in the Figure. This is sufficiently close to zero to confirm that the predicted times do indeed relate to the moment when the eclipses were expected to start. If the predicted times related to the end of the eclipse, then the mean error would be increased by about 3 hours, the average duration of the eclipses. If it were the middle of the eclipse that was intended, then the mean error would still be increased by about $1\frac{1}{2}$ hours.

The average accuracy of the category A eclipses is about 1.31 hours. There is no evidence for any improvement in the accuracy of these predictions down the centuries, mirroring the result found for the observed timings. This suggests that the same methods of making the predictions were used for the early predictions as for the later ones. I shall discuss the implications of this result in Section 2.8 below.

Returning to the category B predictions in Table 2.7, the predicted and computed local times have once more been given. However, these predictions relate to penumbral eclipses, and these events only have virtual contacts. Therefore, the computed time relates to the moment when the moon made its closest approach to the Earth's umbral shadow. The error in the predicted times are shown in Figure 2.13. Unsurprisingly, there is a much greater scatter in these times than there was for the category A predictions. The mean accuracy of these times is about 2.86 hours, more than twice as poor as that of the category A predictions.

Moving on now to the timed solar eclipse predictions, there are 74 records preserved in Late Babylonian history that give a fully preserved time when the eclipse

Date	Description	Category	Predicted LT (h)	Computed LT (h)
-730 Apr 9	Omitted at 60° after sunrise	A	9.72	11.72
-667 May 2	Omitted at 40° after sunrise	A	8.04	9.33
-667 Oct 25	Omitted at 30° before sunset	A	15.64	14.90
-649 May 13	Omitted at 60° before sunset	A	14.86	16.75
-590 Sep 15	Omitted with sunrise	A	5.68	6.02
-572 Sep 25	Omitted at 35° before sunset	A	15.80	13.77
-525 Mar 24	... at 25° before sunset	A	16.35	13.80
-414 Mar 26	Omitted at 12° before sunset	A	17.28	13.47
-409 Jun 28	... at 70° after sunrise	A	9.50	9.72
-408 Nov 11	Omitted at 80° after sunrise	B	11.80	4.14
-395 Mar 26	Omitted at 10° before sunset	A	17.41	13.12
-379 Oct 22	... at 20° after sunrise	F	9.67	-
-378 Oct 11	Omitted at 12° after sunrise	B	6.95	9.36
-356 Feb 14	Omitted at 40° before sunset	B	14.77	14.91
-352 May 28	... at 7° before sunrise	A	4.52	5.83
-334 Dec 3	Omitted at 60° before sunset	A	13.09	12.36
-291 Aug 11	Omitted at 27° after sunrise	B	6.97	6.43
-278 Jun 19	... at 18° before sunset	A	17.95	18.19
-278 Nov 15	Omitted at 45° after sunset	F	20.29	-
-248 Apr 19	Omitted at 39° before sunset	B	15.89	9.11
-248 Oct 13	Omitted at 30° after sunset	B	19.78	21.93
-246 Sep 22	Omitted at 16° after sunrise	A	6.91	7.29
-232 Dec 14	Omitted at 74° after sunrise	A	11.91	10.67
-225 Feb 6	Omitted at 30° before sunset	A	15.34	12.65
-214 Jan 5	... at 58° after sunrise	A	10.54	9.88
-194 Jun 20	Omitted at 15° before sunset	B	18.15	18.99
-194 Nov 16	Omitted at 45° after sunset	B	20.26	23.83
-193 May 11	Omitted at 94° after sunrise	A	11.46	10.98
-191 Apr 19	... at 2° after sunset	A	18.63	17.91
-185 Jun 11	Omitted at 48° before sunset	A	15.92	13.27
-172 Mar 21	Omitted at 47° after sunrise	B	9.11	7.81
-169 Feb 16	Omitted at 31° before sunset	A	15.42	14.41
-169 Aug 13	Not seen at 4° before sunrise	A	4.94	5.60
-168 Jan 7	Omitted at 7° before sunrise	B	6.49	3.24
-162 Sep 23	Omitted at 48° after sunrise	A	9.07	9.04
-161 Feb 18	Omitted at 31° before sunset	B	15.45	8.39
-161 Aug 14	Omitted at 25° before sunset	B	17.10	17.11
-160 Feb 7	Omitted at 10° after sunrise	A	7.32	10.16
-158 Jul 12	Omitted at 58° after sunrise	B	8.75	8.09
-140 Jul 22	Omitted at 34° before sunset	A	16.76	14.43
-140 Dec 17	Omitted at 78° after sunset	B	22.20	1.94
-139 Jun 12	Omitted at 65° after sunrise	A	9.20	8.99
-137 May 22	Not seen at 35° before sunset	A	16.62	16.29
-136 Oct 5	Omitted at 79° before sunrise	B	0.81	2.77
-133 Sep 3	... at 30° after sunset	A	20.46	20.48
-132 Jan 29	Omitted at 92° before sunrise	B	0.62	6.51
-131 Jan 17	... at 60° after sunset	A	21.12	20.55
-122 Aug 2	... at 76° after sunset	A	20.49	23.99
-122 Dec 29	... at 6° before sunrise	B	6.59	8.73
-110 May 24	... at 61° before sunrise	A	0.95	0.35
-110 Nov 16	Omitted at 71° after sunrise	A	11.46	11.25
-106 Mar 11	Omitted at 62° after sunrise	A	10.27	9.73
-86 Aug 24	Omitted at 30° after sunrise	A	7.34	10.52
-76 Feb 9	Omitted at 76° after sunrise	A	11.68	11.77
-75 Jul 24	... at 8° before sunrise	A	4.47	5.17
-62 May 3	... at 9° after sunrise	A	5.89	4.49

Table 2.7: Timed lunar eclipse predictions.

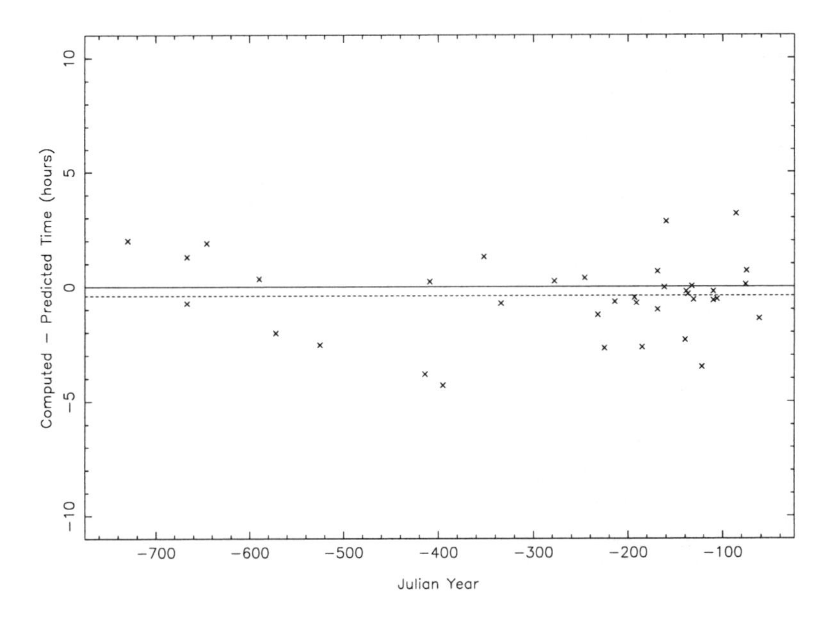

Figure 2.12: The error in the predicted time of the category A lunar eclipses.

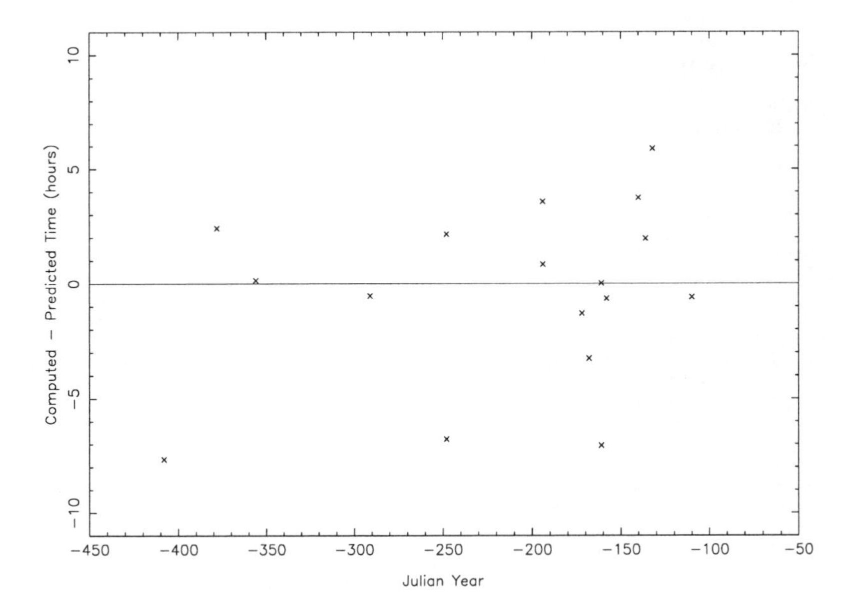

Figure 2.13: The error in the predicted time of the category B lunar eclipses.

Date	Description	Category	Predicted LT (h)	Computed LT (h)
-473 Nov 25	... at 40° after sunrise	B	9.47	16.42
-472 May 20	... at 30° before sunrise	B	3.09	3.06
-471 May 9	... at 70° before sunset	B	14.09	16.96
-458 Aug 12	... at 40° after sunset	B	16.15	22.99
-455 Aug 2	... at 50° before sunset	A	15.62	11.37
-357 Sep 5	Omitted at 50° before sunrise	A	2.21	0.64
-355 Feb 18	Omitted at 46° before sunrise	A	5.73	3.53
-332 Oct 27	Omitted at 30° after sunrise	B	8.43	7.77
-330 Oct 5	Omitted at 1° after sunset	A	18.01	22.86
-302 Sep 25	Omitted at 78° after sunset	B	23.30	0.52
-291 Aug 25	Not seen at 25° before sunset	A	16.94	19.61
-266 Oct 17	... at 57° before sunset	B	13.93	18.00
-255 Mar 24	Pass by at 5° after sunset	B	18.40	1.05
-246 Sep 7	Not seen at 74° after sunrise	B	14.21	10.88
-241 Jun 15	To be watched for at 51° before sunset	A	15.74	11.12
-232 Nov 30	Not seen at 44° after sunrise	B	9.82	0.02
-230 May 15	Omitted at 5° after sunset	A	19.19	20.57
-229 May 5	To be watched for at 54° after sunrise	A	8.88	8.90
-228 Mar 25	Omitted at 28° after sunset	B	19.96	3.27
-226 Mar 3	... at 10° after sunrise	B	6.96	6.50
-225 Jul 17	Omitted at 73° before sunset	B	14.21	11.13
-218 Apr 3	... at 87° after sunset	B	0.03	17.73
-211 May 15	Not seen at 35° before sunset	A	16.54	19.96
-209 Sep 18	Not seen at sunset	A	18.22	19.64
-206 Jan 22	Omitted at 58° after sunset	B	21.03	6.63
-206 Jul 17	Omitted at 49° after sunset	A	22.35	19.44
-205 Jan 11	Omitted at 20° after sunset	A	18.39	22.32
-204 Jun 25	To be watched for at 87° before sunset	B	13.36	18.12
-200 Apr 13	Omitted at 58° before sunrise	B	1.73	23.10
-199 Mar 4	To be watched for at 45° before sunset	B	14.74	13.03
-194 Nov 29	Omitted at 91° after sunset	A	23.19	0.56
-193 May 26	Omitted at 77° after sunset	A	0.12	0.24
-189 Sep 7	Not seen at 28° before sunset	B	16.53	18.24
-188 Feb 2	Omitted at 78° before sunset	B	12.08	18.44
-185 Nov 20	Omitted at 65° before sunrise	B	2.44	22.56
-182 Jun 7	... 60° before sunrise	B	1.41	4.09
-183 May 6	To be watched for at 8° after sunrise	B	5.78	3.71
-168 Jan 22	Omitted at 93° after sunset	A	23.36	21.90
-164 Oct 29	Omitted at 64° after sunset	B	21.78	22.75
-162 Sep 8	Omitted at 30° before sunrise	B	3.61	20.74
-161 Mar 5	Not seen at 30° after sunrise	B	8.25	9.17
-161 Aug 28	Omitted at 16° after sunset	B	19.63	20.92
-154 Oct 10	To be watched for at 35° after sunrise	A	8.49	3.76
-144 Sep 19	Not seen at 59° after sunrise	B	9.74	9.48
-143 Sep 8	Omitted at 39° before sunrise	A	3.02	1.74
-139 Jan 1	Omitted at 87° before sunrise	A	1.19	1.12
-139 Jun 27	Omitted at 48° after sunset	A	22.36	22.00
-136 Oct 20	Not seen at 81° before sunset	B	12.14	14.47
-132 Aug 7	Omitted at 9° after sunset	B	19.45	23.43
-131 Feb 1	Omitted at 21° after sunset	B	18.69	15.14
-128 Nov 20	... at 45° before sunset	A	14.20	16.25
-127 Apr 16	... at 7° after sunset	B	18.93	20.40
-124 Sep 7	Not seen at 60° before sunset	A	14.38	18.25
-122 Feb 7	... at 23° after sunrise	A	8.17	5.83
-122 Jul 19	To be watched for at 71° before sunset	B	14.33	17.51
-121 Jan 12	To be watched for at 44° after sunrise	B	9.80	9.48

Table 2.8: Timed solar eclipse predictions.

Date	Description	Category	Predicted LT (h)	Computed LT (h)
-119 Nov 11	To be watched for at 79° before sunset	A	12.05	13.50
-118 May 7	Omitted at 13° before sunrise	B	4.36	2.85
-118 Oct 31	Omitted at 44° after sunset	A	20.42	22.72
-116 Mar 16	Omitted at 54° after sunset	B	21.56	1.23
-110 Jun 7	To be watched for at 62° before sunset	B	14.97	5.91
-110 Dec 2	Omitted at 87° before sunrise	B	1.11	0.55
-106 Mar 27	Omitted at 81° before sunrise	B	0.30	0.27
-102 Jul 8	To be watched for at 30° before sunset	A	17.13	18.42
-86 Feb 14	Omitted at 76° before sunrise	A	1.46	23.24
-84 Jan 23	Not seen at 63° after sunrise	B	11.02	14.38
-79 Sep 20	Not seen at 10° after sunrise	B	6.50	7.12
-77 Mar 6	To be watched for at 75° after sunrise	A	11.22	11.30
-62 May 18	Not seen at 30° after sunrise	A	7.09	6.80
-40 Feb 15	Not seen at 57° before sunset	B	21.28	18.01
-11 Jul 21	To be watched for at 50° before sunset	B	15.70	12.98
-6 Apr 29	To be watched for at 47.5° before sunset	A	15.50	16.00
-6 Oct 22	To be watched for at sunrise	A	6.37	7.76
+37 Jan 5	To be watched for at 75° before sunset	B	12.04	9.28

Table 2.8 (cont.): Timed solar eclipse predictions.

was expected to occur. These are listed in Table 2.8. As with the lunar predictions, it is possible to divide the solar eclipses in Table 2.8 into different categories: solar eclipses that were visible at Babylon, or would have been visible there if the sun was above the horizon at the time of the eclipse (i.e., the eclipse was visible at the latitude but not necessarily the longitude of Babylon) (A), and eclipses that would not be visible at Babylon under any circumstances (i.e., the eclipse passed either completely to the north or to the south of Babylon) (B). Of the 74 predictions in this sample, 30 (41%) fall into category A, and 44 (59%) into B.

Considering first the category A predictions of solar eclipses, both the predicted and the computed local times are given in Table 2.8. The computed times have once more been determined assuming that the prediction relates to the moment of first contact. The error in the predicted times are shown in Figure 2.14. Once more, the mean error of the prediction is shown by the dashed line in the Figure. This mean error is equal to about +0.10 hours, which is very close to zero providing further proof that the predicted times relate to first contact. The mean accuracy of these predictions is about 2.01 hours.

For the category B predictions, the computed times relate to the moment when the path of the eclipse makes its closest approach to Babylon. The errors in the predicted times of these eclipses are shown in Figure 2.15. Clearly there is a very great scatter in the errors of these times, ranging from close to zero to over 9 hours. The mean accuracy of the category B predictions is about 3.55 hours.

In both categories of the solar eclipse predictions there does not appear to be any change in the accuracy of the times down the centuries. This was also found for the lunar eclipse predictions. Furthermore, there is no evidence for any difference in the accuracy of the predicted times given in the Almanacs and Normal Star Almanacs,

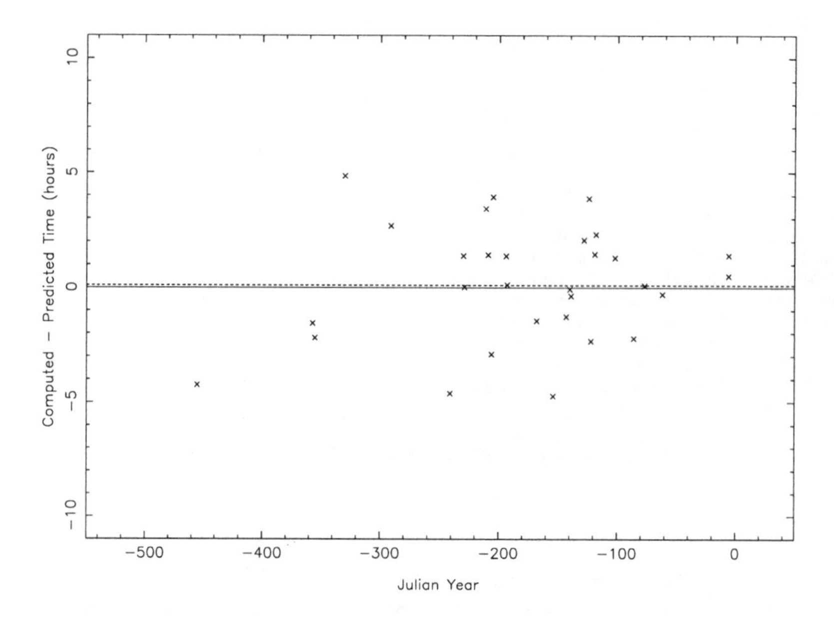

Figure 2.14: The error in the predicted time of the category A solar eclipses.

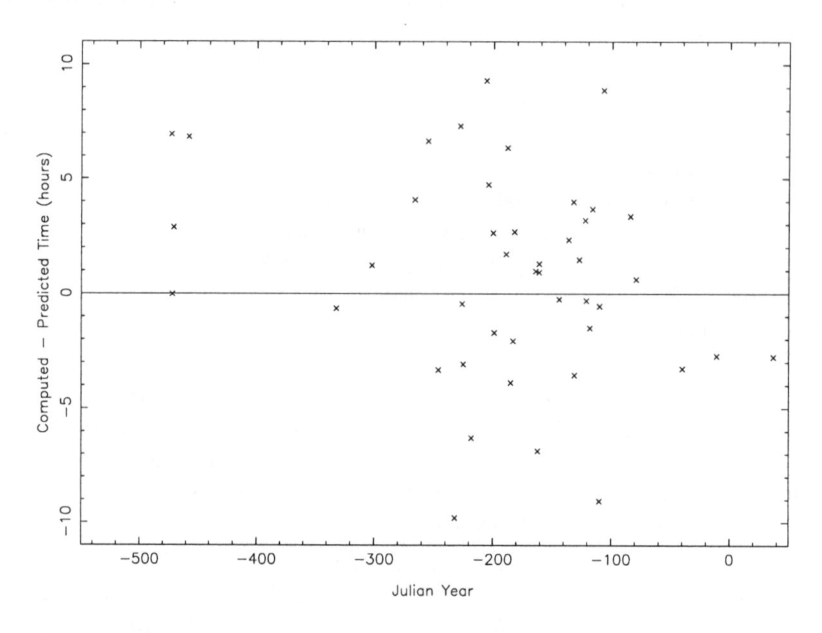

Figure 2.15: The error in the predicted time of the category B solar eclipses.

Date	Description	Predicted LT (h)	Computed LT (h)
-590 Mar 22	At 30° before sunset	15.98	16.29
-562 Sep 5	At 35° before sunset	16.13	17.14
-366 Mar 6	At 10° before sunset	17.08	17.32
-189 Aug 23	At 30° before sunset	16.63	15.68
-170 Aug 23	At 42° before sunset	15.83	15.91
-135 Sep 24	At 30° before sunset	16.09	16.78
-66 Jan 19	At 16° before sunset	16.07	15.07
-65 Jan 8	At 30° before sunset	15.04	14.35
-65 Dec 28	At 6° before sunset	16.60	16.37

Table 2.9: Times of first contact for lunar eclipses which rose eclipsed.

and those given in the Diaries, Goal-Year Texts, and Eclipse Texts. This suggests that all of the eclipse times were predicted in advance, and then if the eclipse was not observed, the predicted time was placed in the Diary. Certainly, predictions accurate to a couple of hours would have at least given the Babylonian astronomers a rough idea of when to watch for an eclipse.

As I have mentioned, the times that the Babylonian astronomers predicted for an eclipse relate to the moment that they expected that the eclipse would begin. In Steele & Stephenson (1997) I suggested that the reason for this might be that if one were preparing for a ritual to be performed at the time of an eclipse, one would wish to know when the eclipse would begin to ensure that the ritual could be begun as soon as the eclipse started. This has since been confirmed by the description of the eclipse ritual on BM 134701, published by Brown & Linsenn (1999). This text reveals that the ritual kettle-drum should be played from the beginning of the eclipse until it began to recover, that is, the end of its maximal phase. Thus it seems natural that the Babylonian astronomers were attempting to predict the beginning, rather than, say, the mid-point, of an eclipse.

Finally, let me discuss the small number of lunar eclipses that were observed to rise eclipsed, and for which an estimate of the time that the eclipse started is given. These are listed in Table 2.9. The errors in the predicted first contact times are shown in Figure 2.16. The mean accuracy of these times is about 0.58 hours. This is significantly better than the accuracy of the purely predicted lunar eclipses discussed above, suggesting that these times were not predicted using the same scheme. It is more likely that the times were instead estimated from the observed later phases of the eclipses.

2.8 Methods of Eclipse Prediction

It would seem that more or less as soon as the Mesopotamian astronomers began to observe solar and lunar eclipses, they also attempted to predict these events in advance.[104] In Old Babylonian omen literature eclipses are forecast by, for example,

[104] For a more detailed study of eclipse prediction in Mesopotamia, see Steele (1999a).

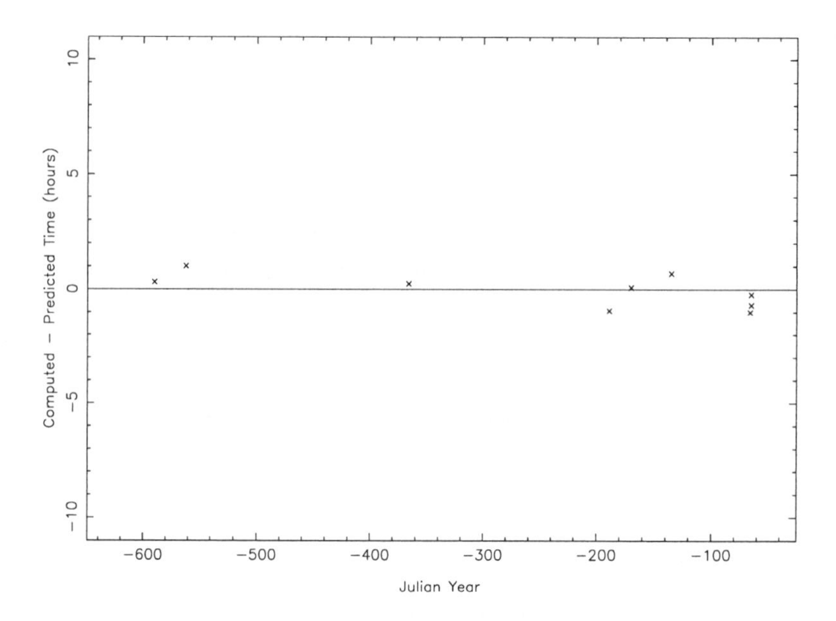

Figure 2.16: The error in the predicted time of first contact for lunar eclipses which rose eclipsed over the Late Babylonian period.

liver- and oil-divination. Some astronomical omens such as halos, the untimely appearance of the new moon, and fog also predict eclipses.[105] Moreover, tablet 20 of *Enūma Anu Enlil* makes frequent reference to predicting eclipses on the basis of lunar observations made on the day of the moon's last visibility:

> "[Observe his (i.e., the moon's)] last [visibility] on the 28th of Nisannu, [and on] the 14th day of Ajaru [you will predict] an eclipse. [The day of last visibility will sh]ow you [the eclipse.]"
>
> [*Enūma Anu Enlil* 20; trans. Rochberg-Halton (1988: 188)]

In the Assyrian and Babylonian Letters and Reports some eclipses were being foretold by ominous events such as these. In other cases, however, it is clear that attempts were being made to predict eclipses astronomically. For example, the Report ARAK 502 and others indicate that lunar and solar eclipses were generally expected to take place within a month of one another:

> "An eclipse of the moon and sun in Sivan will take place. These signs are of bad fortune for Akkad, for the kings of Westland and of Akkad; and now, in this month of Kislev, an eclipse will take place."
>
> [ARAK 502, Rev. 8–10; trans. Hunger (1992: 279)]

[105] Koch-Westenholz (1995: 105).

This Report appears to have been written in Month IX of a year. During this month an eclipse was expected to take place. Six months later, in Month III of the following year, both a lunar and a solar eclipse were expected. This Report also suggests, therefore, that the six month interval between eclipse possibilities (i.e., dates when an eclipse is possible but not necessarily seen) had also been identified. In some texts two successive months were considered as possible dates for the eclipse:

> "Concerning the watch of the sun about which the king, my lord, wrote to me, it is (indeed) the month for a watch of the sun. We will keep the watch twice, on the 28th of Marchesvan and the 28th of Kislev. Thus we will keep the watch of the sun for 2 months."
>
> [LABS 45, Rev. 1–10; trans. Parpola (1993: 1–34)]

This suggests that the scholar who wrote this Letter had realized that not only are eclipses separated by 6, 12, 18 etc. months possible, but also at 5, 11, 17 etc. months. In other words, eclipse possibilities are separated not only by 6n months (where n is a relatively small integer), but also by 6n-1 months. The realization that eclipse possibilities can occur at a five month interval is also implied by the Report ARAK 4 which notes that an observation of an eclipse was made "at an unappointed time" (*ina la mi-na-ti-šú*). A commentary on the omen explains the meaning of "unappointed time":

> "If the moon is eclipsed at an unappointed time, (it means that either) the six months have not yet passed (since the preceding eclipse), or, alternatively, an eclipse occurs on the 12th or 13th day."
>
> [*ACh. Sin* 3, 26[106]]

In the ARAK 4 report, the eclipse was seen on the 14th day of the month. It must therefore have taken place one month earlier than expected for it to have been at an unappointed time.

From these examples it would seem that the scholars were not always sure as to when a five month interval was needed. In general, if an eclipse had just been seen, then a six month interval was called for, but when no eclipse was seen at a given eclipse possibility, an eclipse only five months after this eclipse possibility also needed to be considered. One way to eliminate this problem is to use eclipse periods, such as the Saros of 223 synodic months, after which the distribution of five and six month intervals between eclipse possibilities repeats. Indeed Parpola (1983: 51) writes that "in addition to the simple rule of thumb that the moon may be eclipsed every sixth synodic month, the scholars of this period certainly had recognized the 47 month period and probably also the 18 year Saros." However, from the preserved texts I do not believe it is at present possible to make this claim. To do so, one would need to try to find patterns in the dates of eclipse predictions, and, unfortunately, it is often not possible to date those texts concerned. It may well be that these eclipse periods were in use, but we do not have the evidence to be certain.

[106] See Kugler & Schaumberger (1935: 251), and Rochberg-Halton (1988: 41) whose translation I have largely followed.

After determining the months of the eclipse possibilities, the scholars then tried to decide upon the day of the eclipse and whether it would be seen or would pass by when the luminary was below the horizon. Occasionally, they would even try to predict the watch in which the eclipse would occur. It would appear that the day and approximate time of the expected eclipse was determined by observing whether the moon was seen with the sun on the previous few days. For example, the report ARAK 42 notes that:

> "The moon will be seen [together with] the sun in Elul on the 15th day, it will let [the eclip]se pass by. (Break) [The moon will] be seen together with the sun, it will let [the eclipse] pass by, it will not make (it). From Nabû-ahhe-eriba. [Elu]l, 13th day."
>
> [ARAK 42; trans. Hunger (1992: 26)]

It is easy to see how such observations may be used to make a rough estimate of the time of an eclipse. In the days leading up to an eclipse, the moon's latitude is close to zero. Therefore, the time interval between the sun and moon crossing the horizon is, to a first approximation, dependent upon the difference in longitude of the two luminaries.[107] Thus, as an eclipse occurs when the difference in longitude is either 0° or 180°, if this time interval is great then the moment of syzygy is far off and the eclipse may occur during the following day, whereas if it is small then the syzygy will occur during that night. Perhaps it is to some method such as this that the instruction to observe the last visibility of the moon in *Enūma Anu Enlil* tablet 20 is referring to.

Moving on to the Babylonian NMAT texts, it is immediately clear that more advanced methods of eclipse predictions were being used than in the Assyrian texts. At first sight, and rather confusingly, these superior schemes appear to antedate the Assyrian texts. For example, the earliest firmly dated prediction, from 731 BC, contains the confident statement that an eclipse would be omitted at 60° after sunrise. However, these very early predictions are contained in texts, such as the large compilation LBAT *1414 etc., which were compiled at a much later date. It is possible, although by no means certain, that these very early predictions were back-calculated at a later time. The earliest predictions in a preserved Astronomical Diary date from 652 BC. Unfortunately, this Diary is badly damaged, and it is not known whether either of the two predictions it contains originally included the expected time of the eclipse. However, the eclipse predictions contained on LBAT *1420 only note that an eclipse was omitted during the night. This tablet contains a list of consecutive eclipse observations and predictions from at least 604 BC to 575 BC. It is likely to have been compiled only a short time after 575 BC, much earlier than LBAT *1414 etc., and may for that reason contain accounts abstracted from the Diaries in a more verbatim fashion.

LBAT *1420 does, however, imply that the Babylonians were using the Saros period to predict their eclipses by at least 600 BC. The Babylonian astronomers realized that the Saros not only allowed them to predict eclipses as occurring 223 months after

[107] More generally, this time interval is a very complicated function dependent upon the moon's longitude and latitude, its velocity, and several other factors. See Neugebauer (1957: 107–110) and Brack-Bernsen & Schmidt (1994).

a visible eclipse, but also that it provided a means for determining whether eclipse possibilities were separated by six or only five months. The exact method by which the Babylonians derived the Saros relationships is not known, but it was probably close to what follows.

Beginning in the middle of the eighth century BC it would seem that the Babylonian astronomers had a more or less complete record of observed eclipses. Britton (1989) has shown that by a fairly basic analysis of the observational record, simple schemes for arranging eclipse possibilities could be identified. One of these was that within 223 lunar months there were 38 eclipse possibilities. The Babylonians named this period of 223 months "18 MU.MEŠ" ("18 years") since it is close to 18 years; today it is known as the Saros. The Babylonian astronomers realized that the arrangement of eclipse possibilities in this 223 month period repeats over and over.

Within each Saros period of 223 synodic months, the 38 eclipse possibilities are divided into 33 which were separated by six month intervals, and the remaining 5 by five month intervals. Using the simple rule that the five month intervals should be distributed as evenly as possible, one finds that the eclipse possibilities separated by five months should be arranged so that they split up those separated by six months into groups of seven or eight eclipse possibilities. In other words, if the first eclipse of the 38 eclipses in a Saros period comes five months after the preceding eclipse possibility, then it will be followed by seven eclipses (nos. 2–8) each of which is six months after the preceding eclipse, then an eclipse (no. 9) at five months, six (nos. 10–15) at six months, another (no. 16) at five months, seven (nos. 17–23) at six months, another (no. 24) at five months, six at six months (nos. 25–30), one more (no. 31) at five months, and then finally seven (nos. 32–38) at six month intervals. Thus, the 38 eclipse possibilities are divided into five groups, each of which begins with an eclipse possibility five months after the preceding eclipse, containing eight, seven, eight, seven, and eight eclipses respectively. This is often written as 8–7–8–7–8.[108]

The next problem faced by the Babylonian astronomers was to decide where to begin the scheme they had derived for distributing the eclipse possibilities within the Saros period. Their solution can be immediately seen from the tablet LBAT *1420. The eclipse predictions in this text fit into the expected 8–7–8–7–8 grouping of eclipse possibilities. This same grouping is evident in the preserved parts of the tablets LBAT *1414, LBAT 1415 + 1417 + 1417, and LBAT *1419. As mentioned in Section 2.5 above, Walker (1997) has suggested that these three tablets formed part of a series containing records of eclipses stretching from 747 BC to 315 BC. These sources imply that the first eclipse in the 8–7–8–7–8 distribution was in the Saros series containing the eclipse on 6 February 747 BC.[109] In all probability, it was this very eclipse that should be taken as the start of the distribution — although it is possible that this Saros scheme was projected back onto this eclipse — since this is the probable date of the

[108] Of course, the definition of the beginning of the Saros period is arbitrary, and the distribution could equally well be 7–8–7–8–8, 8–7–8–8–7, 7–8–8–7–8 or 8–8–7–8–7. The last distribution is that found by Aaboe (1972) from a theoretical analysis of equally spaced eclipse possibilities.

[109] In other words, with an eclipse possibility that was a multiple of 223 synodic months before or after the eclipse 6 February 747 BC.

beginning of the large compilation on LBAT *1414, etc. This date was probably chosen because the five month intervals for the first few cycles are placed at the first visible eclipse at an interval of 6n-1 from the previously visible eclipse.

When the distribution of eclipse possibilities within the Saros implied by these texts is compared with other preserved NMAT texts, it accounts for all of the eclipse possibilities found in these sources down to about 250 BC, suggesting that this exact scheme was used until this time.[110] However, by the end of the fourth century BC this scheme was beginning to fail. Small, but probably noticeable, eclipses were beginning to be visible that had not been predicted in advance.[111] Thus the Babylonian astronomers began to formulate revisions to the way in which they used their Saros scheme. Basically, these revisions amounted to changing the date of the first eclipse in the 8-7-8-7-8 distribution.

An example of an alternate scheme derived around this time is probably the distribution found on the so called "Saros Canon" (LBAT 1428) and two similar texts (LBAT *1422 + *1423 + *1424 and LBAT *1425).[112] The Saros Canon is an example of what I have termed a theoretical text. It contains only the dates of lunar eclipse possibilities — arranged in Saros cycles — and, when complete, probably covered the period from 527 BC to 257 BC. The distribution of eclipse possibilities on the Saros Canon is slightly different to that found on LBAT *1414 etc.[113] However, the scheme actually introduced in about 250 BC is slightly different again. This suggests that a number of different schemes were formulated before one was actually adopted. Over the next two centuries it seems that two or three further revisions were made to the way in which eclipse possibilities were distributed within the Saros. There is also evidence that similar schemes were used for predicting solar eclipses. Furthermore, at least from about 250 BC onwards, it seems that the solar and lunar schemes were intimately linked together so that the solar distribution always started four eclipse possibilities earlier than its lunar counterpart.[114]

In addition to using the Saros to determine the date of predicted eclipses, the Babylonians probably also used it to estimate the expected time of the event. The evidence for this was found in Section 2.7 above, namely that the predicted eclipse times do not show any significant improvement in accuracy over the whole of the Late Babylonian period, and they relate to the moment that the Babylonian astronomers expected the eclipse to begin. The first statement immediately suggests that the same methods of eclipse prediction were used throughout the Late Babylonian period, irrespective of the development of mathematical astronomy, as represented by the ACT material from

[110]Steele (1999a).

[111]Unfortunately, the Diaries for the months on which these eclipses would have been recorded are either missing or damaged and so we do not know exactly which unpredicted eclipses were seen.

[112]These texts have been published by Aaboe, Britton, Henderson, Neugebauer, & Sachs (1991).

[113]The suggestion by Britton (1993) that the distribution of eclipse possibilities on the Saros Canon represents the scheme that was used to make the predictions of eclipses for the Diaries, and, consequently, that there must have been a reform to the way the Saros was applied from the earlier scheme represented on LBAT *1414 etc. in about 527 BC, can be discounted as the Saros Canon would not predict all of the eclipse possibilities found in the NMAT sources during the period it covers.

[114]For a detailed discussion of these various schemes, see Steele (1999a).

the Seleucid period. Furthermore, the fact that the predictions are for the start of the eclipses immediately suggests that they were made with the Saros Cycle, for this is the only short period eclipse cycle in which eclipses recur that have similar magnitudes, durations, and other circumstances. Thus, by adding one Saros period to the time of the beginning of an eclipse, one will obtain the approximate time of the beginning of another eclipse. The method is complicated, however, by the fact that the Babylonians measured the time of an eclipse with respect to sunrise or sunset. Thus they would have had to make due allowance for the change in the length of daylight throughout the year when making their predictions. There are a number of methods they could have used to do this, each based upon either the 2:1 or the 3:2 ratio for the longest to the shortest day discussed above, but we do not know which one they used in making their eclipse predictions (or even if the exact same scheme was used in making all of their predictions).

By the Seleucid period, the Babylonian astronomers had developed the mathematical astronomy of the ACT texts. ACT lunar theory comes in two forms named by Neugebauer (1955) System A and System B.[115] He defined these two systems on the basis of the way they calculate the longitude of the syzygies (column B of a lunar ephemeris). In System A, the longitude of the moon at each successive syzygy increases by a function that is equal to one value (30° per month) for part of the zodiac and a second value (28;7,30° per month) for the other part of the zodiac; this is known as a step function. By contrast, in System B, the longitude of the moon increases between syzygies by a value that varies linearly between a maximum (30;1,59,0° per month) and a minimum (28;10,39,40° per month) with a period that represents the length of the year; variations of this kind are known as linear zigzag functions. In fact, the two systems are different in almost all of the functions they employ and the parameters they are based upon.

Both System A and System B ephemerides list, for either successive oppositions or conjunctions, the various functions that are needed to compute the date and time of the syzygy, the magnitude of any potential eclipses, and the moon's visibility. Many tablets will contain an ephemeris covering one year's worth of new moons on the obverse, and the corresponding full moons ephemeris on the reverse.

In a typical System A ephemeris, the fourth column after the date, known as column E, describes the variation in latitude of the moon's centre. Column E is dependent upon the elongation of the moon from the ascending node, which is itself assumed to move in retrograde by a constant amount of 1;33,55,30° per month.[116] Eclipse possibilities are defined as being the syzygy at which the absolute value of the lunar latitude is closest to zero. The next column, Ψ, is dependent upon column E and can be interpreted as the magnitude of a predicted eclipse. System B, however, does not have a column E representing the lunar latitude. Instead, a column Ψ' or Ψ'' is calculated directly. The astronomical meaning of this column is largely analogous to column Ψ of System A, in that it can be interpreted as the magnitude of an eclipse. However, it is

[115] I follow the terminology established by Neugebauer (1955) in discussing these texts.
[116] Aaboe & Henderson (1975).

calculated for all syzygies irrespective of whether it is an eclipse possibility. Eclipses are predicted whenever Ψ' or Ψ'' falls within a particular range.[117]

In both System A and System B, a column M is calculated which gives the time of syzygy with respect to either sunset (System A), or sunset or sunrise (System B). Column M is obtained using columns G and J, where these columns represent adjustments to be applied to the length of the synodic month to take into account the variable motion of the moon and sun. Although the astronomical significance of columns G and J is common to both systems, they are calculated quite differently. In System A, G is a function of column Φ, the excess in the length of the Saros over 6585 days (assuming solar contribution is a maximum). J is a simple step function whose value is dependent on whether the longitude of the syzygy is in the fast or the slow arc. In system B, however, G is a linear zigzag function, and J is given by the differences between successive lines in column H, which is itself a zigzag function. Summing columns G and J gives the length of the synodic month over 29 days, from which the time of syzygy can be obtained by adding this value onto the time of the preceding syzygy and adjusting for the change in the length of daylight.

Systems A or B could, therefore, be used to predict the occurrence, magnitude, and time of eclipses. However, it is clear that the eclipse predictions found in the Diaries and related texts were not made using these methods. For example, a number of eclipse predictions for which we have records in NMAT texts would not have been predicted by the ACT methods. Instead, eclipse would have been predicted for either the preceding or following month. Furthermore, the times of the eclipse predictions found in the NMAT texts are significantly different from those that are given by Systems A and B. This is unsurprising, of course, since System A and B predict the moment of syzygy which corresponds to the maximum phase of the eclipse, and, as we have seen, the predicted times given in the NMAT texts relate to the moment when the eclipse was expected to begin.

This brings us to the interesting question of why the ACT methods for predicting eclipses were not adopted — a question for which the answer is far from clear. In the light of our knowledge of the individuals who wrote the various texts,[118] we cannot now postulate, as once would have been the case, that the astronomers who compiled the Diaries simply did not have sufficient understanding of the working of the mathematical astronomy to use it on a day by day basis. Part of the answer may simply be that there was no need to use the more complicated ACT methods. If the reason that the predictions were being made was simply to guide the astronomers as to when to observe, then the older methods using the Saros cycle were perfectly adequate. Furthermore, if the predictions were related to preparing for an eclipse ritual, then the ACT texts did not provide the required information (i.e., an estimate of the time that the eclipse, and hence the ritual, should begin). We also have to consider whether the ACT schemes were actually more successful than the early methods of eclipse prediction. Naïvely, given their greater complexity, we might think so, but, as I have shown

[117] For a detailed discussion of the methods of predicting eclipses using Systems A and B, see Neugebauer (1945, 1955: 41–85, 1975: 474–539).

[118] See Section 2.2 above.

elsewhere,[119] this turns out not to have been the case, especially when one considers the inaccuracies in the clocks the Babylonians were using.

These are doubtless only some of the reasons why the eclipe predictions in the Diaries were not made using the ACT schemes. It should be noted, for example, that there is very little evidence for any of the predictions of planetary phenomena in the Diaries coming from the ACT texts.[120] The whole question of the relationship between the various types of text requires much more study before any firm conclusions can be drawn.

[119] See Steele (1999a).
[120] Sachs & Hunger (1998: 25).

Chapter 3

The Greco-Roman World

3.1 Introduction

In 331 BC, Alexander the Great defeated the Persian army and brought Egypt under Greek rule. Throughout the Ptolemaic and Roman periods, Alexandria acted as a focal point for scholarly research in the areas surrounding the Mediterranean. Until recently, modern authors have commonly portrayed this scene as an idealistic sanctum, where the ancients amused themselves with scientific theories bearing no relation to their application. However, with the discovery of a large number of papyri from other parts of Egypt, this view has had to be altered. For these new texts have shown that, in addition to the theorizing and philosophizing taking place among the elite in Alexandria, practical astronomy was being undertaken elsewhere.[1]

The main goal of this practical astronomy was astrological. Judging from the number of horoscopes preserved from the first four or five centuries AD, there must have been a considerable number of practicing astrologers throughout the Greco-Roman World. Furthermore, most of the astronomical papyri found (principally from Oxyrhynchus), contain just the type of material that would be needed to calculate the phenomena found in a horoscope. Interestingly, this practical astronomy seems to be based upon two different sources: imported and adapted Babylonian methods, and native Greek astronomy (mainly Ptolemy's). We should not find the use of astronomy for astrology too surprising of course. After all, Ptolemy, in addition to writing the theoretical *Almagest*, did also compile the *Handy Tables*, which provided a simple means to calculate celestial phenomena with no theoretical explanation, and an astrological work, now known as the *Tetrabiblos*.

Greek astrology can be divided into various categories including horoscopic astrology, in which one attempts to determine an individual's fate by considering the positions of the seven heavenly bodies at the date and time of his or her birth, and general predictions for nations and cities. The interpretation of eclipses fell almost

[1] Jones (1994).

exclusively into this latter category. In his *Tetrabiblos*, Ptolemy writes:

> "... the next task (is) to deal briefly with the procedure of the predictions, and first with those concerned with general conditions of countries and cities. The method of inquiry will be as follows: The first and most potent cause of such events lies in the conjunctions and oppositions of the sun and moon at eclipse and the movements of stars at that time."
>
> [*Tetrabiblos*, II, 4; trans. Robbins (1940: 161)]

Due to this ominous significance of eclipses, they are frequently recorded in Greek and Latin literary and historical works, often associated with important events such as battles. However, these accounts generally lack technical precision — they never include the precisely measured time of the eclipse — and can often not be dated without ambiguity, so I will not discuss them here.[2] Instead, I will focus on the few detailed observations made by astronomers that are preserved. These come from only three sources: the Demotic papyrus P. Berlin 13147 + 13146; Ptolemy's *Almagest*; and Theon of Alexandria's commentary on the *Almagest*. In the remainder of this chapter I will consider each of these sources in turn.

3.2 The Demotic Eclipse Text P. Berlin 13147 + 13146

The Demotic papyrus P. Berlin 13147 + 13146 was excavated by Rubensohn in 1904 at Abusir al-Malak in Middle Egypt.[3] The recto contains reports of 23 (or possibly 24) successive eclipse observations or predictions from the first century BC. On the verso there is a text concerning the dates of solstices and equinoxes for the years 73 BC to 70 BC.

Most of the eclipse reports concern predictions of lunar eclipses. For these predicted events, the following information is recorded: the year, month, and day; the interval in months since the last eclipse (always either 6 or 5 months); and the sign of the zodiac in which the moon is located at the moment of the eclipse. Occasionally, however, details of an observation of the eclipse instead of the prediction are given, although this is not the case for all of the eclipses that were theoretically visible. The observational accounts contain, in addition to the information listed above, the time of the eclipse, the part of the moon's disk that was covered, and the visibility of planets during the eclipse. The content of the text, therefore, is similar to some of the Babylonian Eclipse Texts which list successive eclipse observations and predictions.[4]

It has been shown by Neugebauer that the dates of the eclipses in this text are given in the Egyptian calendar (in which years contain twelve 30 day months followed by 5 extra days), and correspond to the period 85 BC to 73 BC.[5] The epoch of the year

[2]For discussions of these eclipse records, often conducted with reference to determining the changes in the Earth's rate of rotation, see, for example, Newcomb (1878), Ginzel (1899), Fotheringham (1920a, 1920b), Newton (1970), and Stephenson (1997b). The latter also gives an overview of earlier works.

[3]P. Berlin 13147 + 13146 has been published by Neugebauer, Parker, & Zauzich (1981) (recto) and Parker & Zauzich (1981) (verso).

[4]For example, LBAT *1414, etc. See Section 2.5 above.

[5]For details of Neugebauer's dating, see Neugebauer, Parker, & Zauzich (1981: 320–323).

Date	Recorded Time	Recorded LT (h)	Computed LT (h)
-84 Dec 28*	-	-	-
-83 Jun 23	-	-	-
-83 Dec 17	-	-	-
-82 May 14	-	-	-
-82 Nov 7	-	-	-
-81 May 3*	9th hour of night	2.67	22.43
-81 Oct 27*	2nd hour of night	19.74	19.43
-80 Apr 21*	-	-	-
-80 Oct 16	-	-	-
-79 Apr 11*	-	-	-
-79 Oct 5*	-	-	-
-78 Mar 5	-	-	-
-78 Aug 26	-	-	-
-77 Feb 19*	6th hour of night	24.00	22.69
-77 Aug 15	-	-	-
-76 Feb 9	-	-	-
-76 Aug 3	-	-	-
-75 Jan 28*	9th hour of night	3.33	20.17
-75 Jun 24*	-	-	-
-75 Dec 19	-	-	-
-74 Jun 14	-	-	-
-74 Dec 8*	-	-	-
-73 Jun 4*	9th hour of night	2.53	3.66

Table 3.1: Eclipse Records on P. Berlin 13147 + 13146

numbers, 102 BC, has since been found to pertain to the fourth Callippic Period.[6] The first column of Table 3.1 lists the Julian dates of the various eclipses reported in this text. On the date of the 19th eclipse in the list, 24 June 76 BC, the moon was actually at too great an elongation from the lunar node to form an eclipse. This led Neugebauer to "correct" this entry to the following month, 24 July 76 BC, when an eclipse was possible. On this latter date a partial eclipse should have been visible in Middle Egypt; however, the moon set shortly after the start of the eclipse, and so it may well have been missed by observers who would not necessarily have been looking for an eclipse on that night. As the text explicitly gives the former date, noting in addition that the eclipse prediction was 5 months after the preceding eclipse possibility, it is better not to correct the text on the basis of this event. Instead we should accept that the theory being used to obtain the dates of the eclipse possibilities did not always result in predictions on dates when the moon was closest to the node. However, this should not necessarily be viewed as a failing of the Egyptian methods of prediction. All of the lunar eclipses visible in Middle Egypt over the period of our text, with the exception of the one questionable event on 24 July 76 BC, were indeed predicted in advance. But what were these methods of eclipse prediction? Unfortunately, we have only a small amount of information upon which to base an answer to this question, and so what follows must by necessity be somewhat tentative.

We can probably discount the use of some form of mathematical astronomy to make these predictions. If mathematical astronomy had been used, then we would

[6] Jones (2000).

have expected that the eclipse predictions would contain more information than simply the sign of the zodiac in which the moon was located at the time of the eclipse. Mathematical astronomy almost inevitably leads one to calculate the moon longitude to a high level of precision, and we would then have to wonder why the author of this text chose to round these to the nearest zodiacal sign. Also, one would expect that the time of the eclipse, its magnitude, etc., would have been calculated, but these are never recorded for the predictions found in the text. It seems more likely that the predictions were based on some eclipse cycle. Once a pattern for placing the five and six month intervals between eclipses within one cycle had been established, this pattern could then be repeated to obtain the dates of all of the eclipse possibilities within subsequent cycles.

Various cycles were known in antiquity. As we have seen, the so called Saros cycle of 223 synodic months, equal to about $6585\frac{1}{3}$ days, was used to predict eclipses in Babylon.[7] This cycle is also described by Ptolemy in the *Almagest* (IV, 2), together with a cycle found by tripling the Saros to obtain a period of 669 synodic months, closely equal to an integer number of days, 19756, called the Exeligmos. Both of these, he says, were used by "the even more ancient astronomers."[8] Ptolemy's commentators also referred to these periods as having been used by "the ancients."[9] However, it is not known to whom these authors are referring. Certainly, they mean astronomers earlier than Hipparchus (2nd century BC), but it may be that Ptolemy (and so also his commentators) did not have any particular individuals in mind. Instead he may simply be suggesting that the Saros was fairly widely known from well before his time.

We cannot rule out, therefore, that the Saros was used to make the predictions in the present text. In the first century BC, it is likely that some aspects of Babylonian astronomy had already found their way into Egypt. Indeed, as early as the third or fourth century BC, eclipse omens were apparently being brought to Egypt from Babylon. These were then adapted to make them relevant to an Egyptian audience, for example, by altering the countries named in the apodoses to the locality of Egypt, and then collected together, presumably on papyrus of which we have a copy from the second or third century AD.[10] It is quite plausible to suppose, therefore, that the Egyptian astronomers could have learned the Saros from the Babylonians, and used it to make their eclipse predictions.

In Babylonian practice, the 38 eclipse possibilities within the 223 synodic month Saros was customarily split up into five groups of either 7 or 8 eclipses. The first eclipse possibility in each group was at an interval of five synodic months from the preceding eclipse, the others all being at six month intervals. The groups were then arranged so that the two containing only seven eclipses were never next to each other.[11]

[7] See Section 2.8 above.

[8] *Almagest* (IV, 2); trans. Toomer (1984: 175).

[9] See, for example, the anonymous commentary found in Par. gr. 2841 and Par. gr. 2415, edited by Jones (1990).

[10] Parker (1959).

[11] Thus, 8-7-8-7-8, 7-8-7-8-8, 8-7-8-8-7, etc. See Section 2.8 above.

Finally, one had to decide when to begin the sequence. At least in the early period, the Babylonians decided that for the first cycle, the first eclipse seen 6n-1 months after another visible eclipse should begin a group. Eventually, however, the scheme has to be reset as unpredicted eclipses begin to occur. The Egyptian astronomers would have been faced by the same problem: when do we begin our Saros cycle? The dates of the eclipse predictions in the first text show that they did not use the rule of making the first eclipse seen at an interval of 6n-1 months after another eclipse the first in a group. Neither did they simply adopt the arrangement used by the Babylonians at that time.[12] Interestingly, however, the Egyptian layout is the same as that which would be given by an earlier Babylonian arrangement. In Babylon, this had been replaced in around 100 BC, about 20 years before the beginning of our text. However, this does not prove that the Egyptians had necessarily copied this earlier Babylonian scheme; only a handful of different arrangements are possible, and so there is always a reasonable chance that two schemes developed independently will turn out to be identical.

Another eclipse cycle known in Babylon, and almost certainly in pre-Roman Egypt, is the period of 135 synodic months.[13] Within this period there are 23 eclipse possibilities: 20 at six month intervals and the remaining 3 at five month intervals. These fall naturally into three groups, two containing eight eclipses and the other only seven. Once again, by repeating the arrangement after one cycle of 135 months, it is possible to predict the dates of eclipse possibilities. It may be significant that our text probably contains 23 eclipses — exactly the length of this cycle.[14] The eclipse possibilities are arranged in four groups containing 3, 8, 7, and 5, dates respectively. Could the first and fourth groups be the end and beginning of a group respectively? If so, we would have the expected 8-8-7 distribution within the 135 month period. Of course, a similar argument could be made that these were part of an incomplete 8-7-8-7-8 arrangement within the 223 month Saros period.

It is impossible to decide which of the 223 and 135 month eclipse periods was used to make the eclipse predictions in this text. In favour of the 223 month Saros is a Greek papyrus from Oxyrhynchus, P. Oxy. 4137,[15] which appears to be part of a list of lunar eclipse predictions from the middle of the first century AD. The preserved fragment contains entries for 10 December 56 AD and 5 June 57 AD. Although considerably more detailed than the Demotic text (in P. Oxy. 4137 each entry probably contained information concerning the duration, magnitude, location, etc. of the eclipse), they may both be examples of a standard variety of text. The predictions on the Oxyrhynchus papyrus were most probably made using the Saros cycle, or perhaps

[12] See Steele (1999a) for this arrangement. The Babylonian scheme would result in predictions on 13 June 83 BC and 24 July 76 BC, both one month later than found in our Egyptian text.

[13] This period was certainly known in the Roman period. The Demotic text P. Carls. 31, published by Neugebauer & Parker (1969: 241–243) where they tentatively date it on paleographical grounds to the second century AD, has been shown by Aaboe (1972) to contain a theoretical function based upon the 135 month eclipse period.

[14] Neugebauer, Parker, & Zauzich (1981: 322) count 24, but they assume that the traces at the beginning of the first line are part of an earlier entry. However, this seems unlikely as there are no other cases of these preserved words ending an account in this text.

[15] P. Oxy. 4137 has been translated and discussed by Jones (1999: v.1 87–94, v.2 16–17).

its triple, the Exeligmos, since no other cycles allow the magnitude and duration of the eclipse to be calculated. However, this does not prove that the predictions on the earlier Demotic text were also made using the Saros.

Once the months of the eclipse possibilities had been found (by whichever method), the dates of the oppositions were probably obtained from a simple numerical scheme. Such schemes are well known from Roman Egypt. For example, P. Carls. 9, written sometime after AD 145, presents a 25-year cycle for determining dates of conjunctions.[16] Neugebauer, Parker, & Zauzich (1981: 325) have shown that projecting this scheme back to the text's date, and adding 14 days to roughly convert from conjunctions to oppositions, yields dates that are in general agreement with our text. Since the Egyptian calendar only differs from the tropical year by about one-quarter of a day, the moon's longitude — which is only given in the text to the nearest zodiacal sign — could easily have been determined from the dates of the eclipse possibilities. As mentioned, the roughness of the quoted longitudes seems to preclude the use of methods which yield precise longitudes, such as the so-called "Standard Lunar Scheme."[17]

Let me now return to the records of the eclipses themselves. On 11 of the 23 dates given in the text, modern computations show that eclipses should have been visible in Middle Egypt.[18] These are indicated by a star after the date in column 1 of Table 3.1. In the majority of these instances there are details given recorded in the text that must have come from observation: the entrance angle of the shadow, the visibility of planets, etc. Somewhat confusingly, however, these appear to be recorded in a mixture of the future and past tenses, rather than simply the past tense as we would normally expect. Either these were predicted in some way (from an eclipse 18 or 54 years earlier?), or the scribe mixed up the tenses (either knowingly or unknowingly).[19] I will suspect the latter.

Times are preserved in five of the records of observed eclipses. These are all given in seasonal hours — a twelfth part of the night — which was presumably the common practice in Egypt.[20] The times are reproduced in column 2 of Table 3.1. In column 3 I have converted these to local times; column 4 gives the computed local times. I have assumed that these times relate to the moment of mid-eclipse, although this is not stated in the text.

From the Table it is evident that there is a considerable range in the errors of the timings. Three of the eclipses, 27 October 82 BC, 19 February 78 BC, and 4 June 74 BC, are in error by -0.31, -1.31, and $+1.13$ hours, respectively. However, the other

[16] P. Carls. 9 has been translated and discussed by Neugebauer & Parker (1969: 220-225).

[17] Jones (1997). The Standard Lunar Scheme is in any case only attested in the Roman period.

[18] I am here assuming that the text was written in Abusir al-Malak (latitude $+29.25°$, longitude $-31.08°$), and that any observations were made from there. This seems to be a reasonable assumption, but even if the "observer" was based elsewhere in Egypt — say in Alexandria — this would make little difference to the visibility and time of the computed eclipses as most potential sites are located within a thin strip of land running more or less north-south along the course of the Nile.

[19] Cf. Neugebauer, Parker, & Zauzich's (1981: 312) comment that "as befits the nature of (the scribe's) text his style is often elliptical."

[20] For example, the early Alexandrian eclipses reported by Ptolemy in the *Almagest* are timed in seasonal hours. See Section 3.3.2 below.

two times, for 3 May 82 BC and 28 January 76 BC, have errors of −4.24 and −7.16 hours respectively. Errors of this size surely cannot be errors in measurement. It is far more likely that these times were wrongly recorded by the scribe. This conclusion is supported by the fact that both of these times, together with the time of the eclipse on 4 June 74 BC, are recorded as being at the "9th hour (of night)." It would be very unusual indeed if three out of five measurements, all related to different events occurring, as is clear from the computed times, at effectively random times, would yield the same hour of night.

3.3 Eclipse Observations in Ptolemy's *Almagest*

Ptolemy's great astronomical treatise, usually known as the *Almagest*, has probably aroused more interest among historians of astronomy than any other work from before the Renaissance. Known in Greek as the *Mathematike Syntaxis*, it was written in Alexandria in the middle of the second century AD.[21] The *Almagest* is a comprehensive treatise of mathematical astronomy, containing, in the words of Pedersen (1974: 11), a "brilliant exposition of everything achieved by Ptolemy himself and by the most remarkable of his predecessors among the Greek astronomers." For over one thousand years after its composition it acted as the standard textbook of astronomy and had a profound influence on both Arabic and European astronomical thinking until as late as the sixteenth century AD. In addition to the *Almagest*, Ptolemy wrote a number of other works including the *Planetary Hypotheses* and the *Handy Tables*, which give a general background to the astronomy of the *Almagest* (occasionally incorporating some minor changes) and simplified rules for its use, the *Geography*, containing the information required to construct a world map, and the *Tetrabiblos*.[22] During the Middle Ages, it was for this last work that Ptolemy was most widely known. It is largely a work of astrology, detailing how the astronomical calculations made using the *Almagest* can be used to deduce the influences of the various heavenly bodies on the Earth.

Although primarily a work of astronomical theory, a number of observations are quoted in the *Almagest* as illustrative examples of the methods that Ptolemy used in calculating the parameters for his theories.[23] These include observations of lunar eclipses, occultations, equinoxes, and a catalogue of over one thousand stars. However, the reliability and authenticity of Ptolemy's observations have been questioned since the nineteenth century,[24] and, in his recent book *The Crime of Claudius Ptolemy*, Robert Newton concludes that "Ptolemy is not the greatest astronomer in antiquity, but he is something still more unusual: He is the most successful fraud in the history of

[21] Toomer (1984: 1) suggests it was written sometime between AD 150 and AD 161.

[22] Pedersen (1974: 391–407) gives a brief overview of Ptolemy's other works.

[23] For details of Ptolemy's theories, see Pedersen (1974) and Neugebauer (1975: 21–261). A list of the observations in the *Almagest* is given by Pedersen (1974: 408–422).

[24] See Britton (1992: ix–x) for an historical overview.

science."[25] However, Newton's statistical arguments for most of the observations in the *Almagest* being fraudulent have been heavily criticised by Swerdlow (1979). The most recent evaluation of Ptolemy's solar and lunar observations is by Britton (1992). He concludes that the observations described by Ptolemy were only a small sample of those that he used in deriving the parameters for his theories.

The astronomical observations reported by Ptolemy in his *Almagest* fall into three distinct groups: those said to have been observed in Babylon, those observed by early Greek (or Egyptian?) astronomers, and those observed by Ptolemy and his contemporaries. It would appear that Hipparchus was the source of most of the observations in the first two categories, as Ptolemy himself notes:

> "Hence it was, I think, that Hipparchus, being a great lover of truth ... especially because he did not yet have in his possession such a groundwork of resources in the form of accurate observations from earlier times as he himself has provided to us, although he investigated the theories of the sun and the moon, and, to the best of his ability, demonstrated with every means at his command that they are represented by uniform circular motions, did not even make a beginning in establishing theories for the five planets, not at least in his writings which have come down to us. All that he did was to make a compilation of the planetary observations arranged in a more useful way, and to show by means of these that the phenomena were not in agreement with the hypotheses of the astronomers of that time."
>
> [*Almagest*, ix, 2; trans. Toomer (1984: 421)]

Although only planetary observations are discussed here, it seems reasonable to suppose the Hipparchus also made a compilation of lunar observations. In a notoriously ambiguous passage, Pliny remarks that:

> "... the courses of both stars (the sun and moon) for 600 years were prophesied by Hipparchus, whose work embraced the calendar of the nations and the situations of places and aspects of the peoples — his method being, on the evidence of his contemporaries, none other than full partnership in the designs of nature."
>
> [*Historia Naturalis*, ii, 9; trans. Rackham (1937: 203)]

It has often been assumed that the "courses of both stars" refers to eclipses, and that this passage indicates that Hipparchus predicted eclipses for a period of 600 years. This would be a monumental — and pointless (especially since Hipparchus admits to possessing a lunar model with only preliminary parameters) — task, leading Neugebauer (1975: 320–321) to suggest that the passage refers to Hipparchus collecting together *observations* of eclipses covering the previous 600 years. If this interpretation were correct then we would have a ready source from which Ptolemy could have taken the eclipse observations he needed for deriving his parameters. However, Goldstein & Bowen (1995) have claimed that Pliny might not be discussing eclipse observations or predictions, but rather that he is attributing a scheme to Hipparchus for determining the daily progress of the moon with a period of 600 years. Whichever

[25] Newton (1977: 379).

interpretation is correct, it seems from the statement by Ptolemy that we can at least assume that Hipparchus compiled a list of some eclipses from the period before his time, and that this was Ptolemy's main source.

The eclipse observations in the *Almagest* are used to deduce some of the basic parameters for Ptolemy's lunar model. He explains that he uses lunar eclipse observations because:

> "... these are the only observations which allow one to determine the lunar position precisely; all others, whether they are taken from passages [of the moon] near fixed stars, or from [sightings with] instruments, or from solar eclipses, can contain a considerable error due to parallax."
>
> [*Almagest*, iv, 1; trans. Toomer (1984: 173)]

As a result, Ptolemy describes no observations of solar eclipses, but only 18 lunar obscurations.[26]

Rather than quote the records in full here, the observed details of each eclipse are summarized in Table 3.2, together with a reference to the page of the full translation of the record in Toomer (1984). Before discussing any of these observations in detail, it is necessary to make a few general remarks about the nature of the records.

Ptolemy used his own chronology throughout the *Almagest*. This commenced with the beginning of the reign of Nabonassar (equivalent to 26 February 747 BC), for, as he says, "that is the era beginning from which the ancient observations are, on the whole, preserved down to our own time."[27] He also uses the Egyptian year of twelve 30-day months followed by 5 extra days,[28] and, when discussing observations made during the night, often gives the date of both the preceding and following day to avoid confusion.[29] Most of the times quoted by Ptolemy are in seasonal hours. Ptolemy's first step in using these times is to convert them to equinoctial hours. He uses Alexandria as his meridian and so in his analysis of the lunar eclipse observations reduces all times to Alexandrian local time.

Let me now consider each of the three sets of eclipse observations — the Babylonian, the Early Greek, and the Later Greek — in turn.

3.3.1 Babylonian Observations

The earliest ten lunar eclipses reported in the *Almagest* are all said to have been observed in Babylon. According to Ptolemy, Babylon is $\frac{5}{6}$ of an hour, or 12.5°, to the east of Alexandria. The true difference in longitude is about 14.5°. This discrepancy reflects one of the greatest problems faced by the ancient Greek astronomers: the determination of geographical longitudes. In his *Geography*, written about a decade

[26]Two further eclipse observations by Hipparchus on 21 April 146 BC and 21 March 135 BC are mentioned at *Almagest* iii, 1. However, no details of these observations are given.

[27]*Almagest*, iii, 7; trans. Toomer (1984: 166)

[28]For a recent discussion of Ptolemy's dating system, and the royal canon on which it is based, see Depuydt (1995).

[29]This seems to have been standard practice among Greco-Roman astronomers and astrologers. See Toomer (1984: 12).

Date	Location	Observation	Page in Toomer
-720 Mar 19	Babylon	The eclipse began, it says, well over an hour after moonrise, and was total	191
-719 Mar 8	Babylon	The [maximum] obscuration, it says, was 3 digits from the south exactly at midnight	191–192
-719 Sep 1	Babylon	The eclipse began, it says, after moonrise, and the [maximum] obscuration was more than half [the disk] from the north	192
-620 Apr 21	Babylon	... at the end of the eleventh hour in Babylon, the moon began to be eclipsed; the maximum obscuration was $\frac{1}{4}$ of the diameter from the south	253
-522 Jul 16	Babylon	1 [equinoctial] hour before midnight at Babylon, the moon was eclipsed half its diameter from the north	253
-501 Nov 19	Babylon	... when $6\frac{1}{3}$ equinoctial hours of the night had passed; at this eclipse the moon was, again, obscured from the south $\frac{1}{4}$ of its diameter	208
-490 Apr 25	Babylon	... at the middle of the sixth hour [of night]. It is reported that at this eclipse the moon was obscured 2 digits from the south	206–207
-382 Dec 23	Babylon	... a small section of the moon's disk was eclipsed from the summer rising-point when half an hour of night was remaining	211–212
-381 Jun 18	Babylon	[the moon] was eclipsed from the summer rising-point when the first hour [of night] was well advanced	212
-381 Dec 12	Babylon	[the moon] was totally eclipsed, beginning from the summer rising-point, after 4 hours [of night] had passed	213
-200 Sep 22	Alexandria	In this eclipse the moon began to be obscured half an hour before it rose, and its full light was restored in the middle of the third hour [of night].	214
-199 Mar 19	Alexandria	... it began when $5\frac{1}{3}$ hours of night had passed, and was total	214
-199 Sep 12	Alexandria	... it began when $6\frac{2}{3}$ hours of night had passed, and was total	215
-173 May 1	Alexandria	... from the beginning of the eighth hour till the end of the tenth hour in Alexandria, there was an eclipse of the moon which reached a maximum obscuration of 7 digits from the north	283
-140 Jan 27	Rhodes	... at the beginning of the fifth hour [of night] in Rhodes, the moon began to be eclipsed; the maximum obscuration was 3 digits from the south	284
+125 Apr 5	Alexandria	... $3\frac{3}{5}$ equinoctial hours before midnight. At this eclipse too the moon was obscured $\frac{1}{6}$ of its diameter from the south	206
+133 May 6	Alexandria	We computed the exact time of mid-eclipse as $\frac{3}{4}$ of an hour before midnight. It was total.	198

Table 3.2: Eclipse observations in Ptolemy's *Almagest*.

Date	Location	Observation	Page in Toomer
+134 Oct 20	Alexandria	We computed that mid-eclipse occurred 1 equinoctial hour before midnight. [The moon] was eclipsed $\frac{5}{6}$ of its diameter from the north	198
+136 May 5	Alexandria	We computed that mid-eclipse occurred 4 equinoctial hours after midnight. [The moon] was eclipsed half of its diameter from the north	198

Table 3.2 (cont.): Eclipse observations in Ptolemy's *Almagest.*

after the *Almagest*, Ptolemy placed Babylon $1\frac{1}{4}$ hours ($18\frac{1}{2}^{\circ}$) east of Alexandria. The probable reason for this discrepancy is that Ptolemy used a different value for the size of the Earth's circumference in the two works: the Eratosthenic value of 180,000 stades in the *Almagest* and Marinus of Tyre's value of 240,000 stades in the *Geography.* Thus an estimate of the east-west distance between Babylon and Alexandria derived from land measurements would convert into a different number of degrees on the two globes.[30]

There has been much debate over the years concerning the source for the Babylonian observations in the *Almagest.* As I have suggested above, Ptolemy probably obtained them from Hipparchus. But how did Hipparchus come to possess them? Did he travel to Babylon himself and persuade a Babylonian astronomer to translate the records for him, as has been suggested by Toomer (1988: 359)? Or was knowledge of Babylonian astronomy widespread among Greco-Roman astronomers? Certainly, Babylonian mathematical astronomy must have been known fairly widely. The Oxyrhynchus papyri show that astrologers both used and understood Babylonian ACT methods, and there is nothing to suggest that the Oxyrhynchus material would differ from that which would have been found in any other medium size city had the conditions for survival of papyri been as good. But the transmission of actual observations is a different matter. To be of any value in determining, say, some parameter of a lunar theory, not just any lunar observation will do. Eclipses, of course, are the most useful, but even then what one really needs are eclipses fulfilling certain conditions. Furthermore, these conditions are in part dependent upon the theory that one is developing. Thus, a Greek astronomer wanting to use Babylonian eclipse observations would probably have to obtain a long run of records, from which he could select certain ones at a later date.

The Babylonian records described by Ptolemy range in date from 721 BC to 382 BC. Even if these represent the earliest and latest reports available to him, Ptolemy must have had access to records covering a period about 350 years. Furthermore, this list of eclipses must have been fairly complete. It has generally been supposed that the original source for these records was the Babylonian Astronomical Diaries. However, there are good reasons for doubting that this was the direct source from which

[30]See Schnabel (1930: 218–219).

Hipparchus made his compilation. A typical Diary covers a period of six months, during which there will be one, or occasionally two, lunar eclipse possibilities (i.e., observations or predictions). In compiling a collection spanning 350 years, therefore, one would have to consult about 700 tablets. It seems unlikely that a Babylonian astronomer would read through all of these tablets, even if they were all preserved, and then explain them at the request of a visiting Greek. There is, however, another ready-made source which would have been of much greater use to Hipparchus: the large compilation of eclipse records, preserved in part on LBAT *1414, LBAT 1415 + 1416 + 1417, and LBAT *1419.[31]

As I have discussed in Section 2.5 above, when complete this large compilation probably comprised eight tablets and covered the period from 747 BC to 315 BC. The texts form a large matrix with each cell representing an eclipse possibility, successive cells in a row separated by either five or six months, and each column separated by one Saros from the preceding column. A direct translation of these texts, preserving their layout, would have solved many problems for Hipparchus. First of all, only eight texts would have to be consulted. Second, several chronographical questions would be answered. For instance, a more or less complete list of the length of each king's reign can be taken directly from the dates of the eclipses recorded in the compilation.[32] By the time they had reached the *Almagest*, the dates of the Babylonian observations had been converted from the Babylonian luni-solar calendar to the Egyptian calendar and the era Nabonassar. Hitherto, people have usually assumed that this would require abstracting the length of each Babylonian month from the Diaries,[33] but, as Alexander Jones recently pointed out to me, this would hardly be be practicable, particularly for the very early records. Fortunately, however, the structure of the large compilation allows one to avoid this problem. Assuming one knew the equivalent dates of the eclipses in the final column of the compilation (333–315 BC), which seems quite possible since Babylon was under Greek rule by that period and so there must surely have been documents detailing the concordance of the native calendar with the Greek calendar, one could obtain the dates of all the other entries in the compilation by using the fact that columns are separated by one Saros of about $6585\frac{1}{3}$ days. Although not exact, using the Saros in this way would allow the number of days between any two observations in the compilation to be calculated with scarcely any error.

It is probably significant that all of the Babylonian eclipse records described by Ptolemy come from the period covered by this large compilation. It may also be worthy of note that the only planetary observations which probably come from Babylon that Ptolemy uses are from the third century BC. As mentioned above, by this period Babylon was under Greek rule and we would expect that the Greeks would have an understanding of the native calendar in use at that time. Furthermore, the use of the Metonic cycle to determine intercalations in Babylon would have eased conversion of

[31] Toomer (1988: 359) and Walker (1997: 21) have both hinted that the Eclipse Texts could have been Hipparchus' source.

[32] In modern times, Parker & Dubberstein (1956) made extensive use of the Eclipse Texts in establishing Babylonian chronology.

[33] See, for example, Toomer (1988: 356), Depuydt (1995: 103), and Steele (1999b).

Babylonian dates to the Callippic calendar.

Let me now compare a record of an eclipse seen in Babylon as described by Ptolemy, with one from a cuneiform source.

> "The first eclipse we used is the one observed in Babylon in the thirty-first year of Darius I, Tybi 3/4 in the Egyptian calendar, at the middle of the sixth hour [of night]. It is reported that at this eclipse the moon was obscured 2 digits from the south..."
>
> [*Almagest*, iv, 9; trans. Toomer (1984: 206–207)]

> "Month VII, the 13th, in 17° on the east side, all was covered; 28° maximal phase. In 20° it cleared from east to west. Its eclipse was red. Behind the rump of Aries it was eclipsed. During onset, the north wind blew, during clearing, the west wind. At 55° before sunrise."
>
> [LBAT *1419, Obv. VI', 2'–8'; trans. Sachs & Hunger (1999)]

In addition to the different calendars, it should be noted that different units are used to quote the times of eclipse in the two sources. The cuneiform records invariably give the time of an eclipse in *bēru* and UŠ. However, Ptolemy generally gives the time of the Babylonian observations in seasonal hours. How did the times come to be converted into seasonal hours? It would not make sense to assume that Ptolemy himself did this, as his first step in analysing the records is to convert the seasonal hours back into equinoctial hours. This would suggest that the conversion was done either by the Babylonians themselves, or by Hipparchus when he compiled his list of eclipses.

Fotheringham (1932) was of the opinion that the conversion was made by the Babylonians using the so called "Ivory Prism." This is a small fragment of a prism, recovered from Nineveh and now in the British Museum, on which he concluded that a scheme for converting between seasonal hours and *bēru* and UŠ was written.[34] The scheme he claimed to identify was based on a ratio for the length of the longest to the shortest day of 2:1. As noted in Section 2.4, this ratio is very inaccurate for the latitude of Babylon; nevertheless there are examples of it found throughout the Late Babylonian period. However, a ratio of 3:2, which is a much better approximation to the true value, was also in use around this time.

Fotheringham claimed that his interpretation of the Ivory Prism was confirmed by the only record of an eclipse in the *Almagest* that is also preserved on an extant Babylonian tablet (BM 33066).[35] This tablet is an example of what I have termed "Miscellaneous Texts" in my discussion of the Babylonian Non-Mathematical Astronomical Texts (NMAT) in Section 2.2 above.[36] On the obverse of BM 33066 lunar six

[34] Fotheringham's reconstruction of the Ivory Prism has been published by Langdon (1935) and discussed by Smith (1969).

[35] This is not strictly true as the tablet BM 37088 + 37652 contains records of three further eclipses reported by Ptolemy: 23 December 383 BC, 18 June 382 BC, and 12 December 382 BC. However, the portion of the tablet containing these three records is so badly damaged as to render it virtually useless.

[36] BM 33066 was first published, although not fully understood, by Pinches (1888). Working from a copy by Strassmaier (1890: no. 400), Epping (1890) and Kugler (1903) then deciphered the text, and a transliteration and translation was published by Kugler (1907: 64–71). An improved version is given by Sachs & Hunger (1999).

data is recorded for the 7th year of Cambyses (523–522 BC). The reverse is divided into six sections containing the Greek-Letter phenomena of Jupiter, Venus, Saturn, and Mars (the "missing" Mercury phenomena are to be found on the right edge of the tablet, although this was not recognized by Kugler), for the 7th, 8th and (for Mars alone) 9th year of Cambyses; planetary conjunctions for his 7th year; and two lunar eclipses, also from his 7th year. The two eclipses date to 16 July 522 BC and 10 January 521 BC. The former date corresponds to the eclipse reported by Ptolemy in the *Almagest* (v, 14).[37] I quote below translations of both the Babylonian and Ptolemy's account of this eclipse.

> "Year 7, month IV, night of the 14th, $1\frac{2}{3}$ *bēru* after sunset, the moon made a total eclipse, a little remained; the north wind blew."
>
> [BM 33066, Rev. 19–20; trans. Sachs & Hunger (1999)]

> "... in the seventh year of Kambyses, which is in the 225th year from Nabonassar, Phamenoth 17/18 in the Egyptian calendar, 1 hour before midnight at Babylon, the moon was eclipsed half its diameter from the north. Thus the eclipse occurred about $1\frac{5}{6}$ equinoctial hours before midnight at Alexandria."
>
> [*Almagest*, v, 14; trans. Toomer (1984: 253)]

Fotheringham (1932) assumes that the time given in the *Almagest* is 1 seasonal hour before midnight, or 5 seasonal hours after sunset. Using his restoration of the Ivory Prism, he finds that 5 seasonal hours are equal to 1 *bēru* and 20 UŠ, or $1\frac{2}{3}$ *bēru*. However, there are a number of problems with his argument. Toomer (1984: 253) believes that the time given by Ptolemy is in equinoctial hours as Ptolemy merely adds his time difference between Babylon and Alexandria ($\frac{5}{6}$ hour) to obtain a time of "$1\frac{5}{6}$ equinoctial hours before midnight at Alexandria." Furthermore, the time of the eclipse given in the cuneiform record is for the start of the eclipse, whereas Ptolemy has the time of its middle. This would mean that Ptolemy would have had to have mistaken the phase of the eclipse and to have then failed to change from seasonal to equinoctial hours before making his calculations.

It is also unwise to base any conclusions concerning the Babylonian records on this tablet alone, since it does not fall into any of the common categories of text.[38] In particular, it is not certain whether this text contains observations or calculations of the phenomena it records. At least some of the data must be calculated. For instance, the full run of lunar six timings for the 7th year of Cambyses cannot all have been measured; clouds would surely have prevented their observation on at least some occasions. The lunar six data must therefore have been either all calculated, as suggested by Kugler (1907: 61–72), or be a mixture of observation and calculation.[39] There is also debate concerning whether the two lunar eclipses were observed or calculated. Huber (1973: 26) is of the latter opinion, basing his argument on the fact that the first

[37] This was first recognized by Oppert (1891).

[38] However, Fotheringham should not be criticised for this. BM 33066 was one of only a handful of tablets he had available to him. It is pure chance that it has turned out to be a very unusual text.

[39] The probable method by which the lunar six were calculated has recently been uncovered by Brack-Bernsen (1997: 90–129; 1999).

eclipse (quoted above) is said to be almost total, but that modern computation shows that only about half of the moon would have been eclipsed on this date. However, in the Babylonian NMAT texts, predicted eclipses of the moon are usually denoted by AN-KU$_{10}$ *sin*, but this text has the usual terminology for an observed eclipse: *sin* AN-KU$_{10}$. In any event, there is sufficient doubt concerning the interpretation of the Babylonian account of the eclipse on 16 June 523 BC that there is little to be gained by comparing it with that given in the *Almagest*.

Without the evidence from the eclipse of 523 BC there is no support for Fotheringham's conclusion concerning the use of the Ivory prism. It seems very unlikely that the Babylonians would have used this scheme to convert from equinoctial to seasonal time if they translated the Babylonian records for Hipparchus. By this period the ratio of 2:1 for the longest to the shortest night had more or less been abandoned. If the Babylonians had made these conversions then they would surely have used the later 3:2 schemes.

A more likely alternative is that the Babylonian observations came to Hipparchus in their original form and the Greeks converted the timings into seasonal hours. All of the *Almagest* eclipses observed by Greek astronomers earlier than Ptolemy are timed in seasonal hours and so this may have been standard Greek practice. It seems quite possible that Hipparchus himself could have made the conversion into seasonal hours so that his compilation of eclipse records, which presumably included contemporary Greek as well as Babylonian observations, had a uniform style.

Table 3.3 compares the times of the Babylonian eclipse observations reported by Ptolemy with the results of modern computation. In each case I have used Ptolemy's interpretation of the timings,[40] despite the fact that in some cases, for example the eclipses on 1 September 720 BC and 19 November 502 BC, the descriptions of the observations are very vague and open to other interpretations. Where necessary I have made the conversion from seasonal to equinoctial hours using modern computations of the length of the night. Fotheringham (1920a) and others have assumed that Ptolemy was mistaken about the phase of some of the eclipses and have chosen various alternatives. However, it seems better to initially follow Ptolemy in his interpretations, and then to consider other possibilities afterwards.

Let me note at this point that the eclipse on 23 December 383 BC could not have been observed in Babylon. On this date, the moon set before the eclipse began. Therefore, Ptolemy's claim that the eclipse was observed to begin half an hour before sunrise clearly cannot be true. Perhaps the Babylonian astronomers predicted an eclipse for this time and this was mistaken for an observation when it was transmitted to the Greeks. I have therefore ignored this record in the subsequent analysis of the errors in the observed times.

There does not appear to be any evidence for systematic errors in the times of the Babylonian eclipses quoted by Ptolemy; the mean error is +0.04 hours. The typical accuracy of these timings is 0.44 hours. In Chapter 2 I found that the accuracy of

[40]For example, Ptolemy assumes that the eclipse on 19 May 721 BC, which is described as beginning "well over an hour after moonrise," started at $1\frac{1}{2}$ hours after moonrise.

Date	Contact	Local Time (h) Observed	Computed
-720 Mar 19	1	19.20	19.58
-719 Mar 8	M	0.00	23.61
-620 Apr 21	1	4.60	4.33
-522 Jul 16	M	23.00	23.59
-501 Nov 19	M	23.58	0.18
-490 Apr 25	M	23.55	22.77
-382 Dec 23	1	6.42	7.16
-381 Jun 18	1	19.55	19.93
-381 Dec 12	1	21.67	21.37

Table 3.3: Babylonian eclipse times reported by Ptolemy.

the eclipse times preserved in the Late Babylonian Astronomical Texts was dependent upon the length of the time interval measured. The average length of the time interval measured in the eclipses quoted by Ptolemy is about $3\frac{1}{4}$ hours. The accuracy of time intervals of this length recorded in the Late Babylonian Astronomical Texts is just under half an hour. Thus, the Babylonian eclipse timings reported by Ptolemy are of comparable accuracy to those found in the cuneiform record. This not only suggests that there was no significant loss of accuracy when the Babylonian times were converted into seasonal hours,[41] but also adds strength to the argument that the observations are indeed genuine.

As Britton (1992) has shown, it is possible to reduce the errors in the times given by Ptolemy for the Babylonian observations on 19 March 721 BC, 1 September 720 BC, 19 November 502 BC and 18 June 382 BC by assuming a mistake in either Ptolemy's interpretation of the time of the eclipse or of the phase at which the time was measured. The most plausible of these corrections is for the eclipse of 721 BC where Ptolemy gives a time of "well over an hour after moonrise." As the Babylonians customarily made their timings with respect to sunset or sunrise it seems possible that Ptolemy had made a mistake in referring to moonrise. If sunset was assumed then the error would be reduced by 0.22 hours. Another possible correction to Ptolemy's reports is to assume that he was mistaken in his belief that the times were seasonal hours. However, there is no firm evidence in favour of making these corrections.

3.3.2 Early Greek Observations

There are five lunar eclipses observed by Ptolemy's Greek predecessors recorded in the *Almagest*. They range in date from 201 BC to 141 BC, and, with the exception of the final one, they were observed in Alexandria (latitude $+31.22°$, longitude $-29.92°$). The eclipse in 141 BC was observed in Rhodes (latitude $+36.43°$, longitude $-28.23°$) by Hipparchus; the names of the other observers are not given, and it is

[41] This strongly implies that the inaccurate Ivory Prism could not have been used to make these conversions.

Date	Contact	Local Time (h) Observed	Computed
-200 Sep 22	4	20.58	20.46
-199 Mar 19	1	23.33	23.05
-199 Sep 12	1	0.63	0.61
-173 May 1	1	0.89	0.49
-173 May 1	4	3.57	3.11
-140 Jan 27	1	21.69	20.71

Table 3.4: Early Greek eclipse times reported by Ptolemy.

possible that some of these may have been made by Egyptian rather than Greek individuals. Ptolemy's source for these observations was probably once again Hipparchus since he notes that most of them were also used by the earlier astronomer.

Seasonal hours were used for all of the time measurements by the early Greek astronomers given in the *Almagest*. They are known to have been in common use in antiquity, as is shown by the clepsydra constructed by Ctesibius at the beginning of the third century BC. This clock, which is described by Vitruvius in *De Architectura* (ix, 8), was marked with different scales for each of the Egyptian months to make allowance for the changing length of the seasonal hour. If we are to postulate that the early Greek astronomers measured the time of the observations with whichever clock was available to them, then it would seem quite possible that this would measure seasonal hours. Despite the fact that equinoctial hours were used by the astronomers in their theories and so any observations must have been converted into this system, it seems reasonable to suppose that in compiling any lists of observational records, the original observations, in seasonal hours, would have been copied.

It is interesting to note that the record of the eclipse in 201 BC not only gives the time that the eclipse was observed to finish, but also an estimate of the unobserved time it began, this being half an hour before the moon rose. This moment was presumably calculated by timing the later phases of the eclipse and estimating its duration. This was also often the practice of the Babylonian astronomers. They could calculate this time with an accuracy of about half an hour.[42] In this case the Greek astronomer was only 0.25 hours late in his estimate.

The eclipse timings made by the early Greek astronomers are listed in Table 3.4. Unlike the Babylonian observations, there appears to be a systematic error in the times of all of these observations. The mean value of this error is −0.38 hours. The time of the start of the eclipse in 141 BC is almost an hour early. This is noticeably less accurate than the other records in this group which suggests that there may be some problem with the record. Britton (1992: 69) has suggested that the time may relate to the middle of the eclipse rather than the beginning. This would reduce the error to −0.09; however, there is no real justification for making this correction and so it seems better simply to say that this record is somehow corrupt, and to ignore it in calculating

[42] See Section 2.7 above.

the mean error, which now reduces to −0.25. Nevertheless, even after discarding this record there is still appears to be a systematic error in the times of these early Greek observations.

3.3.3 Later Greek Observations

There are four eclipse observations made between AD 125 and AD 136 recorded in the *Almagest*. Those in AD 133, AD 134, and AD 136 are all said by Ptolemy to have been "very carefully observed by us in Alexandria."[43] The eclipse of AD 125 was also observed in Alexandria, but the name of the observer is not given. It is possible that this was also Ptolemy; Toomer (1984: 206) suggests that the mathematician Theon may have been the observer.

All four of these records give the time of the middle of the eclipse in equinoctial hours. Ptolemy gives no details of how he computed the mid-point of the eclipses. Presumably the time of the beginning and end of each eclipse was measured and the mid-point taken. Ptolemy also uses equinoctial hours in his astrological work, the *Tetrabiblos*. For example, when discussing the astrological interpretation of eclipses he states that:

> "For when these data are examined, if it is a solar eclipse, we shall understand that the predicted event lasts as many years as the equinoctial hours which we discover, and if a lunar eclipse, as many months."
>
> [*Tetrabiblos*, ii, 6; trans. Robbins (1940: 167)]

In case the reader uses seasonal hours, however, Ptolemy explains how the conversion to equinoctial hours may be made. Over one hundred years earlier, Manilius, in his Latin astrological poem the *Astronomica* (iii, 218–275), had also stressed the importance of using equinoctial hours in astrology. From a study of horoscopes, however, Neugebauer & van Hoesen (1959: 170) have concluded that seasonal hours were usually used in Greek and Roman astrology until well after Ptolemy's time.

Table 3.5 lists the observed and computed times of the four eclipses. As with the early Greek observations there appears to be some evidence of a systematic error; the mean error in the observed times is −0.23 hours. However, the eclipse in AD 136 is considerably less accurate than the other three eclipses, and without it the systematic error is reduced to −0.07 hours. The typical accuracy of these timings is 0.35 hours, but this is reduced to 0.24 hours if the eclipse of AD 136 is removed. It should be remembered, however, that this is a small sample of data.

[43] *Almagest*, iv, 6; trans. Toomer (1984: 198).

Date	Contact	Local Time (h) Observed	Computed
+125 Apr 5	M	20.40	20.65
+133 May 6	M	23.25	22.86
+134 Oct 20	M	23.00	22.93
+136 May 5	M	4.00	3.30

Table 3.5: Later Greek eclipse times reported by Ptolemy.

3.4 A Solar Eclipse Observation by Theon of Alexandria

Little is known of Theon of Alexandria. It is believed that he was active between about AD 360 and AD 380, and that he was a non-Christian living in Alexandria.[44] Though his own contributions to the development of mathematics and astronomy were fairly modest, he became very influential through his editions of and commentaries on works by authors such as Ptolemy and Euclid. Theon was probably involved in teaching these texts, and it seems possible that he wrote his commentaries for the benefit of his students. Perhaps the best known of Theon's works are his commentaries on Ptolemy's *Almagest* and *Handy Tables*, and a larger treatise that attempts to show how Ptolemy derived the *Handy Tables* from his theories in the *Almagest*.[45]

As part of his commentary on the *Almagest*, Theon compares the circumstances of a solar eclipse that he himself has observed, with those calculated by Ptolemy's methods. He describes the eclipse, which he observed on 16 June 364 AD, as follows:

> "... the time reckoned by civil days and equinoctial hours of the exact ecliptic conjunction which we have discussed, and which took place according to the Egyptian calendar in the 1112th year from the reign of Nabonassar $2\frac{5}{6}$ equinoctial hours after midday on the 24th of Thoth, and according to the Alexandrine calendar reckoned by simple civil days in the 1112th year of the same reign $2\frac{5}{6}$ equal or equinoctial hours after midday on the 22nd of Payni ... And moreover we observed with the greatest certainty the time of the beginning of contact, reckoned by civil and apparent time as $2\frac{5}{6}$ equinoctial hours after midday, and the time of the middle of the eclipse as $3\frac{4}{5}$ hours, and the time of complete restoration as $4\frac{1}{2}$ hours approximately after the said midday of the 22nd of Payni."
>
> [Theon of Alexandria 332; trans. Fotheringham (1920b)]

Table 3.6 lists Theon's observed times and those given by modern computations. It is immediately clear that all of the contact timings are early by just under half an hour.[46] However, despite this systematic error, these three timings are very self-consistent.

[44] For further biographical details, see Toomer (1977).

[45] Jones (1996).

[46] The cause of this systematic error is not known. The simplest explanation is that Theon used a poorly calibrated clock to time the eclipse.

	Local Time (h)	
Contact	Observed	Computed
1	14.83	15.25
M	15.80	16.16
4	16.50	16.99

Table 3.6: Timings of the solar eclipse observed by Theon of Alexandria.

3.5 Accuracy of the Observed Times

The errors in the observed times of the eclipses recorded in the Demotic and Greek sources discussed in this chapter are shown in Figure 3.1.[47] The three groups of eclipse observations recorded in the *Almagest* show distinct differences in their variations from modern computations. Unsurprisingly, given their age, the Babylonian observations appear to be the least accurate of the three groups. However, unlike both the early and later Greek observations, they show negligible systematic error in their timings. Indeed the systematic error in the timing of the early Greek observations is very serious and must lead us to question how representative these records are of the observations being made at this period. The systematic error in the later Greek observations is almost wholly caused by the time of the eclipse in AD 136. This was one of the eclipses said to have been observed by Ptolemy himself, but nevertheless it is necessary to question his own interpretation of it.

Although there are only three Demotic timings and so it is hard to judge how typical their errors are, they do appear to be slightly less accurate than the contemporary observations reported by Ptolemy. This is unsurprising, however, since these give the appearance of only being rough estimates of the time of the eclipse, stated to only the nearest seasonal hour, rather than precise measurements of the kind described by Ptolemy. It seems plausible to suggest that Ptolemy will have been selective in choosing which eclipses to use. Not only will he have picked timings he believed to be accurate, on the basis of their precision, but he may also have considered how well they agree with his, or possibly Hipparchus', theories.

Given that the observations in the *Almagest* represent only a proportion, although not necessarily a particularly small proportion, of those available to Ptolemy, and that he probably used some form of averaging in deriving his parameters,[48] we may wonder what were his criteria for choosing them. Obviously, they had to result in plausible parameters for his theories, but there may have been other considerations. In particular, why did he switch from using Babylonian to Greek eclipse observations after the third century BC? The collection of Babylonian eclipses available to him may well have continued up to the date of its compilation. If Hipparchus was its author, as has been suggested, this would have been in the middle of the second century BC. So it would

[47]Only the errors in three of the Demotic timings are shown here. I have omitted the two large errors which I have attributed to scribal mistakes in Section 3.2 above.

[48]Britton (1992: 151).

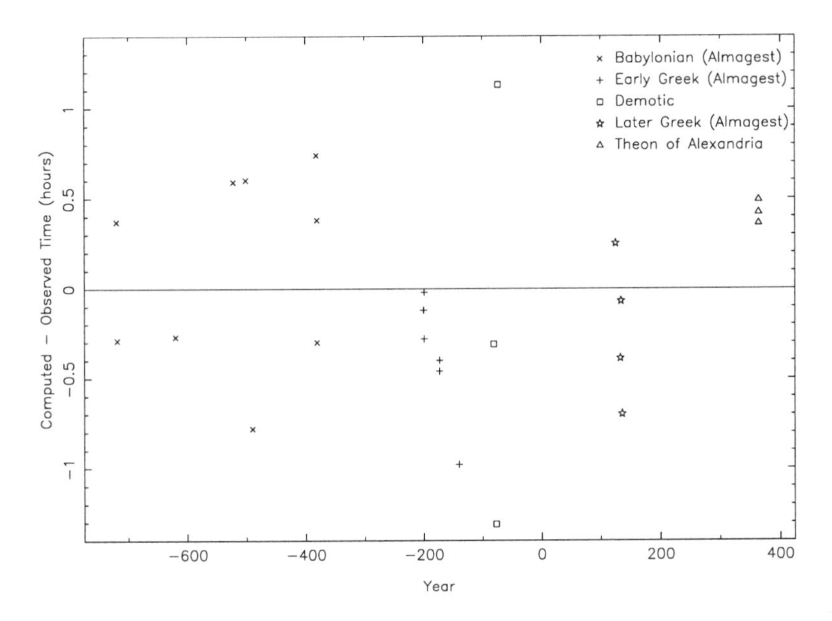

Figure 3.1: Error in the observed eclipse times recorded in Demotic and Greek sources.

appear that Ptolemy preferred to use Greek observations where they were available to him. This may have been for practical reasons: Ptolemy would have known that his correction for the difference in longitude between Babylon and Alexandria was only approximate and that this could seriously affect the derivation of his parameters. Perhaps he was also suspicious of the reliability of the Babylonians as astronomers as he in unlikely to have known how they made their observations, whereas he would probably have the exact details of the practices of the Greek astronomers readily available to him. It is also possible that he believed that his readers might have more faith in Greek observations than those made in a foreign, and now deserted, city. In retrospect, given that the early Greek timings apparently contain a systematic error, this might appear to us to have been a mistake, but in the context of the *Almagest* itself, they served as examples of Ptolemy's methods, and they undoubtedly successfully fulfilled this role.

Chapter 4

The Islamic Near East

4.1 Introduction

At the same time as the so-called European Dark Ages, which, in terms of science, were not as dark as had once been supposed,[1] science and learning were occupying an important place in the society of the neighbouring Islamic Near East. Astronomy, of course, was a major aspect of this science. According to King (1996: 146), "it was not Islam that encouraged the development of astronomy but the richness of Islamic society, a multiracial, highly-literate, tolerant society with one predominant language, Arabic." Islamic astronomy was founded on a mixture of traditional "folk" astronomy and Indian, Persian, and Greek mathematical astronomy, but by the turn of the second millennium it had evolved into a science characteristically its own.[2]

Given the vast number of preserved Islamic astronomical manuscripts,[3] it is surprising how few records of astronomical observations have so far been uncovered. In fact, less than fifty records of eclipses made by Islamic astronomers are currently known, although there are undoubtedly more in sources that have not yet been studied. All of these date from the ninth to the eleventh centuries AD, and were observed by only six different people: Ḥabash, al-Māhānī, al-Battānī, the Banū Amājūr, Ibn Yūnus and al-Bīrūnī. With the exception of those by al-Battānī and al-Bīrūnī, all are contained in the writings of Ibn Yūnus. The Islamic observations have been extensively investigated by a number of authors such as Newcomb (1878), Newton (1972), and, most recently, Said & Stephenson (1997). There is little that can be added to these previous studies, so in this present chapter I will simply give an overview of the available source material. For completeness, I have computed the circumstances of all of the eclipses for comparison with the recorded observations and predictions; however, the results of this analysis vary little from those found by earlier authors.

[1] See, for example, McCluskey (1998) and Chapter 5 below.
[2] King (1996: 144).
[3] King (1996: 143) estimates that there are over 10,000 in libraries throughout the world.

4.2 Timed Eclipse Records from the Islamic Near East

Astronomical records from the Islamic Near East can be found in two main sources: historical chronicles and astronomical treatises. The latter will form the basis for the present investigation as the reports in the chronicles, although relating to a wide variety of celestial observations,[4] generally lack technical details. There are many thousands of manuscripts preserved in collections throughout the world that contain the writings of medieval Islamic astronomers. These include general astronomical and cosmological discussions, commentaries on works such as the *Almagest*, and, perhaps most importantly, a group of texts known as *zījes*.

According to Kennedy (1956: 123), who published a survey of the 125 or so *zījes* then known, a *zīj* "consists essentially of the numerical tables and accompanying explanation sufficient to enable the practicing astronomer, or astrologer, to solve all of the standard problems of his profession." In other words, the *zījes* contain the necessary tables for calculating planetary and stellar positions and eclipses, and for the measurement of time. Sometimes, but not always, the *zījes* contain explanations of the theories that they present and details of the observations on which the tables were based. Although the number of known *zījes* has increased to nearly 200 since Kennedy's survey, nothing has been uncovered to significantly change his findings.[5] However, it should be noted at this point that in many cases all that is known of a *zīj* is its title.

The astronomy of the *zījes* is largely based upon that of Ptolemy. However, by the eighth century AD it had became evident to the Islamic astronomers that some of Ptolemy's parameters were either outdated or inaccurate. It was partly in an attempt to correct these errors that astronomers compiled their *zījes*, which, in some cases, necessitated making careful celestial observations. This is explained by the ninth century Baghdadi astronomer Ḥabash al-Ḥāsib in the introduction to a work whose title is lost but is now commonly referred to as the *Damascus Zīj*:

> "... the astronomers of ancient times, who established the principles of this science, have left nothing for their successors to do with the exception of improvements of exposition, arrangement of material so as to facilitate understanding, correction of certain mistakes on the basis of examples of procedure available in their own texts, or the rectification of errors which were introduced after their own time."
>
> [*Damascus Zīj*, 144–145; trans. Sayili (1960: 80)]

The first organized program of Islamic astronomical observations was instigated by the Abbasid Caliph al-Maʾmūn during the first part of the ninth century AD. First in Baghdad and later in Damascus, al-Maʾmūn built observatories in which he gathered groups of astronomers.[6] They were charged with making observations of the sun and

[4] Including more than one hundred solar and lunar eclipses. See Said, Stephenson, & Rada (1989) and Stephenson & Said (1997).

[5] King (1996).

[6] Sayili (1960: 50–87).

moon, for, according to Ḥabash, it was in the calculation of eclipses that Ptolemy's tables showed the greatest signs of needing to be updated. Some of the results from the observations made in Baghdad were incorporated into the *Mumtaḥan Zīj* by Yaḥyā ibn Abī Manṣūr.[7] In the following centuries, many observatories, both private and royal, were built. It was often within these observatories that astronomers worked, making observations, compiling *zīj*es, and, often, instructing students. Some astronomers did work outside of the observatory structure, however. For example, Ibn Yūnus observed from a variety of sites in Cairo.

Before going any further, it will be useful to outline the Islamic calendar and its relation to the Julian calendar. The Islamic calendar is a strictly lunar calendar — no attempt is made to reconcile the lunar months with the solar year by the use of intercalation. The year consists of twelve months each of which begins with the first sighting of the lunar crescent after the new moon. At all major population centres a watch is kept on the 29th day of every month for the lunar crescent. If it is not seen, either because conjunction has not yet occurred or the moon is obscured by cloud, then an extra day is added to that month. Thus months in the Islamic calendar, like those in the Babylonian calendar discussed in Section 2.3 above, can have either 29 or 30 days. Due to the unpredictability of the sighting of the lunar crescent, it is difficult to calculate the exact details of the Islamic calendar in the past. Thus, to simplify conversion between the Islamic calendar and the western calendar, an idealized Islamic calendar is used. In this system, the first eleven months of the year contain alternatively 30 and 29 days each. The final month contains 29 days in most years, but 30 days in leap years.

Years in the Islamic calendar are numbered from the year that the Prophet Muḥammad fled from Mecca to Medina. This event, called the Hijra, occurred in AD 622. Years reckoned from this epoch are designated *al-Hijra*, which has been latinized to Anno Hegirae (AH). Freeman-Grenville (1977) has constructed tables which allow dates given in the Islamic calendar to be converted to the Julian calendar with relative ease. As these tables are based upon the idealized calendar discussed above, errors of a day are possible in converting between these two calendars. However, for the astronomical records these can generally be eliminated as the astronomers often recorded not only the date of an observation, but also on which of the seven days of the week it was made. Occasionally, the Islamic astronomers recorded the date of their observations in other calendars. For example, al-Bīrūnī used a hybrid system in which the day and month were given in the Persian calendar and the year was given from the era of Nabonassar.[8]

Timed eclipse records are known to be preserved in the writings of only three Islamic astronomers: al-Battānī, Ibn Yūnus and al-Bīrūnī. All three men describe a number of eclipses that they observed, and, in addition, Ibn Yūnus quotes a number of eclipses that were observed by three earlier astronomers: Ḥabash, al-Māhānī and the Banū Amājūr. Many other Islamic astronomers, particularly those based in the

[7] King (1996).

[8] For details of the use and operation of these calendar systems, see Said & Stephenson (1996).

observatories, also observed eclipses.[9] However, their reports have either been lost, or are to be found in works that have not yet been studied.

The Islamic astronomers had two main motives in observing eclipses: testing the accuracy of astronomical tables and determining geographical longitudes. All of the *zīj*es contained tables for predicting eclipses, and, by testing these tables, it was possible to assess the accuracy of the lunar and solar theory upon which they were based. With the exception of Ḥabash, all the astronomers from whom eclipse observations are known made a comparison with the circumstances of the eclipses calculated by means of a *zīj* in some of their records.

The use of simultaneous observations to calculate geographical longitudes has been known since antiquity. For example, Strabo quotes Hipparchus as noting that this was the only method by which longitudes could be precisely determined:

> "Many have testified to the amount of knowledge which this subject (Geography) requires, and Hipparchus, in his Strictures on Eratosthenes, well observes, 'that no one can become really proficient in geography, either as a private individual or as a professor, without an acquaintance with astronomy, and a knowledge of eclipses ... the only means we possess of becoming acquainted with the longitudes of different places is afforded by the eclipses of the sun and moon.' Such are the very words of Hipparchus."
>
> [*Geographica*, I, 1, 12; trans. Hamilton & Falconer (1903: 13)]

Strabo is obviously slightly confused about the method for determining longitude using eclipses. Essentially, one measures the local time of a simultaneous event at two locations, and the difference in time corresponds to the difference in longitude expressed in hours. Later times indicate that one is further to the east. To convert to degrees of longitude, one simply divides by 15. As Strabo writes, eclipses of the moon can be used for this purpose as they are simultaneous events seen at the same moment at any location on the Earth's surface (providing the moon is above the horizon). However, the circumstances of solar eclipses vary from place to place, and so they cannot be used in this way. But in any event, it does not appear that there was a sufficiently well organized system of astronomical observation in antiquity for this method to have been of widespread use,[10] although, it is possible that the comparison of simultaneous observations of lunar eclipses in Babylon and Alexandria or other Greek cities to determine the difference in their longitudes may have been one of the motivations behind the compilation (by Hipparchus?) of historical eclipse observations discussed in Section 3.3 above. Accounts concerning the use of simultaneous eclipse observations to determine longitudinal differences, similar to Strabo's, are found in several Islamic sources.[11]

---

[9]For example, the late fourteenth century Damascene astronomer Ibn al-Shāṭir is known to have written a book entitled *Taʿlīq al-Arṣād* which contained details of how he derived an alternative planetary model to that of Ptolemy from his observations. It seems that a number of eclipse observations were contained in this work. However, all manuscript copies of it have been lost. See Saliba (1987).

[10]Neugebauer (1975: 667).

[11]See, for example, al-Bīrūnī, *Kitāb Taḥdīd Nihāyāt al-Amākin li-Taṣḥīḥ Masāfāt al-Masākin*, 166–168; trans. Ali (1967: 129–130).

In making their observations, the Islamic astronomers usually determined the time of an eclipse by measuring the altitude of either a clock-star or of the eclipsed luminary. In many of the eclipse records the original altitude measurements are reported. Sometimes, the reports contain the local time deduced from these altitudes. Stephenson & Said (1991) have shown that this reduction was usually done with considerable accuracy. Other reports only contain the reduced local time. Both equinoctial and, following ancient Greek practice, seasonal hours are used. Although simple clepsydras were used in the Near East from an early period, it seems that it was not until the eleventh century AD that any significant attempts were made to build a highly accurate device.[12] Furthermore, the earliest evidence for the use of a clepsydra by an astronomer comes from the thirteenth century AD when Maghribī used one at the Marāgha Observatory.[13] Thus it seems clear that water-clocks played no part in the timing of eclipses by early Islamic astronomers.

The instrument used to measure altitudes during the observed eclipses is hardly ever mentioned, but it was in all likelihood a hand-held astrolabe. This device functioned both as an instrument for measuring altitudes and as a analogue computer for converting altitudes into local times.[14] Tables are also known in various works on timekeeping that enable altitude measurements to be converted to local times.[15] However, it is not clear to what extent these tables were used in preference to an astrolabe.

The Islamic astronomers were among the first to distinguish between the true and the apparent beginning of an eclipse. Al-Bīrūnī explains that it is commonly held that the beginning of a lunar eclipse cannot be distinguished until one twelfth of the moon is covered. He then remarks that this "is subject to enquiry and examination. I consider the amount of one digit excessive."[16] Ibn Yūnus occasionally notes both the time when the eclipse was first perceived and his estimate of the true time of first contact. However, in most of the other observations, no distinction is made between these two times. Hence, as it is impossible to determine the delay in perceiving the beginning of the eclipse since this is a physiological factor dependent on the eye of the observer, I will generally consider the times to relate to the moment of true first contact.

In the following sections I will discuss the eclipse observations made by Ḥabash, al-Māhānī, al-Battānī, the Banū Amājūr, Ibn Yūnus and al-Bīrūnī in turn. In doing so I will make use of the translations of these observations by Said & Stephenson (1997). They have corrected a number of scribal errors in the texts which I have, on the whole, accepted without comment. Ḥabash, al-Māhānī, and the Banū Amājūr, made their observations from Baghdad, Ibn Yūnus made his from Cairo, al-Battānī from al-Raqqah and Anṭākyah, and al-Bīrūnī and his colleagues made their observations

[12] Al-Hassan & Hill (1986: 55–59).

[13] Saliba (1986).

[14] For a detailed explanation of the construction and use of an astrolabe, see North (1974). Of the many treatises on the astrolabe by Islamic astronomers the most detailed is by al-Ṣūfi. See Kennedy & Destombes (1966) for a commentary on this text. For a preliminary survey of known astrolabes and their makers, see Mayer (1956). He concludes that most astrolabes were built by the astronomer who intended to use it.

[15] Goldstein (1963); King (1973).

[16] *Kitāb Taḥdīd Nihāyāt al-Amākin li-Taṣḥīḥ Masāfāt al-Masākin*, 168; trans. Ali (1967: 130–131).

City	Latitude (°)	Longitude (°)
al-Raqqah	35.94	−39.02
Anṭākyah	36.20	−36.17
Baghdad	33.34	−44.40
Cairo	30.05	−31.25
Ghaznah	33.55	−68.43
Jurjān	36.83	−54.48
Jurjāniyyah	42.30	−59.16
Nīshāpūr	36.21	−58.83

Table 4.1: Near Eastern observation sites.

from Nīshāpūr, Jurjān, Jurjāniyyah, and Ghaznah. The latitudes and longitudes of these sites are given in Table 4.1. It should be noted that, with the exception of Cairo (Arabic: al-Qāhirah), I have used the Arabic names for these cities.

4.2.1 Ḥabash

Ḥabash al-Ḥāsib worked in Baghdad during the first half of the ninth century AD. In addition to making numerous observations, he compiled several important *zīj*es. These rely heavily on Ptolemy, but were based in part upon Ḥabash's own observations.[17] Ḥabash also wrote on the construction of astrolabes and other astronomical instruments, and made important contributions to the fields of trigonometry and cartography.[18]

According to the description of the surviving manuscripts of Ḥabash's *zīj*es published recently by Debarnot (1987), no eclipse observations are contained in these sources. However, two observations of eclipses by Ḥabash in AD 829 are reported by Ibn Yūnus in his *al-Zīj al-Kabīr al-Ḥākimī*:

> "Aḥmad b. ʿAbd Allāh known as Ḥabash said: 'There was a lunar eclipse after Nowrūz (i.e., Persian new year) in the year 198 of Yazdijerd. (The prediction of the) calculations of (*al-Zīj*) al-Mumtaḥan and of Ptolemy were near to each other ... As for the solar eclipse, which (occurred) in this year at the end of the month of Ramaḍān, all calculations (concerning the eclipse) were in error. The altitude of the sun at the beginning was 7° as they (the astronomers) claim. The eclipse ended when the altitude of the sun was about 24°, as though it was 3 (seasonal) hours of day (i.e., after sunrise).' "
>
> [*al-Zīj al-Kabīr al-Ḥākimī*; trans. Said & Stephenson (1997: 30)]

Unfortunately, no details of the lunar eclipse observation are given. However, the altitude of the sun at both the beginning and end of the solar eclipse, and the reduced time of its end are reported. This eclipse occurred on 30 November 829 AD.

The measured altitudes and times of the eclipse observed by Ḥabash are shown in Table 4.2. For comparison, the values given by modern computations are also given.

[17] Kennedy (1956: 127).

[18] For further biographical details, see Tekeli (1972).

Date	Type	Contact	Object	Altitude (°) Observed	Altitude (°) Computed	Local Time (h) Observed	Local Time (h) Computed
829 Nov 30	Solar	1	Sun	7.00	1.63	7.68	7.07
		4	Sun	24.00	22.82	9.49	9.35

Table 4.2: Eclipse observation made by Ḥabash.

Clearly, there is a considerable error in the measured altitude of the sun when the eclipse began. The cause of this error is not known. It may have been due to a delay by Ḥabash in noticing the beginning of the eclipse; since the sun was rising higher in the sky during the eclipse, this would have resulted in a measured altitude that was too great. However, it seems more likely that this is simply an observational error as an altitude of only 2° is difficult to observe, particularly from within a city.

As there is only one record of a timed eclipse observation by Ḥabash preserved it is not possible to make a reliable estimate of the typical accuracy of his eclipse timings. For reference, however, the mean accuracy of these two timings is about 0.38 hours.

4.2.2 al-Māhānī

Al-Māhānī worked on various problems of astronomy and mathematics during the latter half of the ninth century AD. Based in Baghdad, his main achievements were in mathematics where he wrote, among other works, a commentary on parts of Euclid's *Elements*. Between AD 854 and AD 866 al-Māhānī observed three lunar and one solar eclipse. These, together with a number of conjunctions, are reported by Ibn Yūnus in his *al-Zīj al-Kabīr al-Ḥākimī*. It seems likely that al-Māhānī may have made many further observations, but no source has as yet been found to contain them.[19]

The record of the lunar eclipse on 12 August 854 AD is interesting as al-Māhānī notes that his altitude measurements were converted into local times by the use of an astrolabe:

> "This lunar eclipse was mentioned by al-Māhānī. 'The moon was eclipsed on the night of Sunday 13th of the month of Rabīʿ al-Awwal in the year 240 of al-Hijrah. It was found by observation that the time of the beginning of the eclipse was when the altitude of (the star) *al-dabarān* (Aldebaran: α Tau) was 45;30° in the east. We did not find its times (accurately) except this time (i.e., of the beginning), which was exact and precise. We measured the time of the completion of (the first phase of) the eclipse, which is the time of the beginning of the staying (Arabic: *al-makth*) (in totality) and found it (to be) when the altitude of (the star *al-shiʿrā) al-shāmiyyah* (Procyon: α CMi) was between 22° and 23° in the east. This (latter) measurement is not exact but approximate. We determined the time of the beginning from the altitude of *al-dabarān* by the astrolabe and found it to be 44° (of the celestial sphere) after midnight. The time of beginning was 8° later than its (calculated) time. We (also) determined the time (of the beginning of) the

[19] For further biographical details, see Dold-Samplonius (1974).

Date	Type	Contact	Object	Altitude (°) Observed	Altitude (°) Computed	Local Time (h) Predicted	Local Time (h) Observed	Local Time (h) Computed
854 Feb 16	Lunar	1	-	-	-	-	22.05	22.21
854 Aug 12	Lunar	1	α Tau	45.50	45.32	2.40	2.93	2.91
		2	α CMi	23.00	18.30	-	4.50	4.11
856 Jun 21	Lunar	1	α Tau	9.50	8.02	2.83	3.33	3.33
866 Jun 16	Solar	1	-	-	-	12.06	12.33	12.37
		M	-	-	-	13.40	13.72	13.73
		4	-	-	-	14.72	15.00	14.98

Table 4.3: Eclipse observations made by al-Māhānī.

stay by the astrolabe, taking the altitude of (*al-shiʿrā*) *al-shāmiyyah* as 23° and found it to be $23\frac{1}{2}$ parts (i.e., degrees) of the celestial sphere after the (time of the) beginning (of the eclipse).' "

[*al-Zīj al-Kabīr al-Ḥākimī*; trans. Said & Stephenson (1997: 31)]

As Stephenson & Said (1991) have shown, al-Māhānī reduced his altitude measurements to local times with a high degree of accuracy. In this example, the error in his two reductions is less than 3 minutes of time (or less than 1° of the celestial sphere). Indeed, the error in the reduction of the altitude of α Tau to the local time is less than half a minute.

When recording many of his eclipse observations, including the example quoted above, al-Māhānī compared his observed local times with those calculated by means of tables. Unfortunately it is not known which set of tables al-Māhānī used, but it is unlikely to have been those of his own *zīj*. According to Kennedy (1956: 136) this *zīj* is not extant, but is known only from the writings of al-Fārisī who said that it was compiled by al-Māhānī in about AD 860. As most of al-Māhānī's eclipse observations are earlier than this it seems likely that these observations were among those that he used to compile his *zīj*.

Table 4.3 summarizes al-Māhānī's eclipse observations and predictions. The mean accuracy of the observed times is about 0.09 hours and there is no evidence for any systematic error in the times. For the predicted times, the true accuracy is about 0.38 hours. The observed accuracy of the predicted times is also about 0.38 hours. The accuracy of the predicted times of the solar eclipse on the 16th June 866 AD is slightly better than those of the earlier lunar eclipses. This may suggest that al-Māhānī used his newly compiled *zīj* to make this prediction, but as there are so few eclipses it is not possible to draw any firm conclusions in this regard.

4.2.3 al-Battānī

Al-Battānī is one of the most noted of all Islamic astronomers. He was born in AD 858 in the city of Ḥarrān, but spent most of his working life in the city of al-Raqqah.[20]

[20]For further biographical details, see Hartner (1970).

Date	Type	Contact	Local Time (h) Predicted	Observed	Computed
883 Jul 23	Lunar	M	19.25	20.00	19.70
891 Aug 8	Solar	M	12.00	13.14	12.89
901 Jan 23	Solar	M[1]	-	8.33	8.12
		M[2]	10.50	8.50	8.37
901 Aug 2	Lunar	M[1]	-	3.33	2.86
		M[2]	-	3.58	3.08

1. Observed by al-Battānī in Anṭakya.
2. Observed for al-Battānī in al-Raqqah.

Table 4.4: Eclipse observations made by al-Battānī.

According to Ibn al-Qifṭī, a thirteenth century AD biographer, al-Battānī was "one of the illustrious observers and foremost in geometry, theoretical and practical astronomy, and astrology."[21] In about AD 900 al-Battānī compiled his *zīj* entitled the *Al-zīj al-Ṣabiʾ*. This *zīj* was one of the most influential compiled by an Islamic astronomer. It was largely based on the astronomy of Ptolemy, but showed a considerable improvement over his parameters.[22]

From AD 887 al-Battānī made over forty years worth of observations from his observatory at al-Raqqah. This observatory, built with his own money, was well equipped with astrolabes, a gnomon, a large parallactic ruler, and a mural quadrant with a radius of over one metre. In his *zīj*, al-Battānī reports observations of four eclipses. The first two he observed from his observatory for the purpose of investigating the accuracy of Ptolemy's tables. In AD 901, however, he travelled to Anṭakya to observe two eclipses. He arranged for a colleague to observe the same two eclipses from al-Raqqah so that he could attempt to determine the difference in longitude between the two cities. I quote below a translation of his account of the solar eclipse on 23 January 901 AD:

> "This solar eclipse was observed by us at the city of Anṭakya on the 23rd of (the month of) Kānūn al-Thānī in the year 1212 of *Dhū al-Qarnain* (i.e., Alexander IV), which is the year 1224 after the death of al-Iskandar (i.e., Alexander III, the Great). The middle of the eclipse was about $3\frac{2}{3}$ equal hours before midday. (A little) more than $\frac{1}{2}$ of the sun (i.e., sun's surface) in sight was eclipsed. In this eclipse the sun was at its nearest distance (perigee) and the moon was nearly at its middle distance ... This (same) eclipse was observed by someone on our behalf at the city of al-Raqqah. The middle of the eclipse was (a little) less than $3\frac{1}{2}$ equal hours before midday. A little less than $\frac{2}{3}$ of the sun in view was eclipsed. According to calculation from Ptolemy('s tables), the sun should have been totally eclipsed, and the (time of) the middle of the eclipse was later than the observed time by about two hours. Such a discrepancy is not acceptable."
>
> [*Al-zīj aṣ-Ṣabiʾ*; trans. Said & Stephenson (1997: 43–44)]

[21] *Taʾrīkh al-Ḥukamāʾ*; trans. Hartner (1970: 507–508).
[22] Kennedy (1956: 132–133).

Al-Battānī's eclipse records are summarized in Table 4.4. The mean accuracy of his observed times is about 0.31 hours. Furthermore, all of his times are late; however, there are too few observations to conclude that his timing methods suffered from a systematic error. The accuracy of the times predicted with Ptolemy's tables is also very poor — the true accuracy is about 1.16 hours and the observed accuracy is about 1.30 hours. In light of this, al-Battānī's remark that Ptolemy's tables are "not acceptable" appears to be fully justified.

4.2.4 The Banū Amājūr

The Banū Amājūr is the name given to a group of astronomers who observed in Baghdad, and possibly also Shiraz, between AD 885 and AD 933. The group consisted of Abū al-Qāsim ʿAbd Allāh ibn Amājūr, his son Abū al-Ḥasan ʿAlī ibn Amājūr, and their freed slave Mufliḥ ibn Yūsuf. It is also possible that others, in particular a third member of the Amājūr family, may have collaborated with them on occasions.[23] In addition to their extensive observational programme, the Banū Amājūr compiled five *zīj*es, none of which is extant.[24]

Ibn Yūnus reports observations of a number of eclipses made by the Banū Amājūr between AD 923 and AD 933. The record of the solar eclipse on 11 November 923 AD gives some details of the site in Baghdad from which they made their observations:

> "This solar eclipse was calculated and observed by Abū al-Ḥasan ʿAlī ibn Amājūr from *al-Zīj al-ʿArabī* of Ḥabash. This eclipse was at the conjunction (i.e., new moon) of (the month of) Shaʿbān in the year 311 (AH). We as a group observed (this eclipse) and clearly distinguished it. The estimate of all (observers) for the middle of the eclipse was that it occurred when the altitude of the sun was 8° in the east; its clearance was at $2\frac{1}{5}$ seasonal hours (after sunrise), when the altitude of the sun was 20°. We observed this eclipse at several sites on the 'Ṭārmah.' The estimate of Abū al-Ḥasan for the middle of the eclipse at his house was when the altitude of the sun was 8°, as I estimated myself at my house before he arrived. The magnitude of the eclipse was $\frac{1}{2}$ plus $\frac{1}{4}$ (i.e., $\frac{3}{4}$) of the sun's diameter; the middle of the eclipse, which we estimated when the sun's altitude was 8°, is to be when the elapsed time (after sunrise) was 0;50 seasonal hours, and the (celestial) sphere had revolved (through) 10;40°. (The interval) between the middle of the eclipse and its clearance in this observation was 1;22 seasonal hours; the (corresponding) time was 1;10 equal hours because the sphere revolved (through) 28;9° at the moment of clearance, which is (equivalent to) 1;53 equal hours. The middle would be at 0;43 equal hours. According to calculation from the conjunction tables in the Ḥabash *zīj* the middle was at 0;31 hours and its clearance at 0;44 hours, calculation being in advance of observation."
>
> [*al-Zīj al-Kabīr al-Ḥākimī*; trans. Said & Stephenson (1997: 33–34)]

The "Ṭārmah" referred to by the Banū Amājūr is an elevated platform on the side of a building.[25] Elsewhere it is written that this platform had certain slits in its walls

[23] Sayili (1960: 101–103).

[24] Kennedy (1956).

[25] Said & Stephenson (1997).

Date	Type	Contact	Object	Altitude (°) Observed	Computed	Local Time (h) Predicted	Observed	Computed
923 Jun 1	Lunar	M	-	-	-	-	20.80	20.57
		4	α Cyg	29.50	33.80	-	22.13	21.95
923 Nov 11	Solar	M	Sun	8.00	7.63	7.30	7.51	7.52
		4	Sun	20.00	18.72	7.52	8.70	8.55
925 Apr 11	Lunar	1	α Boo	11.00	34.04	19.07	19.39	19.53
		4	α Lyr	24.00	24.77	22.84	23.12	22.84
927 Sep 13	Lunar	1	α CMa	31.00	33.76	4.12	4.03	4.22
		M	-	-	-	5.21	-	5.04
		4	-	-	-	5.99	-	5.57
928 Aug 18	Solar	1	-	-	-	4.77	-	5.01
		M	-	-	-	5.64	-	5.67
		4	Sun	11.89	11.21	6.39	6.44	6.36
929 Jan 27	Lunar	1	α Boo	18.00	31.50	22.81	22.88	23.89
		2	-	-	-	0.43	-	1.06
		M	-	-	-	0.56	-	1.69
		3	-	-	-	0.93	-	2.12
		4	-	-	-	2.42	-	3.29
933 Nov 4	Lunar	1	α Boo	15.00	15.18	4.18	4.56	4.63
		2	-	-	-	5.43	-	5.77
		3	-	-	-	6.88	-	7.30
		4	-	-	-	8.00	-	8.43

Table 4.5: Eclipse observations made by the Banū Amājūr.

indicating fixed directions. Although there is no direct evidence that this platform was specifically intended to be used for celestial observations, Sayili (1960: 94) has suggested that this building, their house, may have also acted as a private observatory for the Banū Amājūr.

At the end of the eclipse report quoted above, the Banū Amājūr note that there is a difference between their observed times and those predicted using Ḥabash's *zīj*. This appears to be the primary motivation behind the Banū Amājūr's eclipse observations. Presumably they used these and other observations to construct their own *zīj*es. Unfortunately, as these *zīj*es are lost it is not possible to find out whether they gave better agreement with observation than did those of Ḥabash.

Table 4.5 summarizes the Banū Amājūr's eclipse observations and predictions (using the tables of Ḥabash). The mean accuracy of the observed times is about 0.15 hours and there is no evidence for any significant systematic error. The true accuracy of the times predicted using the tables of Ḥabash is about 0.52 hours. This is significantly poorer than the observed accuracy of these predicted times which is about 0.32 hours. This implies that the Banū Amājūr would have overestimated the accuracy of these tables in predicting eclipses.

4.2.5 Ibn Yūnus

Along with al-Battānī, Ibn Yūnus is regarded as one of the greatest astronomers of medieval Islam. He was born and worked in Egypt during the latter half of the tenth and the former half of the eleventh centuries AD. Between AD 977 and AD 1004 he made astronomical observations for the Caliph al-ʿAzīz and his successor, the young Caliph al-Ḥākim. These observations were made in Cairo at a variety of sites.[26] Ibn Yūnus' major work was the *Ḥākimī Zīj*, which contained a very detailed account of his astronomical theory, and proved to be very influential in Egypt and the surrounding area for many centuries to come. He also compiled a set of tables for determining the time of day from solar observations,[27] and even works of poetry.[28] His *zīj* is only extant in fragments, but is known to have contained eighty-one chapters.[29] Parts of this *zīj* have been translated into French by Caussin (1804). The *zīj* is unique among extant examples in containing details of a number of observations made not only by Ibn Yūnus, but also by earlier astronomers such as al-Māhānī and the Banū Amājūr, as discussed above.[30]

Ibn Yūnus records 16 of his own eclipse observations in his *zīj*. These were all observed in Cairo. His earliest observation, dating from 13th December 977 AD, was made from the roof of a mosque:

> "This solar eclipse was in the early morning of Thursday the 28th of the month of Rabīʿ al-ākhir, in the year 367 of al-Hijrah, which is the 22nd of the month of Ādhar in the year 346 of Yazdijerd. We, a group of scholars ... (ten names are given) attended at al-Qarāfah (a district of Cairo) in the Mosque of Abū Jaʿfar Aḥmad ibn Naṣr al-Maghribī to watch this eclipse. Everyone waited for the beginning of this eclipse. It began to be perceived when the altitude of the sun was more than 15° but less than 16°. (Those) present all agreed that about 8 digits of the sun's diameter were eclipsed, that is (a little) less than 7 digits of surface. The sun was completely cleared when its altitude was more than 33° by about $\frac{1}{3}$ of a degree, as estimated by me, and agreed by all those present. The sun and moon in this eclipse were at their nearest distance (from the Earth — i.e., at perigee)."
>
> [*al-Zīj al-Kabīr al-Ḥākimī*; trans. Said & Stephenson (1997: 37)]

On two other occasions Ibn Yūnus states that he observed the eclipse from this Mosque; no observation site is mentioned for the other cases.

In two of his eclipse records, Ibn Yūnus compares the observed times with those calculated by means of Yaḥyā ibn Abī Manṣūr's *Mumtaḥan Zīj*, noting deviations of about half an hour. His observed and computed times are summarized in Table 4.6. The mean accuracy of his observed times is about 0.14 hours with negligible systematic error. The mean true accuracy of the predicted times is about 0.24 hours, and the mean observed accuracy of the predicted times is about 0.31 hours. However, as there are only three predicted times, it is not possible to draw any firm conclusions on the accuracy of the *Mumtaḥan* tables.

---

[26] It has often been claimed that al-Ḥākim built an observatory for Ibn Yūnus in Cairo, but Sayili (1960: 130–156) has shown this to be untrue. Instead he argues that Ibn Yūnus may have built his own private observatory in his house.

[27] King (1973).

[28] King (1976).

[29] Kennedy (1956: 126).

[30] For further biographical details, see King (1976).

Date	Type	Contact	Object	Altitude (°) Observed	Altitude (°) Computed	Local Time (h) Predicted	Local Time (h) Observed	Local Time (h) Computed
977 Dec 13	Solar	1	Sun	15.50	15.79	-	8.40	8.42
		4	Sun	33.33	33.80	-	10.71	10.79
978 Jun 8	Solar	1	Sun	56.00	58.71	-	14.50	14.29
		4	Sun	26.00	24.97	-	16.83	16.91
979 May 14	Lunar	4	-	-	-	-	20.07	20.01
979 May 28	Solar	1	Sun	6.50	7.31	-	18.38	18.30
979 Nov 6	Lunar	1	Moon	64.50	65.09	-	22.41	22.45
		4	Moon	65.00	66.21	-	1.61	1.51
980 May 2	Lunar	1	Moon	40.66	40.24	-	0.84	0.93
		4	-	-	-	-	4.65	4.76
981 Apr 22	Lunar	1	Moon	21.00	19.99	-	3.53	3.63
		4	-	-	-	-	5.15	5.33
981 Oct 16	Lunar	1	Moon	24.00	24.67	4.85	4.45	4.37
983 Mar 2	Lunar	1	Moon	66.00	65.82	23.33	0.00	0.01
		4	Moon	35.83	3.50	2.75	3.42	3.28
985 Jul 20	Solar	1	Sun	23.00	26.70	-	16.95	16.68
		4	Sun	6.00	8.58	-	18.34	18.10
986 Dec 19	Lunar	1	Moon	30.50	27.06	-	4.39	4.65
990 Apr 12	Lunar	1	Moon	38.00	22.37	-	21.81	22.33
		4	-	-	-	-	1.27	1.49
993 Aug 20	Solar	1	Sun	27.00	28.57	-	7.67	7.79
		M	Sun	45.00	43.88	-	9.07	8.99
		4	Sun	60.00	59.86	-	10.34	10.32
1001 Sep 5	Lunar	1	-	-	-	-	20.17	19.87
1002 Mar 1	Lunar	1	α Boo	52.00	53.37	-	23.48	23.59
		1	α Aur	14.00	13.78	-	23.55	23.59
1004 Jan 24	Solar	1	Sun	18.50	19.05	-	15.68	15.63
		M	Sun	5.00	6.59	-	16.88	16.75

Table 4.6: Eclipse observations made by Ibn Yūnus.

4.2.6 al-Bīrūnī

Al-Bīrūnī was something of a polymath. During the first half of the eleventh century AD he wrote more than 146 works, of which only 22 are extant, on subjects ranging from astronomy and geography to history and literature.[31] He travelled widely, both voluntarily and for political reasons, ranging from Baghdad to various parts of India. Al-Bīrūnī's most important works related to astronomy are his *zīj*, *al-Qānūn al-Masʿūdī*,[32] and his *Kitāb Taḥdīd Nihāyāt al-Amākin li-Taṣḥīḥ Masāfāt al-Masākin*,[33] whose main theme is the determination of geographical coordinates.[34]

Al-Bīrūnī records observations of four lunar eclipses in the two works noted above.[35] These were observed in a variety of cities: Jurjān in AD 1003, Jurjāniyyah

[31] For a list of al-Bīrūnī's works, including details of those that are extant, and those that have been published, see Kennedy (1970).

[32] This work is described by Kennedy (1956: 157–159).

[33] This work has been translated into English by Ali (1967).

[34] For further biographical details, see Kennedy (1970).

[35] Al-Bīrūnī also reports an observation of an annular solar eclipse by al-Irānshahrī on 28 July 873 AD. Although no timings of this eclipse are reported, it is of considerable historical interest as according to Ptolemy's eclipse theory (*Almagest*, V, 14), annular eclipse are not possible. See Goldstein (1979).

Date	Type	Contact	Object	Altitude (°) Observed	Computed	Predicted	Local Time (h) Observed	Computed
1003 Feb 19	Lunar	M	-	-	-	-	18.73	18.70
1003 Aug 14	Lunar	M	-	-	-	-	23.60	23.35
1004 Jul 4	Lunar	M	-	-	-	-	2.61	2.90
1019 Sep 17	Lunar	1	α Aug	66.00	66.77	-	2.29	2.39
		1	α CMa	17.00	18.62	-	2.34	2.39
		1	α CMi	21.00	23.99	-	2.32	2.39
		1	α Tau	63.00	63.09	-	2.38	2.39
		4	-	-	-	4.19	-	5.59

Table 4.7: Eclipse observations made by al-Bīrūnī.

in AD 1004, and Ghaznah in AD 1019. I quote below a translation of the first of his observations:

> "This lunar eclipse was on the night of Saturday the 14th of the month of Rabīʿ al-ākhir in the year 393 (of al-Hijrah). I observed the beginning and clearance at Jurjān by the altitude of the (two stars) *al-Shiʿrayān* (i.e., *al-shiʿrā al-yamāniyyah* — Sirius: α CMa — and *al-shiʿrā al-shāmiyyah* — Procyon: α CMi). The moon was eclipsed by $\frac{1}{4}$ of its diameter by estimate. The longitude difference between Jurjān and Ghaznah is 2;21 minutes of day. The middle of the eclipse at it (presumably at Ghaznah) was 10;11 (minutes of day) after midday of Friday, the 6th of the month of Isfandārmadh in the year 1751 of Bukhtinassar (i.e., Nabonassar)"
>
> [*al-Qānūn al-Masʿūdī*; trans. Stephenson & Said (1997: 45)]

The observed local time at Jurjān is reduced by al-Bīrūnī to Ghaznah using a longitude difference of 2;21 degrees of day, where one degree of day corresponds to $\frac{1}{60}$ of a day, or 0.4 equal hours. Al-Bīrūnī's longitude difference (0.94 hours) is very close to the modern value (0.93 hours).[36]

Table 4.7 summarizes al-Bīrūnī's observed eclipse times. The mean accuracy of these times is about 0.10 hours, with no significant systematic error. In reporting the eclipse on 17 September 1019 AD al-Bīrūnī notes that:

> "... some astronomers from Khurāsān predicted that the completion of the clearance would be when $10\frac{1}{4}$ hours of night had elapsed. Since night hours were then nearly equal to daytime hours, because the sun was in the last degrees of Virgo, this would be when 1 plus $\frac{1}{2}$ plus $\frac{1}{4}$ (i.e., $1\frac{3}{4}$) hours of night remained (i.e., before sunrise). It was clear to the sight that the world was lit up, the stars had disappeared, the sun was about to rise, and the moon was about to set behind the mountains which screened it. A small portion of the eclipse (still) remained in its body (i.e., disk) and I was unable to observe it (i.e., the time of clearance) exactly."
>
> [*Kitāb Taḥdīd Nihāyāt al-Amākin li-Taṣḥīḥ Masāfāt al-Masākin*; trans. Said & Stephenson (1997: 47)]

[36] Said & Stephenson (1997).

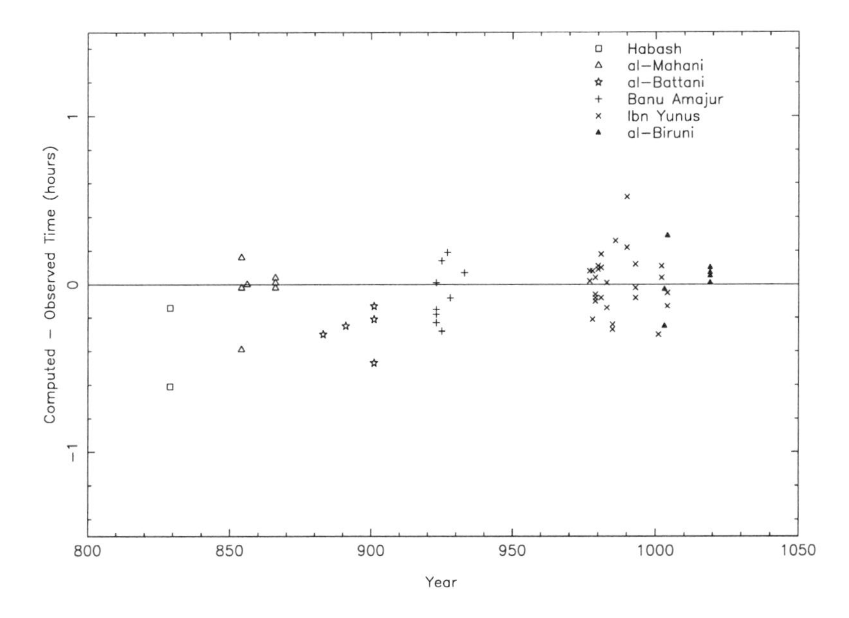

Figure 4.1: Error in the observed eclipse times.

It is clear from Table 4.7, however, that this prediction was very inaccurate — the predicted time is 1.40 hours earlier than computed.

4.3 Accuracy of the Observed and Predicted Times

The errors in the observed times of all of the eclipses presently known to have been observed by medieval Islamic astronomers are shown in Figure 4.1. The typical accuracy of these observations is between about 6 and 10 minutes. As I have discussed above, the timings by Ḥabash and by al-Bāttānī are of slightly poorer accuracy than those of their contemporaries. This conflicts with the view expressed by Ibn al-Qifṭī that al-Bāttānī was one of the foremost observers of his age.

Figure 4.2 shows the true and observed errors in the eclipse times predicted by al-Māhānī, al-Battānī, the Banū Amājūr, Ibn Yūnus and al-Bīrūnī. The prediction of the time of the maximum phase of the solar eclipse on 23 January 901 AD is in error by more than 2 hours, and so has not been included in the Figure. It is clear that the typical accuracy of each group of eclipse predictions is not the same. This is not surprising as they were made using different sets of tables. The accuracy of each group of predictions is shown more clearly in Figure 4.3. The lines in this figure connect points showing the mean accuracy of each set of predictions at their mean date. For reference, the accuracy of each set of observed times is also shown. It should be noted,

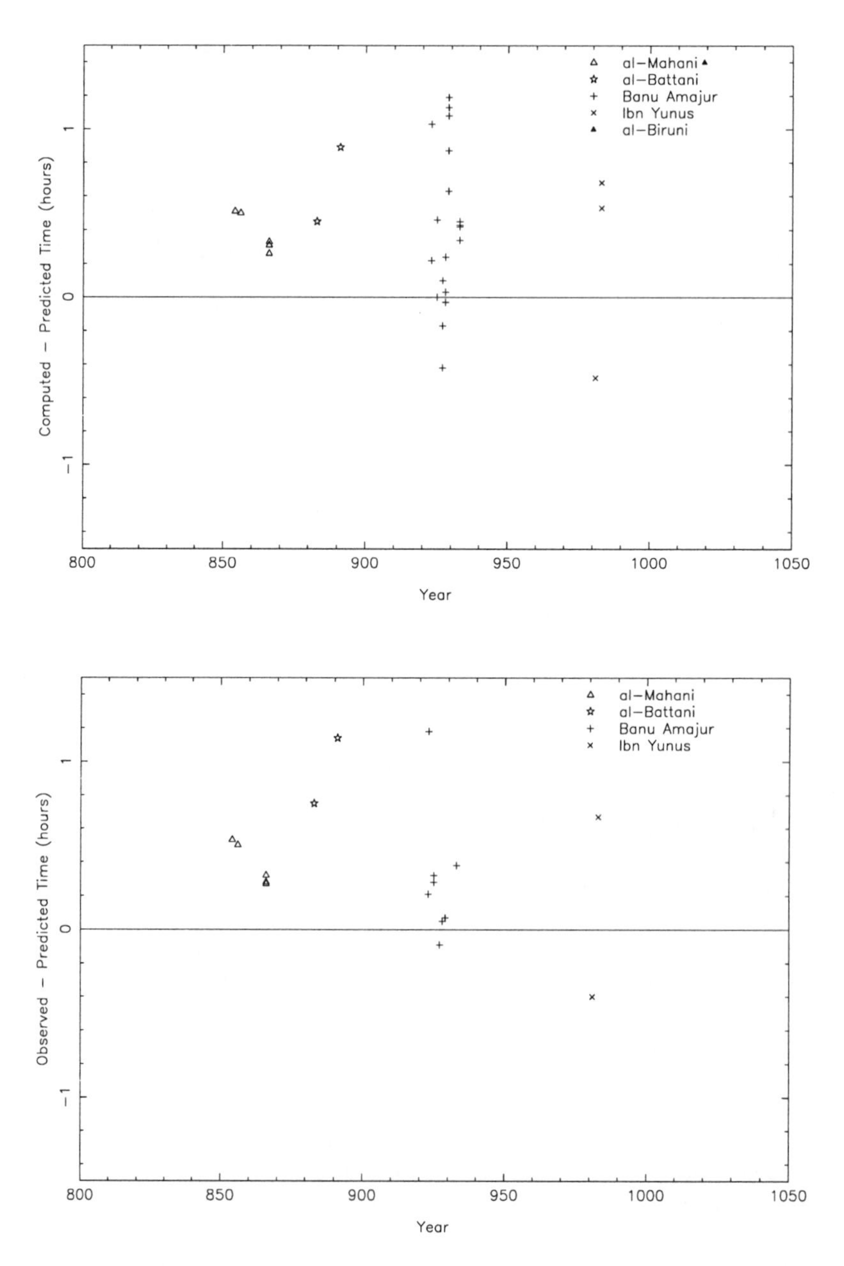

Figure 4.2: The true (above) and observed (below) error in the predicted eclipse times.

however, that the straight lines in the Figure should not be interpreted as implying any form of linear change in accuracy between the points, but merely illustrate the general trend in the accuracy of predictions in this period. The very inaccurate prediction made by "some astronomers from Khurāsān" and reported by al-Bīrūnī has not been included in the Figure as it is only an isolated example.

It is immediately apparent from Figure 4.3 that the predictions made by al-Battānī around the turn of the tenth century AD are greatly inferior to those made by the other astronomers. Al-Battānī used the tables in Ptolemy's *Almagest* to make these predictions. These tables were already more than seven hundred years old by this time, and so it is not surprising that they were giving inaccurate predictions. This was caused by the cumulative effect of small errors in Ptolemy's parameters over many years, and, as I have discussed in Section 4.2 above, it was errors in the prediction of eclipses that were largely responsible for the attempts by the Islamic astronomers to make improvements to Ptolemy's parameters in their *zīj*es.

The tables used by al-Māhānī to make his predictions during the ninth century AD are not known. They are of comparable accuracy to the tables of Ḥabash used by the Banū Amājūr in about AD 960, and so I would suggest that either these tables, or ones based upon them, were used. It is interesting that the *Mumtaḥan Zīj*, used by Ibn Yūnus at the end of the tenth century AD, appears to be slightly more accurate than the *zīj*es of Ḥabash. This *zīj* was compiled by Yaḥyā ibn Abī Manṣūr in around AD 930, some 10 or 20 years before the *zīj*es of Ḥabash. This could suggest one of two possibilities: either the *Mumtaḥan Zīj* was more accurate than the later *zīj*es of Ḥabash, or Ibn Yūnus applied the tables more reliably than the Banū Amājūr.

Overall, the medieval Islamic astronomers of the Near East achieved a considerable level of accuracy in timing eclipses. At no earlier period of history, be it in Mesopotamia, Alexandria, or China, had it been possible for eclipses to be timed with an accuracy of better than 10 minutes. Furthermore, using the *zīj*es that they compiled, the Islmaic astronomers were able to predict the time of an eclipse to an accuracy of better than 20 minutes. Again, this is a considerable achievement.

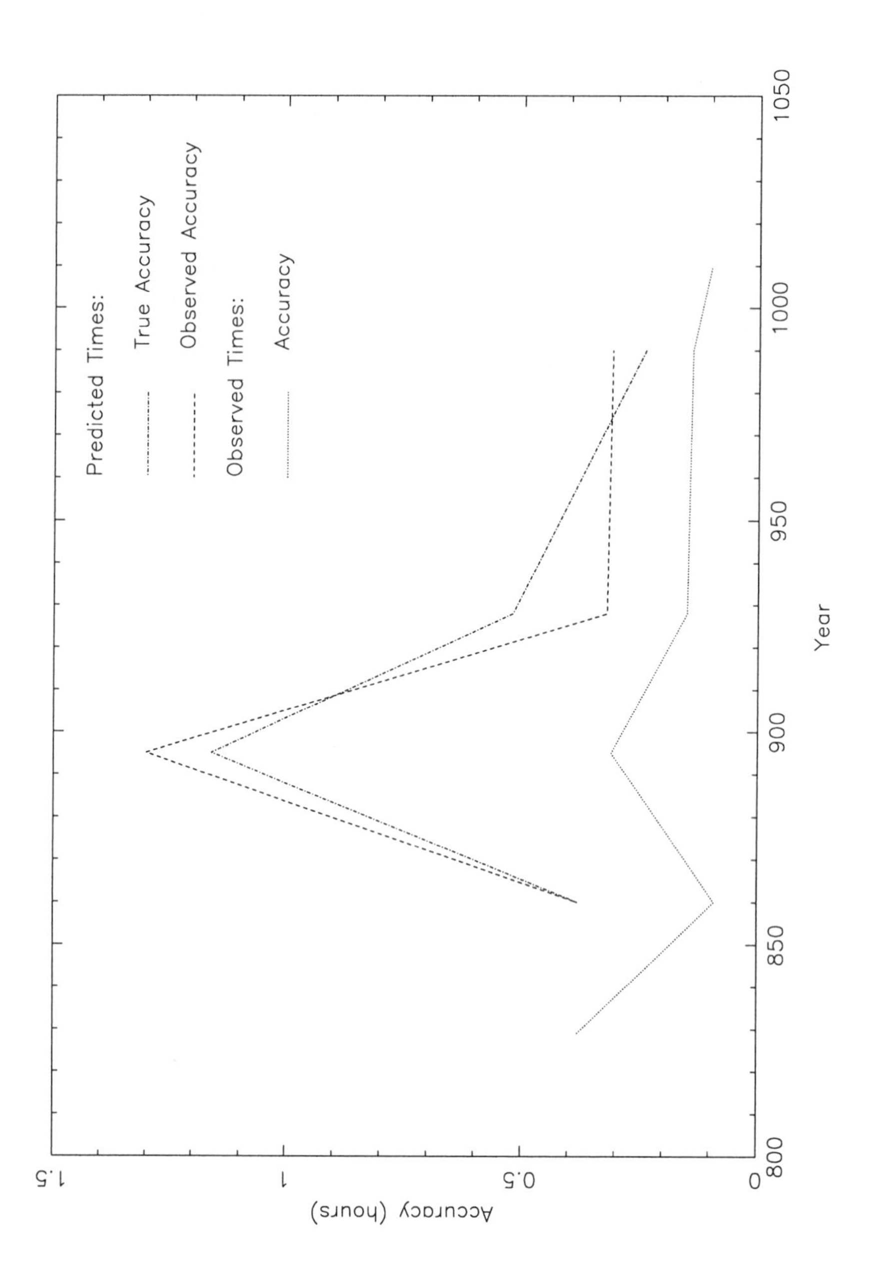

Figure 4.3: Schematic representation of the accuracy of the eclipse times.

Chapter 5

Late Medieval and Renaissance Europe

5.1 Introduction

After the fall of the Greek and Roman Empires, scientific learning in Europe went into a period of decline. By the latter half of the first millennium, the works of Ptolemy and other Greek astronomers had more or less been lost in Western Europe, fortunately to survive either in the original Greek or in Arabic translation in the Near East. But that is not to say that astronomy had no place in Medieval Europe. Instead, a new form of "practical" astronomy developed whose goals were to assist in solving some of the problems, such as determining the date of Easter and the times of prayers in monasteries, of religious and civil life.[1] By the twelfth century AD, however, European interest in science had been rekindled and there began a search to recover ancient scientific texts. This led to the many scientific achievements of the European Renaissance in the fifteenth and sixteenth centuries AD .

The general outline given above applies throughout the whole of Europe with one notable exception: Spain. At the end of the seventh century AD, Ṭāriq b. Ziyād and Mūsā b. Nuṣayr conquered Spain. The northern half of the country was soon recaptured by the Christians, but it was not until AD 1492 that the Muslim strongholds in the south finally fell. In the early part of this period there was a significant amount of contact between the Islamic astronomers based in the Near East, and those in southern Spain. However, due in part to the difficulty of long-range communication, in later periods the most recent astronomical writings were not always available to the Spanish scholars.[2] This resulted in the astronomical developments taking place in Spain gradually becoming less dependent on those being made in the rest of the Islamic World.[3]

[1]McCluskey (1998).
[2]King (1996).
[3]Samsó (1991).

Nevertheless, it was through southern Spain that Islamic astronomy was transmitted first to northern Spain, and then to the rest of Europe, in the first half of the second millennium AD.

There are two main sources of astronomical records in European history: accounts of noticeable celestial events recorded in various historical chronicles, and, from the fourteenth century AD onwards, various treatises and other assorted manuscripts written by astronomers, some of which contain detailed descriptions of astronomical observations. The reports of observations recorded in the chronicles are generally very descriptive, but, on the whole, lack technical precision.[4] They will therefore not be considered further in this study. Of much greater interest are the works written by astronomers. These range from published treatises such as Copernicus' *De Revolutionibus*, to astronomical tables and their explanatory canons, to miscellaneous collections of manuscripts on astronomical subjects.

Unlike most of the eclipse reports recorded in other parts of the world, which were often made by anonymous astronomers or groups of astronomers, all of the reports from Late Medieval and Renaissance Europe were made by identifiable astronomers.[5] Thus, it is possible to directly compare the observations and predictions of eclipses by these different observers to determine their relative accuracy. This will be the main subject of the present chapter. Before this, however, it is necessary to make some introductory remarks on the use of astronomical tables to predict eclipses in Europe, and on the instruments and techniques used by the astronomers in their observations.

5.2 European Astronomical Tables

One of the most important legacies of Islamic astronomy in Europe was the tradition of compiling astronomical tables. The earliest tables found in Europe come from Spain. Examples of these include the *zījes* of al-Khwārizmī and al-Battānī, both of which were imported from the Near East. It was not until the second half of the eleventh century, however, that a set of astronomical tables was actually compiled in Europe. These were the *Toledan Tables*, described by Samsó (1991: 14) as "a hasty adaptation of all the available astronomical material (al-Khwārizmī, al-Battānī and the *Almagest*) to the coordinates of Toledo."[6] The tables themselves were not particularly accurate; nevertheless, they were widely used in Spain, and also parts of Christian Europe, for the next two centuries and are extant in more than 200 manuscripts.

Towards the end of the thirteenth century, King Alfonso X of Castille patronized an important body of scientific work.[7] This included a collection of translations of

[4]A wide spectrum of astronomical events are reported in the chronicles including solar and lunar eclipses, meteors, and comets.

[5]With the exception of some of the eclipse records in the historical chronicles. However, as I have said, these records will not be used in the present study. For details of the eclipse records in the chronicles, see Newton (1972) and Stephenson (1997b).

[6]For a detailed description of the *Toledan Tables*, see Toomer (1968).

[7]Note, however, that Alfonso's patronage was not limited to scientific works; he also sponsored books of historical, legal, and literary studies. For limited biographical details of Alfonso, see Thomas (1970) and

Islamic astronomical works, a collection of treatises on the use of various astronomical instruments together with a star catalogue, entitled *Los libros del saber de astronomia*, and, most famously, a set of astronomical tables known as the *Alfonsine Tables*.[8] All of these works were written in Castilian; however, they are not all extant in their original form. In particular, the *Alfonsine Tables* are now known only in Latin. In the prologue to the tables it is written that they were compiled by two Jewish astronomers, Isaac ben Sid and Jehuda ben Moses Cohen, under the direction of the king.[9]

Appended to all extant copies of the *Alfonsine Tables* are a series of treatises, usually known as "canons," describing how to use the tables, but giving no explanation of the astronomical theories that underlie them. These is no question that these are all original Latin works since they are attributed to specific authors such as Jean de Murs and Jean de Saxe.[10] The *Alfonsine Tables* were quickly distributed throughout Europe where they achieved great popularity. One of their principal uses was in predicting eclipses. For example, Thorndike (1951, 1952, 1957) has uncovered a number of manuscripts containing predictions of eclipses made with these tables for several years in the fourteenth and fifteenth centuries AD. Unfortunately, the meridians used in making these predictions are not known,[11] and so it is not possible to evaluate their quality.

The basic astronomy of the *Alfonsine Tables* was Ptolemaic, although minor modifications were made to a number of the parameters based upon more recent observations. Eclipses were calculated following the procedure set out by Ptolemy: the mean position of the moon at conjunction or opposition is calculated, and if the moon's distance from the node is less than a certain amount (the "eclipse limit," $\approx 12^\circ$), then an eclipse is possible. Once it has been determined that an eclipse will take place, the time and magnitude can be calculated by finding the time and position of true conjunction or opposition, the parallax of the moon (for solar eclipses), and the moon's anomaly in the relevant tables.

Despite the general acceptance of the *Alfonsine Tables* throughout Europe, it did not take long before they began to be criticized. Levi ben Gerson, a French Jew working in Orange, does not mention them specifically, but he may have had them in mind when he gave a general criticism of Ptolemy's lunar model. Based upon his own observations,[12] he proposed an alternative lunar model which attempted to correct for the differences he noted between Ptolemy's model and his observations. He also

---

the references therein.

[8]Procter (1945).

[9]However, Poulle (1988) has argued, although not fully convincingly, that the Latin version of the *Alfonsine Tables* was not a translation of the Castilian original, but a new set of tables compiled in Paris by Jean de Murs in the fourteenth century.

[10]The canon of Jean de Saxe and the *Alfonsine Tables* have been translated into French by Poulle (1984). Extracts have also been translated into English by Thoren & Grant (1974).

[11]Thorndike (1957) suggests Oxford for some of them, but I do not believe that there is sufficient justification to make this claim.

[12]Levi ben Gerson was almost unique in the Medieval World in mainly using his own observations, rather than those of the ancients, to construct his astronomical models. See Goldstein (1972, 1974).

compiled his own set of astronomical tables,[13] but it is not clear what impact they had on other astronomers of the time. They certainly did not rival the *Alfonsine Tables* in popularity.

The reliability of the *Alfonsine Tables* came to be increasingly questioned during the latter half of the fifteenth century. This criticism largely came from the city of Vienna, where Georg Peurbach and his student and associate Regiomontanus worked. Peurbach, a member of the faculty at the University of Vienna and court astrologer to King Ladislaus V, is most famous for his work the *Tabulae eclipsium*, probably completed in AD 1459.[14] These tables were derived entirely from the *Alfonsine Tables* but, due to their arrangement, allowed the circumstances of solar and lunar eclipses to be calculated with greater ease. However, after observing a lunar eclipse on 3 September 1457 AD, Peurbach noted that the time did not agree with that given by the Alfonsine calculations by nearly 10 minutes, and began to formulate improvements. Unfortunately he died in AD 1461 before the completion of this work. However, some fifty years later, Johannes Angelus, also from Vienna, published annual ephemerides that he claimed were based upon Peurbach's corrections to the *Alfonsine Tables*, which he had completed himself.[15] Undoubtedly, Peurbach discussed his misgivings about the *Alfonsine Tables* with Regiomontanus, who by AD 1464 was able to write to the Italian astronomer Giovanni Bianchini declaring that the astronomy of the *Alfonsine Tables* was incorrect.[16] Indeed he went so far as to accuse the astronomers who blindly accepted them of great indolence:

> "... I cannot but wonder at the indolence of the common astronomers of our age who, just as credulous women, receive as something divine and immutable whatever they come upon in books either of tables or their canons, for they believe in writers and make no effort to find the truth."
>
> [MS Nuremberg Cent V app. 56c; trans. Swerdlow (1990: 170–171)]

Regiomontanus intended to reform the astronomy of the *Alfonsine Tables* by means of the application of Ptolemy's methods to new observations. However, he was never to complete this task (he died in AD 1476), and it was not until Copernicus published *De Revolutionibus* in AD 1542 that his intentions were fulfilled. Although the astronomy of *De Revolutionibus* placed the sun at the centre of the planetary system, in its methods it was still largely Ptolemaic, relying on the use of epicycles. In AD 1551 Erasmus Reinhold, a mathematics professor at the University of Wittenberg compiled the *Prutenic Tables* based upon parameters derived from Copernicus' observations. These tables came to replace the *Alfonsine Tables* throughout Europe until the middle of the seventeenth century AD when Kepler's *Rudolphine Tables* finally came to be widely accepted.[17] The *Rudolphine Tables* were compiled by Kepler from his theories of planetary motion derived from Tycho Brahe's observations. Tycho had

[13]These have been edited by Goldstein (1974).

[14]For a biography of Peurbach, see Hellman & Swerdlow (1978).

[15]Dobrzycki & Kremer (1996).

[16]Swerdlow (1990).

[17]Swerdlow (1996).

himself noted significant errors in both the *Alfonsine Tables* and the *Prutenic Tables*. However, although he proposed his own Earth-centred cosmology with the sun carrying the planets revolving around the Earth, he did not publish any tables to replace them.

5.3 Observational Techniques and Instruments

In addition to the tradition of compiling astronomical tables, Islamic astronomy also profoundly influenced the way that astronomers of the Late Medieval period made astronomical observations. In particular, Islamic astronomers had concluded that clepsydras were not reliable enough to make accurate time measurements. It was deemed better to determine the time indirectly using measurements of the altitude of the sun, moon or fixed stars. This practice was inherited by later Medieval European astronomers.

The most important instrument used to measure altitudes was the astrolabe. This device consists of a circular plate, one side of which contains a graduated scale and an alidade.[18] The other side of the plate contained a movable rete over an inscribed base which could be used to convert the altitude measurements to a local time. Thus the astrolabe functioned both as an observational instrument, and as an analogue computing device for converting altitudes to local times.[19] The limiting factor in the reliability of an astrolabe is often the accuracy with which the scale is graduated. Chapman (1983) has made a study of a number of European astrolabes of this period and found that in most cases these scales are typically accurate to about 5 minutes of arc. This is equivalent to better than 1 minute of time. Figure 5.1 shows a drawing of an astrolabe.

Another instrument widely used among European astronomers for measuring altitudes was the quadrant. As its name suggests, this instrument consists of a graduated quarter-circle plate fixed with one edge horizontal but allowed to rotate to any direction of azimuth. An alidade, fixed at what would be the centre of the circle, could then be used to measure the altitude of a star. A drawing of a quadrant is shown in Figure 5.2.

A third instrument that could be used to determine altitudes was the Ptolemaic ruler. This simple instrument consisted of three rulers, two of which were attached with hinges to the ends of the third, which was itself held vertical. The free end of one ruler was then allowed to slide along the length of the other, forming a triangular shape with an inclined base which extended beyond the apex of the triangle. The construction of the Ptolemaic ruler can be seen more clearly from Figure 5.3.

Around the end of the fifteenth century AD, mechanical clocks started to be used by European astronomers. Regiomontanus appears to have been a pioneer among astronomers in using a clock.[20] However, although some other astronomers, most

[18] An alidade is basically a straight edge used for sighting the reference object and then used as a marker on a graduated scale.

[19] For a full discussion of the construction and use of an astrolabe, see North (1974).

[20] Zinner (1990: 138).

Figure 5.1: Drawing of an astrolabe being used to determine the altitude of the sun taken from Johann Stoffler's *Elucidatio Fabricae Ususque Astrolabii* (Lutetiae 1553). [Courtesy: The Burndy Library, Dibner Institute for the History of Science and Technology, MIT]

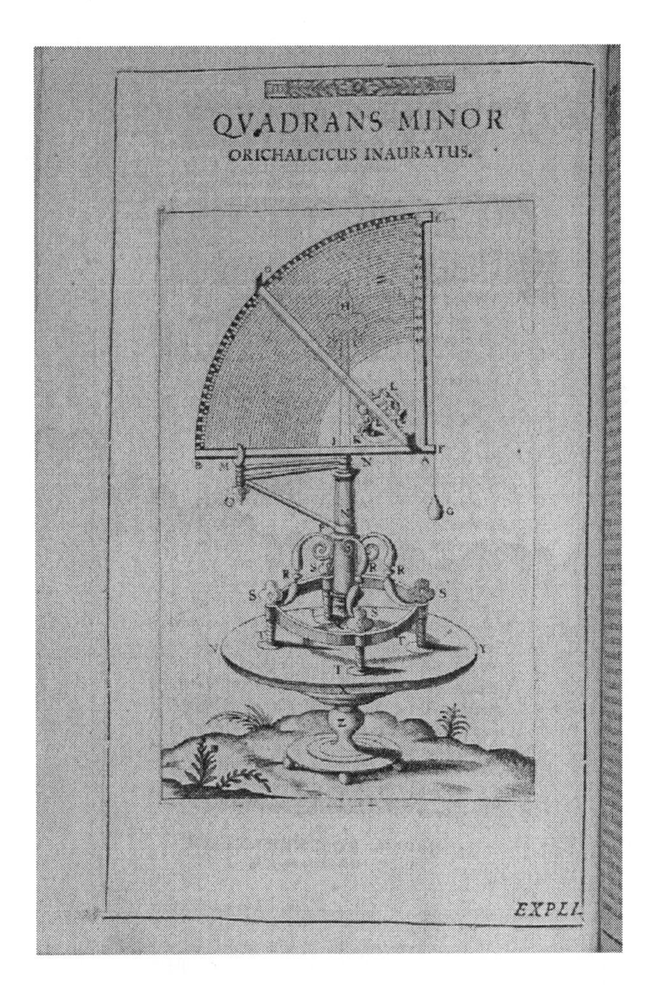

Figure 5.2: Drawing of a quadrent taken from Tycho Brahe's *Astronomiae Instauratae Mechanica* (Wandesburg 1598). [Courtesy: The Burndy Library, Dibner Institute for the History of Science and Technology, MIT]

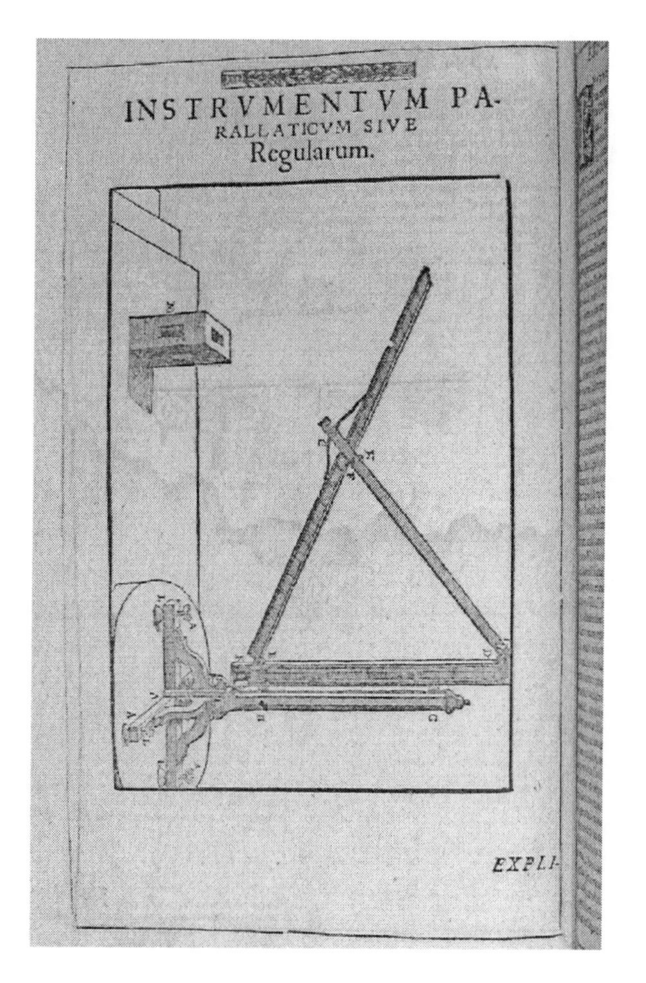

Figure 5.3: Drawing of a Ptolemaic ruler taken from Tycho Brahe's *Astronomiae Instauratae Mechanica* (Wandesburg 1598). [Courtesy: The Burndy Library, Dibner Institute for the History of Science and Technology, MIT]

notably Bernard Walther, used clocks in their observations, many others decided that the available devices were not sufficiently accurate and so continued to determine the time from stellar altitudes.

5.4 Timed Eclipse Records in European History

There would appear to have been very little interest in observing eclipses by European astronomers in the Late Medieval and Renaissance periods. Indeed, before the beginning of the seventeenth century AD, only seven astronomers are known to have made detailed timed observations of eclipses: Isaac ben Sid, Levi ben Gerson, Jean de Murs, Regiomontanus,[21] Bernard Walther, Nicholas Copernicus, and Tycho Brahe. This trend was radically reversed during the seventeenth century AD, in particular after the invention of the telescope, when many astronomers began to make systematic observations of eclipses.[22] It is quite possible, however, that other earlier astronomers did record observations of eclipses in works that have now been lost.

The sources in which the observations of the above mentioned astronomers are found vary widely. Some, for example those by Isaac ben Sid and Jean de Murs, are found in manuscripts that have only recently been found. Others, for example those by Regiomontanus, Walther and Copernicus, are contained in works that were published in the sixteenth century. In particular, the observations by Regiomontanus and Walther are found in a work entitled *Scripta Clarissimi Mathematici M. Ioannis Regiomontani*. This is a collection of short mathematical and astronomical treatises written by Regiomontanus and Peurbach and collected together by Schoener (1544). He also reports many, but not all, of Regiomontanus' and Walther's astronomical observations. However, this compilation, which was made forty years after the last observation it reports, contains a number of scribal errors, as will be noted in Sections 5.4.4 and 5.4.5 below.

Before discussing the observations made by the Late Medieval and Renaissance European astronomers in detail, it is necessary to make two comments. First, the observations were made from various cities throughout Europe. These are listed, together with their geographical latitude and longitude in Table 5.1. The second comment concerns the calendars which the European observers used. With the exception of one observation by Tycho Brahe (in AD 1600) which uses the Gregorian calendar, all of the reports use the Julian calendar. However, there was no consensus through Europe on the date on which the year began. For example, Regiomontanus started the year on 1 January, but Levi ben Gerson chose 1 March as the date of New Year. In some parts of Europe 25 March was used. Throughout this study, however, I shall always use 1 January as the date of New Year. In addition, all Julian dates after the introduction of the Gregorian calendar in AD 1582 will be converted to their Gregorian equivalents.

[21] Some of Regiomontanus' eclipse observations were made jointly with Georg Peurbach.

[22] See, for example, the many eclipse observations collected by Pingré (1901).

City	Latitude (°)	Longitude (°)
Evreux	49.05	−10.18
Frauenberg	54.34	−19.67
Hven	55.92	−12.75
Knudstrup	55.30	−10.87
Melk	48.23	−15.35
Nuremberg	49.45	−11.08
Orange	44.13	−4.80
Prague	50.10	−14.42
Padua	45.40	−11.88
Rome	41.90	−12.48
St. Germain des Pres	48.87	−2.33
Toledo	39.87	+4.03
Vienna	48.20	−16.37
Viterbo	42.40	−12.10

Table 5.1: European Observation Sites

5.4.1 Isaac ben Sid

Isaac ben Sid was one of a number of Jewish scholars at the court of King Alfonso X of Castille during the thirteenth century. Together with Jehuda ben Moses Cohen he was responsible for compiling the *Alfonsine Tables*. As part of the preparation for compiling the tables, the two astronomers made a number of observations of the path of the sun throughout the year, planetary conjunctions, and lunar and solar eclipses.[23] Little is known of the life of ben Sid either before of after the compilation of the *Alfonsine Tables*; ben Moses Cohen is known to have been involved in translating some of the Arabic texts for Alfonso's *Los Libro de las Estrellas*, but ben Sid does not appear to have been involved in this project.

No contemporary record of Isaac ben Sid's astronomical observations is known; however, some forty years later Isaac Israeli of Toledo reported four of ben Sid's eclipse observations in his astronomical work *Jesod Olam*.[24] This work is unavailable to me at present; fortunately, however, the eclipse observations have been described by Goldstein (1979). The three lunar eclipses were observed on 24 December 1265 AD, 19 June 1266 AD, and 13 December 1266 AD. The solar eclipse was observed on 5 August 1263 AD. In each case the observed time of mid-eclipse in Toledo is reported; in addition, for the solar eclipse and the first lunar eclipse, the predicted time of mid-eclipse is given. It is reasonable to suppose that these predicted times were calculated from the *Toledan Tables*. It is not known how Isaac ben Sid determined the observed times of the eclipses, but I would suggest that, following the practice of the Islamic astronomers,[25] he measured the altitude of either the eclipsed luminary or of a clock-star and converted this into a local time.

Isaac ben Sid's eclipse observations are summarized in Table 5.2. The mean ac-

[23] Procter (1945).

[24] Cf. Dreyer (1920) who incorrectly remarks that Isaac Israeli only refers to three eclipse observations.

[25] See Section 4.3 above.

			Local Time (h)		
Date	Type	Contact	Predicted	Observed	Computed
1263 Aug 5	Solar	M	5.05	14.00	14.27
1265 Dec 24	Lunar	M	20.66	3.66	3.42
1266 Jun 19	Lunar	M	-	3.13	3.40
1266 Dec 13	Lunar	M	-	18.62	18.74

Table 5.2: Eclipse observations made by Isaac ben Sid.

curacy of the observed times is about 0.23 hours. It is clear, however, that the times predicted using the *Toledan Tables* are extremely inaccurate — both of the two predicted times are early by more than 7 hours. As there are only two predicted times it is meaningless to try to determine the typical accuracy of the *Toledan Tables* from these records, but it should at least be noted that errors of more than 7 hours are very serious. For example, they are significantly less accurate than contemporary eclipse predictions made by Islamic astronomers.[26]

5.4.2 Levi ben Gerson

Levi ben Gerson lived and worked in Orange,[27] France during the first half of the fourteenth century AD.[28] A Jewish scholar, he was fortunate to live in this part of France for it escaped the expulsion of the Jews from the country by King Philip the Fair in AD 1306. He worked in a number of different fields, writing on mathematics, astronomy, philosophy, and religion, although there is no evidence that he knew either Latin or Arabic; he seems to have worked from Hebrew translations of ancient books. All of his works were written in Hebrew, although some, including his philosophical treatise the *Milḥamot Adonai*, were later translated into Latin.

The *Milḥamot Adonai* ("The Wars of the Lord") is commonly regarded as Levi ben Gerson's greatest work. It comprises six books, of which the fifth is devoted to astronomy. The first draft of the astronomical treatise was completed in AD 1328 but did not reach its final form until AD 1340. Some of the observations that Levi made as part of this work date from AD 1321, and so the final version of the astronomical treatise must represent the culmination of many years of work. The main part of the treatise comprises a set of astronomical tables. Another part of the treatise contains a discussion of an instrument known as the Jacob's Staff. This instrument, which Levi claimed to have invented, is made of two movable pieces of wood in the form of a cross and can be used to measure angular differences between stars.[29]

In constructing his astronomical tables, Levi made extensive use of his own ob-

[26] See Section 4.3 above.

[27] Ben Gerson calls this town *ʿir ha-ezov* in Hebrew, which is translated into Latin as Aurayca. On the identification of Aurayca with Orange, see Goldstein (1974: 19–20).

[28] For a detailed biography of ben Gerson, see Samsó (1973).

[29] For a description of the Jacob's Staff, see Samsó (1973).

			Local Time (h)		
Date	Type	Contact	Predicted	Observed	Computed
1321 Jun 26	Solar	1	4.55	4.41	4.56
		M	-	5.28	5.50
1321 Jul 9	Lunar	1	3.00	-	3.21
1330 Jul 16	Solar	M	16.30	-	16.67
1331 Dec 14	Lunar	4	5.83	-	6.25
1333 May 14	Solar	M	-	15.50	15.49
		4	-	16.63	16.76
1333 Oct 23	Lunar	M	20.12	-	20.54
1334 Apr 19	Lunar	4	23.97	0.00	0.34
1335 Oct 3	Lunar	2	-	2.88	2.98
		3	-	3.65	3.67
		4	5.03	4.73	4.94
1337 Mar 4	Solar	1	7.33	7.29	7.33
		4	9.33	9.30	9.65
1339 Jan 26	Lunar	1	-	4.50	4.42

Table 5.3: Eclipse observations made by Levi ben Gerson.

servations.[30] Amongst them were 10 observations of eclipses. These are recorded in chapters 80 and 100 of the *Milḥamot Adonai*, which have been translated by Goldstein (1979). In many cases Levi reports not only the observed details of the eclipse, but also its circumstances calculated using his tables. As an example, I quote Levi's account of the lunar eclipse on 3 October 1335 AD.

> "The fifth eclipse took place in the year 1335. The end of this eclipse was 1;24 hours before sunrise of 3 October as determined with our accurate instruments. This was the first night to the festival of *Succot*, and there was present with us a noble Christian who wrote down all that we observed. The moon remained in darkness (i.e., totality) 0;44 hours as most accurately determined."
> [*Milḥamot Adonai*, 80, 22–25; trans. Goldstein (1979: 110)]

In chapter 100, Levi also notes that the time of the middle of the eclipse was 15;19 hours after apparent noon, and so the time of 2nd contact was 14;57 hours after apparent noon, and that of 3rd contact was 15;41 hours after apparent noon. He further notes that the time of the end of the eclipse calculated from his tables is 16;51 hours after apparent noon, 6 minutes earlier than observed. It is not known how Levi made his time measurements, but I may once more speculate that they were obtained from altitude measurements.

Levi ben Gerson's eclipse observations and predictions are summarized in Table 5.3. The mean accuracy of the observed times is about 0.16 hours. However, it appears that Levi's observed times are subject to a systematic error; their mean error is about +0.14 hours. The cause of this error is unknown. Perhaps he attempted to correct for the delay in detecting the true moment of an eclipse contact and overestimated the time interval between a true and an apparent contact. The mean true accuracy of Levi's eclipse predictions is about 0.25 hours, but once more these are subject to a

[30]Goldstein (1972).

systematic error, this time of about +0.23 hours. The mean observed accuracy of the predictions, however, is about 0.11 hours, with no significant systematic error. The fact that Levi did not observe any systematic error in his predictions is unsurprising since his tables were based upon his own observations. Therefore, any systematic error in his observations would be carried through to his predictions.

5.4.3 Jean de Murs

Working mainly in Paris and the surrounding areas, Jean de Murs was active in the field of science from about AD 1317 to about AD 1345.[31] Like many European scholars of this period, however, de Murs did not restrict himself to one discipline, but had a range of interests spanning astronomy, mathematics, and music, and during his life wrote a number of books on all of these subjects. In astronomy, he is most widely known as the author of a set of canons for the *Alfonsine Tables* first begun in AD 1321 but not completed until AD 1339.

By AD 1321, Jean de Murs had become a master of arts at the Sorbonne in Paris where he was to continue to work for many years. However, a manuscript uncovered by Beaujouan (1974) at the Escorial Library,[32] has revealed that he continued to travel throughout France during this period.[33] In AD 1333 he traveled to Evreux, near St. Germain, where it would appear he held at least two official positions. Whilst there he observed an eclipse of the sun on 14 May. His observation is recorded in the Escorial Manuscript and has been published in Latin by Beaujouan (1974). In translation, de Murs' account of the eclipse reads:

> "[In the current year of our Lord 1333, on the 14th day of May] ... At Evreux, in the region of St. Germain, 3 brothers and I, in the presence of the Queen of Navarre, observed the beginning of this eclipse. And the altitude of the sun at the initial point of the eclipse was near to 50°, and the altitude at the end of the eclipse was 33°. The digits (*puncta* — i.e., magnitude) of the eclipse, according to our estimate, was 10 digits; a smaller (i.e., more precise) discernment was not possible. And so the beginning of the eclipse was after our midday by 2 hours and 20 minutes."
>
> [Escorial, MS O.II.10, fol. 92]

Jean de Murs then notes that the eclipse occurred about 17 minutes earlier than was predicted by the *Alfonsine Tables*.

By AD 1336, de Murs had returned to the Sorbonne in Paris. The following year he was to observe another eclipse from nearby St. Germain des Pres. This solar eclipse, also reported in the Escorial Manuscript, took place on 3 March:

> "In the year of our Lord 1337, on the 3rd day of March, after the sun rose. That day, in St. Germain des Pres, we saw the beginning of the eclipse of the sun. The sun was at an altitude of 10°, and already a (small) part was eclipsed, from which

[31] For detailed biographical accounts of Jean de Murs, see Poulle (1973) and Gushee (1969).

[32] Escorial, MS O.II.10.

[33] Gushee (1969).

Date	Type	Contact	Object	Altitude (°) Observed	Altitude (°) Computed	Local Time (h) Predicted	Local Time (h) Observed	Local Time (h) Computed
1333 May 14	Solar	1	Sun	50.00	54.44	14.02	14.33	13.76
		4	Sun	33.00	31.20	15.90	16.18	16.21
1337 Mar 3	Solar	1	Sun	9.00	9.55	6.93	7.20	7.26
		4	Sun	29.00	27.69	9.34	9.61	9.43

Table 5.4: Eclipse observations made by Jean de Murs.

> we concluded that the peripheries of the luminaries could have touched (when they were) at an altitude of 9°. Similarly, we saw that the moon left contact with the sun, as much as (it) was possible (to see), (when) the sun was at an altitude of 27° and about 30'. However, at Paris, the Alfonsine Tables placed the beginning of the eclipse of the sun (when) the sun had an altitude of 14° ... It finished, I think, at an altitude of 29° as it was seen at an altitude of $27\frac{1}{2}°$. Therefore, I consider it necessary to quickly correct and make known the errors in placing eclipses by the Alfonsine Tables. And it follows similarly from the preceding eclipse in the year 1333. That one anticipated the tables by about a third of an hour ... The tables place the amount in darkness as 7 digits (*puncta*) of the diameter. However, it was only 5 digits. In this experiment there were 10 of us present and a number had good astrolabes."
>
> [Escorial, MS O.II.10, fol. 93]

It is interesting to note de Murs' attitude towards the inaccuracy of the *Alfonsine Tables*. He found errors of about one-quarter of an hour between the tables and observation to be unacceptable, but when they are compared to the errors observed by earlier Islamic astronomers,[34] this does not seem to be too bad. Nevertheless, de Murs remarks that "it is necessary to quickly correct and make known the errors" in the tables. Perhaps this was to form the stimulus for the revision of his canon to the tables in AD 1339.

After AD 1337, no further eclipse records by de Murs are known. However, this was not the end of his interest in astronomy. In AD 1344, de Murs, together with Firmin de Belleval, was called to Avignon by Pope Clement VI to discuss calendar reform.[35] His suggestions, which were published as a memoir in AD 1345, were not, however, adopted. Later, at some point before AD 1352, de Murs wrote to Clement VI to inform the pope of the favourable conditions for a crusade in AD 1365 signified by a conjunction in that year. This is the latest evidence we have for de Murs' work, and it would appear that he died shortly after this date.

Jean de Murs' eclipse observations are summarized in Table 5.4. With the exception of the time of first contact of the eclipse in AD 1333, the timings are very accurate, having a mean accuracy of about 0.08 hours. Including this timing, the mean accuracy is about 0.21 hours. Furthermore, there is no evidence for any systematic errors in

[34] See Section 4.4 above.
[35] Poulle (1973).

the timing. The mean true accuracy of the times calculated from the *Alfonsine Tables* is about 0.09 hours, which is very impressive at this comparatively early period. The mean observed accuracy is only 0.28 hours, but this is still fairly good and does not seem to justify de Murs' harsh criticism of the tables. It is interesting to note that both of these eclipses were also observed by Levi ben Gerson. With the exception of de Murs' inaccurate first contact timing of the eclipse in AD 1333, the two astronomers reached a comparable level of accuracy in both observing and predicting the times of these eclipses.

5.4.4 Regiomontanus

Johannes Muller of Konigsberg, who later adopted the name Regiomontanus, enrolled in the University of Vienna in AD 1450 at the age of fourteen, and within two years had been awarded his bachelor's degree.[36] Whilst still a student, he was befriended by Georg Peurbach, whom Regiomontanus often refers to as "my teacher," and together they began making a study of astronomy. Their earliest observation for which a record is preserved is of the lunar eclipse on 3 September 1457 AD. This was made at Melk castle, a short distance to the west of Vienna. Because of the low precision of the mechanical clocks of the period, Regiomontanus used an astrolabe to measure the altitude of the star 27 Tau and thus to determine the time when the eclipse started and finished. He then proceeded to deduce the time of the true opposition and compared the observed time with that calculated by the *Alfonsine Tables*, noting a significant error.

On 11 November of that same year, Regiomontanus was appointed to the faculty of the University of Vienna. He remained there for the next four years, and during that time observed another three lunar eclipses. Once more the times of the eclipses were determined by measuring the altitudes of either bright stars or the moon itself, and were then compared with the times given by tables. These tables were probably Peurbach's *Tabulae eclipsium*, which Regiomontanus had copied during this period.

In AD 1460, Cardinal Bessarion, the papal legate to the Holy Roman Empire, visited Vienna. He asked Peurbach to write a commentary or *Epitome* of Ptolemy's *Almagest*, and to accompany him to Rome. Peurbach accepted on the condition that Regiomontanus was to travel with them. However, before they departed, Peurbach fell ill and died, leaving Regiomontanus to complete the writing of the commentary and to travel to Rome alone with Bessarion. This was in the autumn of AD 1461.[37]

Regiomontanus spent the next five years in Italy, where, among other things, he completed writing the *Epitome* of the *Almagest*, entered into his correspondence with Bianchini, and continued making astronomical observations. He observed two lunar and one solar eclipse over the next year, either in Rome itself, or in nearby Viterbo. In AD 1464, Regiomontanus moved to Padua to present a series of lectures. Whilst there, he observed the lunar eclipse on 21 April, once again measuring the altitude of

[36] For a detailed biography of Regiomontanus, see Zinner (1990) and Rosen (1975).
[37] Zinner (1990: 51).

two bright stars at the moment that the eclipse began. He used an instrument called the "great quadrant" to make these measurements. This quadrant evidently allowed Regiomontanus to make much more precise determinations of altitude than his usual instruments, for he quotes the altitudes to the nearest five minutes of arc, instead of to his usual degree or half-degree.

After this brief sojourn in Padua, Regiomontanus returned to Rome, before moving on to Hungary in AD 1467 to join the faculty of the University of Pressburg. There he passed the next four years until he was invited to Nuremberg and given a Royal Commission to make celestial observations. In Nuremberg, he met Bernard Walther (who became his pupil and associate) and set up a print shop to publish scientific works. These were mainly written by himself and Peurbach, and include the *Kalendarium* which contained the ecclesiastical calendar, tables for calculating the position of the sun and moon, and a canon of eclipses for the years 1475 to 1530. He also began making systematic astronomical observations. For this purpose, he built a Jacob's staff and a Ptolemaic ruler;[38] however, it is not clear whether he used either of these instruments during his eclipse observations.

Over the next four years, Regiomontanus undertook a programme of stellar and planetary observations. However, at the end of June 1475 AD, before it was complete, he was summoned to Rome by Pope Sixtus IV to participate in the discussions on the reformation of the calendar,[39] and by the following year he had died, perhaps murdered, at the age of 40.[40]

In all, Regiomontanus observed nine eclipses in Vienna, Rome, Padua, and Nuremberg. These are reported in Schoener's *Scripta Clarissimi Mathematici M. Ioannis Regiomantani*,[41] which was compiled from Regiomontanus' notebooks some sixty-nine years after Regiomontanus' death. Below I give translations of the relevant parts of the report of each observation, taken from Steele & Stephenson (1998b):

- 3 September 1457 AD

> "Master Georg Peurbach and Johannes Regiomontanus observed in Melk, Austria, near Vienna, in the year of our Lord 1457 a total eclipse of the moon at the true opposition in September, namely after sunset on the 3rd day of the month. Moreover, at the beginning of totality the penultimate star of the Pleiades (27 Tau) had an easterly altitude of 22 degrees, and according to calculation the sun was 48 minutes in the 20th degree of Virgo. However, at the end of totality the altitude of this same star was 36 degrees. This

[38] Zinner (1990: 141).

[39] Zinner (1990: 151).

[40] Shortly after Regiomontanus' death, rumours spread that he had been the victim of a terrible crime. The sons of Trebizon, whose translation of the *Almagest* Regiomontanus had criticized, were said to have murdered him. However Zinner (1990: 152) notes that there is little evidence in support of this story, and suggests instead that Regiomontanus was probably the victim of a plague that was epidemic in Rome in that year.

[41] Schoener (1544). Regiomantanus' observations are given in folios 36–43, entitled *Ioannis de Monteregio, Georgii Peurbachii, Bernardi Waltheri, ac aliorum, Eclipsium, Comentarum, Planetarum ac Fixarum observationes*.

observation was at Melk castle in Austria, which is distant from Vienna 11 German miles towards the west. From these two altitudes of the said star, the actual time of mid-eclipse can be calculated ..."

Note: There follows a long calculation in which the time of true opposition is deduced as 11 hours 6 minutes, which compares with the time calculated from the *Alfonsine Tables* of 11 hours 14 minutes.

- 3 July 1460 AD

  "There was a partial eclipse of the moon during the night which followed the 3rd day of July; its beginning was exactly 7 hours 16 minutes after midday. Moreover, the middle (was) at 8 hours 13 minutes and the end occurred at 9 hours 10 minutes; (it was) 2;56 ecliptic digits. This (was) according to the tables for the meridian of Vienna. However, I myself observed the middle of this eclipse in the sky, and it seemed to be eclipsed rather more than four digits. Moreover at the end I measured the altitude of the moon as 15 degrees 18 minutes. Also present was Georg, my teacher."

- 27 December 1460 AD

  "In the same year there was a total eclipse of the moon at the true opposition of the luminary, which was the 27th day of December, in which by observation at the start of the eclipse the star which is called *Alramech* (α Boo) had an altitude in the east of 7 degrees. At the beginning of totality the altitude was 17 degrees and at the end of totality the altitude was 28 degrees. At the beginning of the eclipse the moon was seen on a great circle passing through the head of the preceding Twin (α Gem) and the bright (star) of Canis Minor (α CMi). However, at the end it was above a circle passing through the head of the following Twin (β Gem) and Canis Minor. The observers were Georg Peurbach and Johannes Regiomontanus in the town of Vienna."

- 22 June 1461 AD

  "There was a total eclipse of the moon at the opposition of the luminary, which was the 22nd day of June. Moreover, at the beginning of totality the altitude of the Flying Vulture (α Aql) was 26 degrees; the moon was then noted at an altitude of 6 degrees 30 minutes. At the end of the whole eclipse, the altitude of the Vulture (α Aql) was 47 degrees 30 minutes. Master Johannes Regiomontanus noted the altitude of the moon as 17 degrees 30 minutes in the town of Vienna. It was therefore probable that the opposition of the luminary occurred one hour and 21 minutes after midnight. By calculation using tables, this occurred one hour and 20 minutes after midnight, a difference of one minute."

  Note: It seems probable that the time of opposition Regiomontanus determined both from his observations and from the *Alfonsine Tables* should be before midnight rather than after midnight. From his altitude measurements, he determined that the eclipse became total just before 9 pm, and that the eclipse ended after 11:30 pm. A time of opposition before 11 pm therefore seems more likely.

- 17 December 1461 AD

"At the start of the following night on the 17th day of December, the moon rose eclipsed by 10 digits of its diameter. Indeed I merely noted 8 (digits). Moreover, from the Alphonsine computations the end of the eclipse occurred at 1 hour and 56 minutes after sunset. At this same end of the eclipse the altitude of the star *Alhaioth* (α Aur) in the east was 38 degrees 30 minutes, whereas (the altitude of) the star *Aldebaran* (α Tau) was 29 degrees in the east. The location of the sun according to computation was 5;24 (degrees) in Capricorn. This was in the city of Rome, whose latitude is 42 degrees 2 minutes, although some place it as 41 degrees 50 minutes..."

Note: The text contains a printing error, giving the altitude of *Alhaioth* as 38 minutes 30 seconds instead of 38 degrees 30 minutes.

- 11 June 1462 AD

"On the night which followed the 11th of June, it happened that there was a partial eclipse of the moon at 15 hours 15 minutes after midday. Moreover, it was eclipsed 6;34 digits according to the tables. I observed the eclipse in Viterbo near Rome, which is believed to be to the east [sic] of Vienna by 4 degrees and a little more according to geography. I could not, however, note the start nor the end on account of obscuring clouds. However, in the middle, the Flying Vulture (α Aql) had an altitude of 51 degrees in the west. I judged that it was eclipsed about 7 digits."

- 21 November 1462 AD

"On the 21st day of November, I observed an eclipse of the sun about midday. I did not observe the start of the eclipse, but when I caught sight of it, it seemed that two digits of the sun were eclipsed from the south side. The sun then had an altitude of $26\frac{1}{2}$ degrees and was due south. Furthermore, at the end of the eclipse, which I carefully noted, the sun had an altitude of 24;36. The degrees of azimuth of the sun towards the west of the meridian were 16 degrees 15 minutes. But, as far as I was able to conjecture, it seemed that a third of the time of the whole eclipse had passed from the beginning of the eclipse up to the instant of its first observation. For, a little before (my) first observation, which was precisely at midday, I observed the sun not yet eclipsed. All from Viterbo near Rome."

- 21 April 1464 AD

"There was a total eclipse of the moon, namely at its opposition, which was on the 21st of April at fully 12 hours and 59 minutes after midday, in equal hours at the meridian of the city of Padua, whose latitude is said to be 45 degrees 24 minutes. The true place of the sun was 10 degrees 52 minutes in Taurus from the Alfonsine calculations. Moreover, the moon was in opposition. The true argument of latitude of the moon was 5 degrees 25 minutes 23 seconds; the northerly latitude of the moon at the middle of the eclipse was 0 degrees 24 minutes and 5 seconds.

Start of eclipse	11 h 15 m
Start of totality	12 h 33 m
Middle of eclipse	12 h 59 m
End of totality	13 h 25 m
End of eclipse	14 h 43 m
Semi-duration of day	7 h 5 m
Total duration of the eclipse	3 h 28 m

All according to the Alfonsine parameters.

At the beginning of this eclipse I found that the altitude of (the star) at the Heart of the Scorpion (α Sco) was 12 degrees 45 minutes in the east. At the same time also the altitude of α Hyd was 9 degrees 40 minutes by the great quadrant, all (measured) as carefully as possible."

- 2 June 1471 AD

"On the night of the 2nd of June there was an eclipse of the moon, at the start of which the Heart of the Scorpion (α Sco) had an altitude in the west of 14 degrees 15 minutes. Further, the Dolphin or Moscida of Pegasus (β Peg) had an altitude in the east of 22 degrees 30 minutes. Four digits appeared to be obscured. Afterwards the moon appeared to fill up again. The true end did not appear on account of intervening clouds at Nuremberg."

Regiomontanus' eclipse observations are summarized in Table 5.5. Clearly he achieved considerable success in observing his eclipses. The mean accuracy of his altitude measurements is about 0.80°, which corresponds to a mean accuracy in time of about 0.12 hours, or just over 7 minutes. As Regiomontanus generally quotes his altitude measurements to the nearest half degree, it would appear that he was observing at a level of accuracy close to the precision of the instruments that he used. Furthermore, there is no evidence for any systematic errors in his measurements; the mean error in his altitudes is about $-0.15°$ which is well within the level of precision of his instruments.

Between AD 1472 and AD 1475, Regiomontanus made a series of observations of the meridian zenith distance of the sun with his Ptolemaic ruler. Newton (1982) has shown that these observations typically have an accuracy of better than about 2 minutes of arc. This is a noticably better level of accuracy than Regiomontanus achieved during his eclipse observations. However, this instrument was, of course, being used here for meridian observations. Eclipse contacts would have to be observed rapidly at a wide variety of azimuths and often when the lunar or solar altitude was changing at an appreciable rate. These circumstances would have presented practical difficulties. Furthermore, when observing an eclipse there is the additional problem of defining the exact moment of contact with the unaided eye.

As noted above, one of the main reasons Regiomontanus observed eclipses was to compare his measurements with the calculations given by the *Alfonsine Tables*. From Table 5.5 it is clear that there is a considerable difference between the Alfonsine times

				Altitude (°)		Local Time (h)		
Date	Type	Contact	Object	Observed	Computed	Predicted	Observed	Computed
1457 Sep 3	Lunar	2	27 Tau	22.00	22.97	-	22.48	22.58
		M	-	-	-	23.23	23.10	23.17
		3	27 Tau	36.00	34.68	-	23.89	23.75
1460 Jul 3	Lunar	1	-	-	-	19.27	-	19.97
		M	-	-	-	20.22	-	21.03
		4	Moon	15.30	15.04	21.17	22.14	22.09
1460 Dec 27	Lunar	1	α Boo	7.00	5.01	-	23.71	23.48
		2	α Boo	17.00	14.73	0.78	-	0.54
		3	α Boo	28.00	26.83	1.89	-	1.76
1461 Jun 22	Lunar	2	α Aql	26.00	27.02	-	20.78	20.85
		2	Moon	6.50	6.56	-	20.84	20.85
		M	-	-	-	22.67	22.65	21.72
		4	α Aql	47.50	47.35	-	23.72	23.68
		4	Moon	17.50	17.69	-	23.52	23.68
1461 Dec 17	Lunar	4	α Aur	38.50	39.29	18.50	17.35	17.43
		4	α Tau	29.00	28.16	18.50	17.41	17.43
1462 Jun 12	Lunar	M	α Aql	51.00	51.49	3.25	2.81	2.72
1462 Nov 21	Solar	4	Sun	24.60	24.07	-	12.94	13.12
1464 Apr 21	Lunar	1	α Sco	12.75	13.99	23.25	22.94	23.16
		1	α Hyd	9.66	10.01	23.25	23.20	23.16
		2	-	-	-	0.55	-	0.36
		M	-	-	-	0.98	-	0.79
		3	-	-	-	1.42	-	2.23
		4	-	-	-	2.72	-	2.42
1471 Jun 2	Lunar	1	α Sco	14.25	15.05	-	0.00	23.72
		1	β Peg	22.50	22.43	-	23.73	23.72

Table 5.5: Eclipse observations made by Regiomontanus.

on the one hand and both Regiomontanus' observed times and the times given by modern computation on the other. The mean true accuracy of the predicted times is about 0.50 hours, and the mean observed accuracy is about 0.55 hours.

5.4.5 Bernard Walther

Bernard Walther does not figure largely in histories of astronomy.[42] He is often regarded as Regiomontanus' student and possible patron, but nothing more. This may be because he seems to have made no contribution to the development of astronomical theories, but rather to have merely been an observer. However, this is to understate his importance, for he left a legacy of more than thirty years' worth of careful observations — observations which would be extensively used by Copernicus, Tycho, and Kepler in testing their planetary theories.[43]

When Regiomontanus came to Nuremberg in AD 1471 he met Walther who had settled there four years earlier. First as teacher and student, and then as collaborators, the two men began their programme of systematic observations of the Heavens. After

[42] Walther is not even given his own entry in the *Dictionary of Scientific Biography*. For some biographical details, see deB. Beaver (1970). Regarding this article, however, note the cautionary remarks in note 2 of Kremer (1980). For a more recent biographical study, see Eirich (1987).

[43] Kremer (1981).

Regiomontanus' death in Rome, Walther continued to observe in Nuremberg until only 16 days before his own death on 19 June 1504 AD.[44]

Between AD 1478 and AD 1504, Walther observed four solar and two lunar eclipses. In AD 1501 he had two extra south-facing windows built in his house from which it would appear he made his observations. The location of his earlier observation sites in Nuremberg are not known. Like Regiomontanus, Walther often determined the time of an eclipse by measuring the altitude of the luminary, using either a Ptolemaic ruler, or an armillary sphere. However, in AD 1487, he used a mechanical clock to measure the time of the eclipse on 8 February. In two of his observations, those of AD 1485 and AD 1497, Walther does not state the method he has used to determine the time of the eclipse, but as he is known to have made use of his clock after AD 1484, it seems most likely that he would have used it on these occasions.

Walther's astronomical observations were published in Schoener's *Scripta Clarissimi Mathematici M. Ioannis Regiomantani*.[45] I give below translations of the relevant parts of his eclipse observations, once more taken from Steele & Stephenson (1998b):

- 29 July 1478 AD

  > "July 29. At about the first hour after midday, namely when the sun's altitude was $54\frac{1}{2}$ degrees, a solar eclipse began. Further, it ended when the altitude of the sun reached $41\frac{1}{2}$ degrees."

- 16 March 1485 AD

  > "On the 16th day of March there was an eclipse of the sun, beginning 3 hours and 26 minutes after midday; the end was at 5 hours and 28 minutes (after midday) and about 12 points (= digits) were obscured..."

- 9 February 1487 AD

  > "On the 8th of February, an eclipse of the moon occurred at about the second hour after midnight, in the morning. When the moon began to be eclipsed, the sun had a depression of 29 degrees. When it was the middle of the eclipse, the true time as indicated by the clock was 3 hours and 45 minutes. At the end of total obscuration, the sun had a depression of 24 degrees, while the clock read 4 hours 18 minutes. The end of the eclipse was at 5 hours 20 minutes after midnight. The times were checked by altitudes..."
  >
  > Note: Clearly the statement that the sun had a depression of 29° at the start of the eclipses is a misprint in the text as the computed value of the solar depression is about 44° at this instant. Presumably the solar depressions were simply obtained from the negative value of the lunar altitude.

[44] Zinner (1990: 146–147).

[45] Schoener (1544). Walther's observations are given in folios 44–60, entitled *Observationes factae per doctissimum virum Bernardum Waltherum Norimbergae*.

				Altitude (°)			Local Time (h)	
Date	Type	Contact	Object	Observed	Computed	Predicted	Observed	Computed
29 Jul 1478	Solar	1	Sun	54.50	55.76	-	13.09	12.78
		4	Sun	41.50	41.00	-	14.95	15.01
16 Mar 1485	Solar	1	-	-	-	-	15.43	15.47
		4	-	-	-	-	17.47	17.50
9 Feb 1487	Lunar	M	Clock	-	-	-	3.75	3.81
		3	Sun	-24.00	-24.61	-	4.44	4.38
		3	Clock	-	-	-	4.30	4.38
		4	Clock	-	-	-	5.33	5.50
10 Oct 1493	Solar	4	Zenith	-	-	-	16.40	16.49
29 Jul 1497	Solar	4	-	-	-	15.40	15.40	15.66
1 Mar 1504	Lunar	4	Zenith	-	-	3.37	3.11	2.98

Table 5.6: Eclipse observations made by Bernard Walther.

- 10 October 1493 AD

  "On the 10th October there was an eclipse of the sun, the start of which was between the first and second hours after midday. Its start was poorly observed, but at its end the zenith was in the first degree (*puncto*) of Capricorn, namely 4 hours and 24 minutes after midday."

- 29 July 1497 AD

  "On the 29th of July there was an eclipse of the sun. I did not observe the start of this eclipse but the end, which was about 3 hours 24 minutes after midday, agreed with its calculated value."

- 1 March 1504 AD

  "On the morning of the 1st of March I observed an eclipse of the moon, but I could see neither the start of the eclipse, nor totality, nor even the end of totality on account of clouds. However, I saw the end of the eclipse more completely, and I observed accurately through the Earth's shadow with an armillary; and I found that the zenith was 10 degrees in Virgo. However, according to the previous observations, the sun at the instant was 20 degrees 7 minutes in Pisces. The R.A. at the start was therefore 80 degrees 55 minutes in Capricorn. However, the R.A. of the zenith was 307 degrees 35 minutes. Whereby the end of the eclipse was 3 hours, 6 minutes and 40 seconds after midnight. Calculation gives 3 hours 22 minutes."

Walther's measurements of the eclipses are summarized in Table 5.6. From the table it is evident that the typical accuracy of Walther's eclipse timings is about 0.12 hours, the same value as Regiomontanus' measurements. This is interesting as Walther preferred to measure the position of the sun or to use a clock, rather than to measure the altitude of a bright star as Regiomontanus had done. But it would seem that the same level of accuracy was typically achieved no matter how the time was determined.

However, as the clock used by Walther was probably regulated by comparison with observations of stellar or solar altitudes, this is perhaps not surprising. On the three occasions when Walther did measure altitudes during the eclipses, he achieved a level of accuracy almost identical to that of Regiomontanus. Over the same period that Walther observed the eclipses described above, he also made more than 700 measurements of the meridian zenith distance of the sun. These measurements, which Kremer (1983) has shown were in all but one case made with the Ptolemaic ruler, are accurate to about one minute of arc.[46] As with Regiomontanus' timings, it would appear that the various problems of determining the exact moment of an eclipse contact may have been the limiting factor in Walther's measurements of the time of an eclipse.

Bernard Walther's two calculations of the time of an eclipse using the *Alfonsine Tables* are of a comparable accuracy to those made by Regiomontanus. It is interesting to compare Walther's eclipse predictions with those contained in Regiomontanus' *Kalendarium* (see Figure 5.4). The end of the solar eclipse on 29 July 1497 was said by Walther to have been observed at "about 3 hours 24 minutes after midday, (which) agreed with its calculated value." According the *Kalendarium*, the moment of conjunction was 3 hours 2 minutes after midday and the duration of the eclipse was 36 minutes. Thus, the end of the eclipse was calculated by Regiomontanus to be 2 hours hours 38 minutes after midday.

The end of the lunar eclipse on 1 March 1504 was said by Walther to have been calculated to take place 3 hours 22 minutes after midnight. In the *Kalendarium*, the moment of opposition is given as 1 hour 36 minutes after midnight (written as 13 hours 36 minutes after midday on 29 February), and the duration as 1 hour 46 minutes. Thus, Regiomontanus calculated the end of the eclipse to be at 2 hours 29 minutes after midnight. Evidently, therefore, Walther did not simply take his calculated eclipse times from Regiomontanus's *Kalendarium*. Presumably he calculated them himself. Incidentally, it was the prediction of the eclipse on 1 March 1504 by the *Kalendarium* that was used by Christopher Columbus in an attempt to determine the longitude of Jamaica.[47]

5.4.6 Nicholas Copernicus

Nicholas Copernicus has become justly famous as one of the foremost astronomers of the European Renaissance.[48] He was born in AD 1473 at Torún in Poland to a prosperous merchant family, and following his father's death in AD 1483, was raised by his uncle, Lucas Watzenrode. In AD 1491 Copernicus entered the University of Cracow, where he developed an interest in astronomy. Five years later he was elected, through the influence of his uncle who had became the bishop of Ermland in AD 1489, as a canon of the Cathedral of Frauenberg. Officially to study canon law at the University of Bologna, but also to pursue his interest in astronomy, Copernicus travelled to Italy in AD 1496. In AD 1500, he visited Rome to lecture "on mathematics

[46] Newton (1982).

[47] Zinner (1990: 122).

[48] For a detailed discussion of Copernicus and his work, see Swerdlow & Neugebauer (1984).

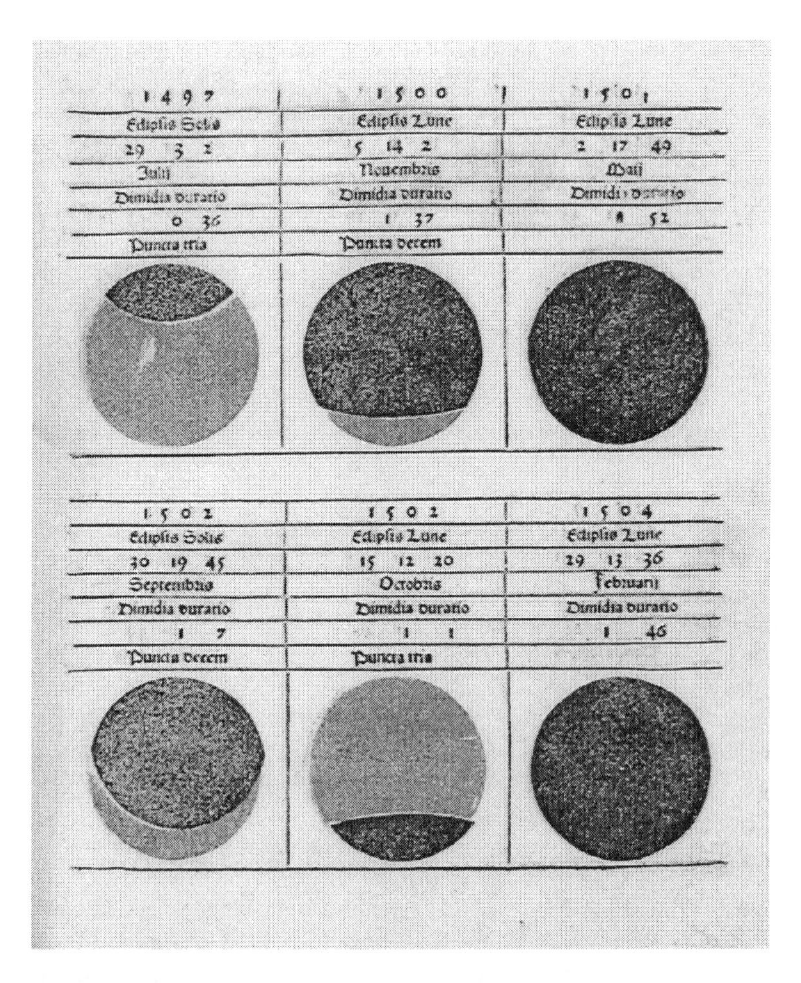

1497	1500	1501
Eclipsis Solis	Eclipsis Lune	Eclipsis Lune
29 3 2	5 14 2	2 17 49
Julii	Novembris	Maii
Dimidia duratio	Dimidia duratio	Dimidia duratio
0 36	1 37	1 52
Puncta tria	Puncta decem	

1502	1502	1504
Eclipsis Solis	Eclipsis Lune	Eclipsis Lune
30 19 45	15 12 20	29 13 36
Septembris	Octobris	Februarii
Dimidia duratio	Dimidia duratio	Dimidia duratio
1 7	1 1	1 46
Puncta decem	Puncta tria	

Figure 5.4: Eclipses for the years AD 1497–1504 from Regiomontanus' *Kalendarium* (Vienna 1474). [Courtesy: The Burndy Library, Dibner Institute for the History of Science and Technology, MIT]

before a large audience of students and a throng of great men and experts in this branch of knowledge."[49] Whilst there, Copernicus observed the lunar eclipse on 6 November 1500 AD, noting the time of the eclipse in his manuscript copy of Regiomontanus' *Ephemerides*.

After gaining permission from his chapter to continue his studies in Italy, Copernicus enrolled in the University of Padua in AD 1501 to study medicine. He returned shortly afterwards to Frauenberg where he was to spend the remainder of his days in the service of his chapter. He also continued to make astronomical observations. Around this time, he wrote the first draft of his *Commentariolus*, outlining his heliocentric world system. This he circulated privately to friends in order to gain an idea of the reception his theory would receive. The response cannot have been too encouraging, for, although he commenced writing his *De Revolutionibus*, which described his world system in great detail, he was reluctant to publish it. It was not until Georg Joachim Rheticus, a professor of mathematics at the University of Wittenberg, visited Copernicus that he began to consider publication.

In AD 1540, Copernicus gave Rheticus permission to write an account of the *De Revolutionibus* to try to stimulate interest in his work. This Rheticus did in his *Narratio Prima*, and its reception encouraged Copernicus to complete the *De Revolutionibus*. By May of AD 1542 the book was ready for printing in Nuremberg. However, he fell ill later in that same year and only received a copy of his book on 24 May 1543 AD, the day of his death.

According to Rheticus in his *Narratio Prima*, Copernicus was a careful and diligent observer who, "for nearly 40 years in Italy and here in Frauenberg ... observed eclipses and the motion of the sun."[50] Unfortunately, however, Copernicus' original observational record is lost, and only those observations he cited in his published works are extant. In his *De Revolutionibus* Copernicus records observations of five lunar eclipses that he himself observed. Following Ptolemy, he uses these observations, together with those recorded in Ptolemy's *Almagest*, to determine various parameters for his theories. As he notes, lunar eclipses are particularly useful for this purpose since they are not affected by parallax:

> "For that reason (i.e., parallax) its (i.e, the moon's) position cannot be observed by means of astrolabes or any other instruments. But the benevolence of nature supplies human needs even in this area, so that it is located all the more certainly by its eclipses than by the use of instruments, and without any suspicion of error. For the rest of the universe being untouched and full of daylight, it is accepted that the darkness is merely the shadow of the Earth, which extends to a conical shape, and ends in a point. The moon, entering this, is dimmed, and when its position is in the centre of the darkness, it is supposed that it has reached a point opposite to the sun. However the eclipses of the sun, which are caused by the interposition of the moon, do not present a certain proof of the moon's position. For a conjunction of the sun and moon comes within our observation at that time, although with reference to the centre of the earth it has already passed or has not yet occurred,

---

[49] Rheticus, *Narratio Prima*; trans. Rosen (1959: 111).

[50] *Narratio Prima*; trans. Rosen (1959: 125).

			Local Time (h)	
Date	Type	Contact	Observed	Computed
1500 Nov 6	Lunar	M	2.00	1.83
1509 Jun 2	Lunar	M	23.75	23.77
1511 Oct 6	Lunar	1	22.88	22.62
		4	2.33	1.78
1522 Sep 5	Lunar	1	23.60	23.46
		M	1.33	1.19
1523 Aug 25	Lunar	1	2.80	2.39

Table 5.7: Eclipse observations made by Nicholas Copernicus.

on account of the parallax already mentioned as the cause of the displacement. Consequently the same eclipse of the sun is not observed to be equal in magnitude and duration in all countries, nor following the same pattern. However, in lunar eclipses no such obstacle occurs, but they are alike everywhere."

[*De Revolutionibus*, iv, 3; trans. Duncan (1976: 190)]

The five lunar eclipses recorded in the *De Revolutionibus* were observed on 6 November 1500 AD, 2 June 1509 AD, 6 October 1511 AD, 5 September 1522 AD, and 25 August 1523 AD. The first eclipse was observed in Rome during Copernicus' lecture visit. The remaining eclipses were observed in Frauenberg. Copernicus notes that Frauenberg is at the same longitude as Cracow; in truth, Frauenberg is about $\frac{1}{4}^{\circ}$ to the west. As an example of Copernicus' eclipse records, I quote below a translation of the account of the observation of the eclipse in AD 1511:

"This being Ptolemy's method, let us follow it and proceed to another trinity of eclipses, which were also very carefully observed by ourselves. The first was in the year of Christ 1511, when six days of the month of October had passed; and the moon began to be in eclipse one hour and an eighth before midnight, and returned to complete visibility at two hours and a third after midnight. Thus the middle of the eclipse was $\frac{7}{12}$ of an hour after the middle of the night which followed the morning of the seventh of October (the Nones) ..."

[*De Revolutionibus*, iv, 5; trans. Duncan (1976: 198–199)]

It is not known how Copernicus made his time measurements. He may have either used a clock or have measured the altitude of the moon or a clock-star at the time of the eclipse.

Copernicus' eclipse observations are summarized in Table 5.7. It is immediately clear from this Table that, with the exception of the observation in AD 1509, all of Copernicus' observed times are slightly late. This suggests that Copernicus did not detect the apparent moment of an eclipse contact until some time after the true moment of this contact, and, furthermore, that he made no allowance for the difference between true and apparent contacts. The mean accuracy of these observations is about 0.21 hours, which is quite poor in comparison with that of his predecessors, Regiomontanus and Walther.

5.4.7 Tycho Brahe

Born into a noble family in AD 1546, Tycho Brahe spent most of his adult life engaged in astronomical observation.[51] Tycho was brought up by his paternal uncle who ensured that, from the age of seven, he was educated in Latin and the preparatory subjects. Between AD 1559 and AD 1562 he attended the Lutheran University of Copenhagen where he studied the *quadrivium*.[52] By AD 1560, Tycho had developed an interest in astronomy, and, following his sighting of a partial solar eclipse in August of that year, began making astronomical observations. The young Tycho's growing interest in science was a worry to his uncle, who decided in AD 1562 to send him to the University of Leipzig to study law. However, Tycho's interest could not be overcome, and he continued to make astronomical observations in secret.[53]

After he left Leipzig in AD 1565, Tycho spent the next eight years travelling throughout northern Europe, visiting various men who were interested in astronomy, and lecturing on astronomy in a number of universities. Then, on 11 November 1572 AD, Tycho noticed a new star shining brightly in the constellation of Cassiopeia.[54] This observation was to result in Tycho abandoning the notion of the fixed stars as being immutable objects, and, together with his later cometary observations, led him to reject the Aristotelian cosmology of the crystalline spheres. In AD 1573, Tycho published a short tract on the new star, which, along with the details of his observations, also included a section of the star's astrological significance, and predictions for a lunar eclipse on 8 December of that year. These predictions he had made himself by adapting the *Prutenic Tables* of Reinhold. The following year, Tycho observed this same eclipse, noting with considerable satisfaction the accuracy of his earlier predictions:

> "I carefully observed the eclipse from the Senate House in Knudstrup, for the sky was sufficiently clear. I had the help of my beloved sister Sophie Brahe, at that time a girl of 14, who was a willing and competent assistant. Indeed, by means of the ungilded brass Quadrant, with which a single degree could sufficiently readily be measured, we saw that the moon entered the shadow and when (only) a small part was covered, the Head of the Lower Twin (β Gem) was at an altitude of $14\frac{1}{2}^{\circ}$, so that the star was scarcely 14° above the horizon at the beginning. Our value was higher than the same star's height given by earlier calculations of 13°. But this was only a small difference with the observation. At the beginning of totality, we observed that the same star had an altitude of $22\frac{2}{3}^{\circ}$. Our earlier calculations put this at $21\frac{1}{2}^{\circ}$. That is also only a moderate difference. At the beginning of the departure and shining forth from the shadow, Canis Minor (α CMi) was observed to have an altitude of $17\frac{5}{6}^{\circ}$. But the earlier (calculated) value was 17°, which is hardly any noticeable difference. Near the very end of the whole eclipse, no stars were observed, for from the time when about half of the moon had left the Earth's shadow, all the way until the complete end and beyond, thin clouds veiled

[51] For detailed biographical information, see Hellman (1970), Dreyer (1890), and Thoren (1990).

[52] Arithmetic, geometry, music, and astronomy.

[53] Hellman (1970).

[54] Clark & Stephenson (1977: 172–190).

> the sky, and the moon was not discernible. On the other hand, when its disk reached an altitude of around 50° 25', its whole light was seen to be recovered, as much as could be discerned through the gaps in the clouds. However, the final observation was not sufficiently accurate, in part because the moon was in clouds, as mentioned, and partly because it was approaching the south and so the altitude scarcely changed with time. The aforementioned observations agree sufficiently well with the earlier (calculation). For the last altitude observation is only about 1° higher, which is tolerable. Indeed, perhaps as little as 5 minutes of time lacked (in comparison with calculation) at the Royal City of Knudstropium Haffnia, or rather more assuming a place with a longitude to the east, so this requires virtually no correction. And, for myself, I cannot but be very surprised that even at this youthful age of 26 years, and without many accurate observations of the motion of the luminaries, I was able to get such accurate results. Indeed nothing was changed, and my observations of the heavens were fully reported in the book in which annual observations were recorded; they were easily written by my sister using the light of a lantern."
>
> [*Opera Omnia*, x, 38–39]

Tycho had good reason to be proud of the accuracy of his predictions. For example, the computed altitude of β Geminorum at the start of the eclipse is 13.19°, very close to his predicted 13°, and at second contact it is 20.72°, once more close to his predicted $21\frac{1}{2}^{\circ}$.

In AD 1575, Tycho visited William IV in Kassel. William was also very interested in astronomy, and it was possibly on his recommendation that in the following year, King Frederick II offered Tycho the island of Hven and asked him to construct an astronomical observatory there. Within a few months Tycho had established himself on the island, and construction began on his new home and observatory, which he named Uraniborg. He equipped this observatory with a wide range of astronomical instruments, although many of these would be replaced in the following years in his search for ever greater accuracy.[55] Tycho later published a description of the instruments in the Uraniborg observatory entitled *Astronomiae Instauratae Mechanica*.[56]

Over the next twenty years on Hven, Tycho made many thousands of observations of stars, planets, the sun and moon, and eclipses. These observations, in particular those of the planet Mars, were later to be extensively used by Kepler in formulating his laws of planetary motion. However, very few of these observations were published at the time, and it was not until J. L. E. Dreyer collected and published transcriptions of all of Tycho's known manuscripts in *Tychonis Brahe Dani Opera Omnia* that the full extent of Tycho's observational activity became known.[57] It should be noted that in recording the date of his observations, Tycho continued to use the Julian calendar up until AD 1599, seventeen years after it had officially been replaced by the Gregorian calendar in Denmark. The reason for this is not known, for he does not seem to have had any practical or theological objections to the Gregorian calendar.[58] It is possible

[55] Thoren (1973).

[56] This has been translated into English by Raeder, Strömgren, & Strömgren (1946).

[57] Four volumes of this collection are filled with Tycho's observations: Dreyer (1923, 1924, 1925, 1926).

[58] Hellman (1970).

that continuity of dates made comparison of observation and calculation easier.

Among Tycho's many observations made on Hven are more than thirty of solar and lunar eclipses. However, his observational accounts of many of these eclipses are difficult to evaluate — most consist simply of sketches of the eclipsed luminary made at specific times during the eclipse, rather than measurements of the time of the various eclipse contacts. Nevertheless, there are at least ten eclipse observations which do include timings of the contacts.[59] In many cases it is not known how these timings were made. Sometimes Tycho measured the altitude of a clock star using either his giant mural quadrant or a smaller device, at other times he used a mechanical clock. Indeed it is likely that in many cases he used more than one method. For example, he timed the beginning of clearing of the lunar eclipse on 26 September 1577 AD both by using a clock and by measuring the azimuth of the star α Orion:

> "I observed a total eclipse of the moon; the beginning and middle of which could not be seen on account of clouds. However the beginning of clearing was at 1 hour 56 minutes after midnight according to a clock corrected by the sun which showed single minutes; moreover (at this time) the bright star in the right shoulder of Orion (α Ori) was carefully measured to be $47\frac{2}{3}^{\circ}$ along the horizon according to careful observation. When a third of the moon had recovered its light, the same star had moved to a direction of $32\frac{5}{6}^{\circ}$ from the meridian along the horizon towards the east. From observation of the star in Orion, the time the shadow left was 2 hours 10 minutes after midnight, but in this I place little faith. Note: the observation by the shoulder of Orion is not very good. In fact errors of more than 2° may be on account of a not very well derived meridian line, and also may be caused by the wood of the instrument. I prefer to trust the clock."
>
> [*Opera Omnia*, x, 50]

In years following the death of King Frederick II, Tycho gradually lost his popularity in court, particularly with the young King Christian IV who did not see the value of Tycho's astronomical observations.[60] By AD 1597, Tycho's unpopularity had reached such a level that he decided to leave Denmark. During the next three years he once again travelled through northern Europe, looking for a patron interested in his work. Finally, in AD 1600, he settled in Prague where he was granted some financial support from the emperor. He began to observe again — his observations during this period include the solar eclipse on 10 July 1600 AD — but died the following year after a short illness.

The eclipse observations made by Tycho from which a contact time can be derived are summarized in Table 5.8. Clearly, Tycho's observations, in particular those made on Hven, are very accurate. The mean accuracy of the whole set of timings is about 0.09 hours, improving to about 0.07 hours when the eclipse in AD 1573 (observed in Knudstrup) and the eclipse in AD 1600 (observed in Prague) are ignored. Tycho often expressed considerable confidence in the accuracy of his instruments. This confidence

[59] The choice of these observations is based upon whether Tycho's sketches seem to relate to the moment of a contact or not, and so is inevitably rather subjective. I have tried to be as cautious as possible, rejecting some observations where it was questionable whether the sketch was of a contact or not.

[60] Hellman (1970).

Date	Type	Contact	Local Time (h) Observed	Computed
1573 Dec 8	Lunar	1	18.33	18.22
		2	19.45	19.21
		3	21.00	20.78
1577 Apr 2	Lunar	1	19.08	19.18
		2	20.17	20.16
		3	21.67	21.72
		4	22.75	22.70
1577 Sep 26	Lunar	3	1.93	2.09
1578 Sep 15	Lunar	1	0.41	0.44
		2	2.17	2.19
1579 Feb 25	Solar	1	16.73	16.91
		4	18.77	18.80
1581 Jan 19	Lunar	1	19.98	20.10
		2	21.30	21.28
		3	22.67	22.71
1584 May 10	Solar	1	5.03	5.07
		4	6.35	6.38
1588 Feb 26	Solar	1	13.50	13.48
		4	14.87	14.98
1588 Mar 12	Lunar	1	1.26	1.28
		2	2.51	2.35
		3	3.67	3.74
1590 Jul 31	Solar	1	6.83	6.81
		4	9.00	9.05
1591 Jul 20	Solar	1	14.90	14.72
		4	16.39	16.34
1600 Jul 10	Solar	1	12.72	12.49
		4	14.66	14.76

Table 5.8: Eclipse observations made by Tycho Brahe.

has been shown by Wesley (1978) to have been fully justified — his angular measurements of stellar positions are typically accurate to better than 1 minute of arc. This would typically translate into an error in the time of an observation of only a fraction of a minute. Thus, it would seem that the accuracy of Tycho's eclipse timings was not limited by his ability to measure the altitude of a clock-star, if he was using this method, but rather by the difficulties of determining the exact moment of an eclipse contact.

5.5 Accuracy of the Observed and Predicted Times

The errors in the observed time of all of the eclipses observed by Later Medieval and Renaissance European astronomers are shown in Figure 5.5. The typical accuracy of these observations ranges from about 6 to 12 minutes. Unsurprisingly, given his reputation, Tycho's observations are the most accurate of the seven sets of data. However, the observations by Regiomontanus and Walther are only marginally less accurate, implying that these two astronomers also took great care over their observations.

Figure 5.6 shows the true and computed errors in the eclipse times predicted by

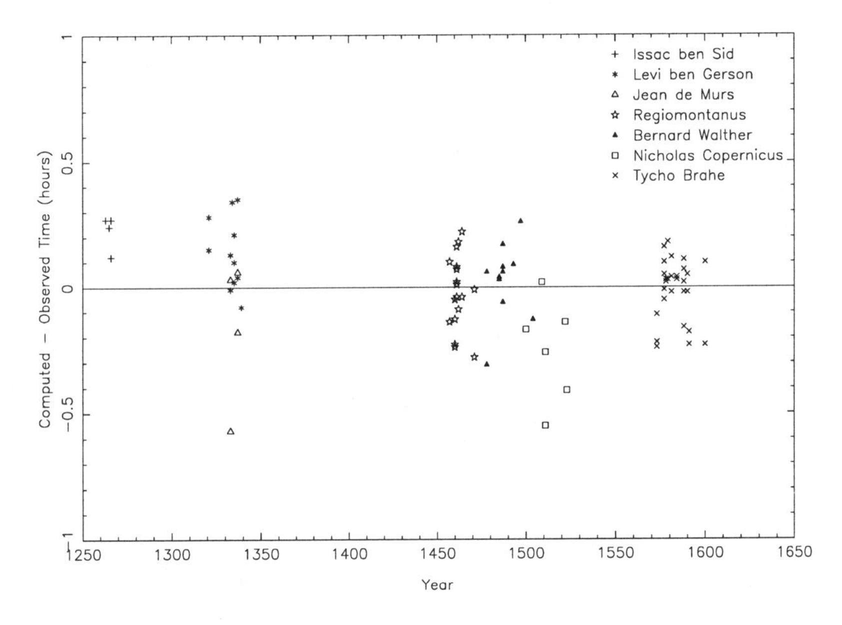

Figure 5.5: The error in the observed eclipse times.

Levi ben Gerson, Jean de Murs, Regiomontanus and Bernard Walther. The predictions made by Isaac ben Sid using the *Toledan Tables* are not included since they are of significantly poorer quality than the others — perhaps as much as six times less accurate. Thus Isaac ben Sid had good reason to compile the *Alfonsine Tables* as a replacement. The typical accuracy of the various sets of observations and predictions is shown schematically in Figure 5.7. Once again I must stress that this type of diagram simply shows the general trend in the accuracy of predictions down the centuries; the lines should not be taken to imply any form of linear increase, or indeed decrease, between any two points.

The *Alfonsine Tables* were used by Jean de Murs, Regiomontanus and Bernard Walther in making their predictions. It is clear from Figure 5.7 that de Murs' predictions are significantly more successful than those by Regiomontanus and Walther. This may have been caused by a number of factors. De Murs' predictions were made at a time much nearer to the epoch of the tables; by the time Regiomontanus and Walther made theirs, there may have been a considerable cumulative error in the calculations due to small inaccuracies in the parameters used. Another reason may be the problems of determining the geographical coordinates of the sites for which the predictions were made. For example, Regiomontanus himself notes in the report of the eclipse on 17 December 1461 AD, that the latitude of Rome is not agreed upon by all scholars. Errors in longitude were probably more serious. Kremer & Dobrzycki

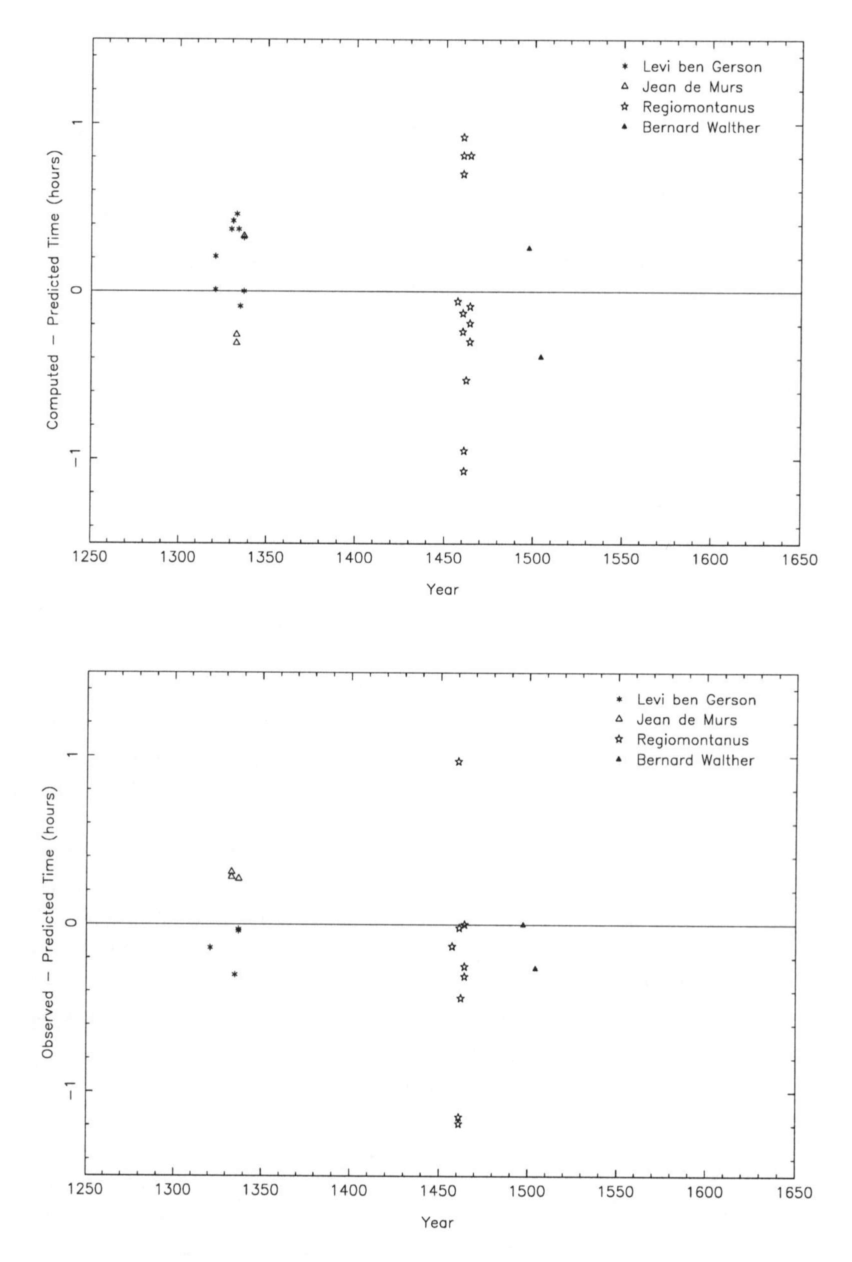

Figure 5.6: The true (above) and observed (below) errors in the predicted eclipse times.

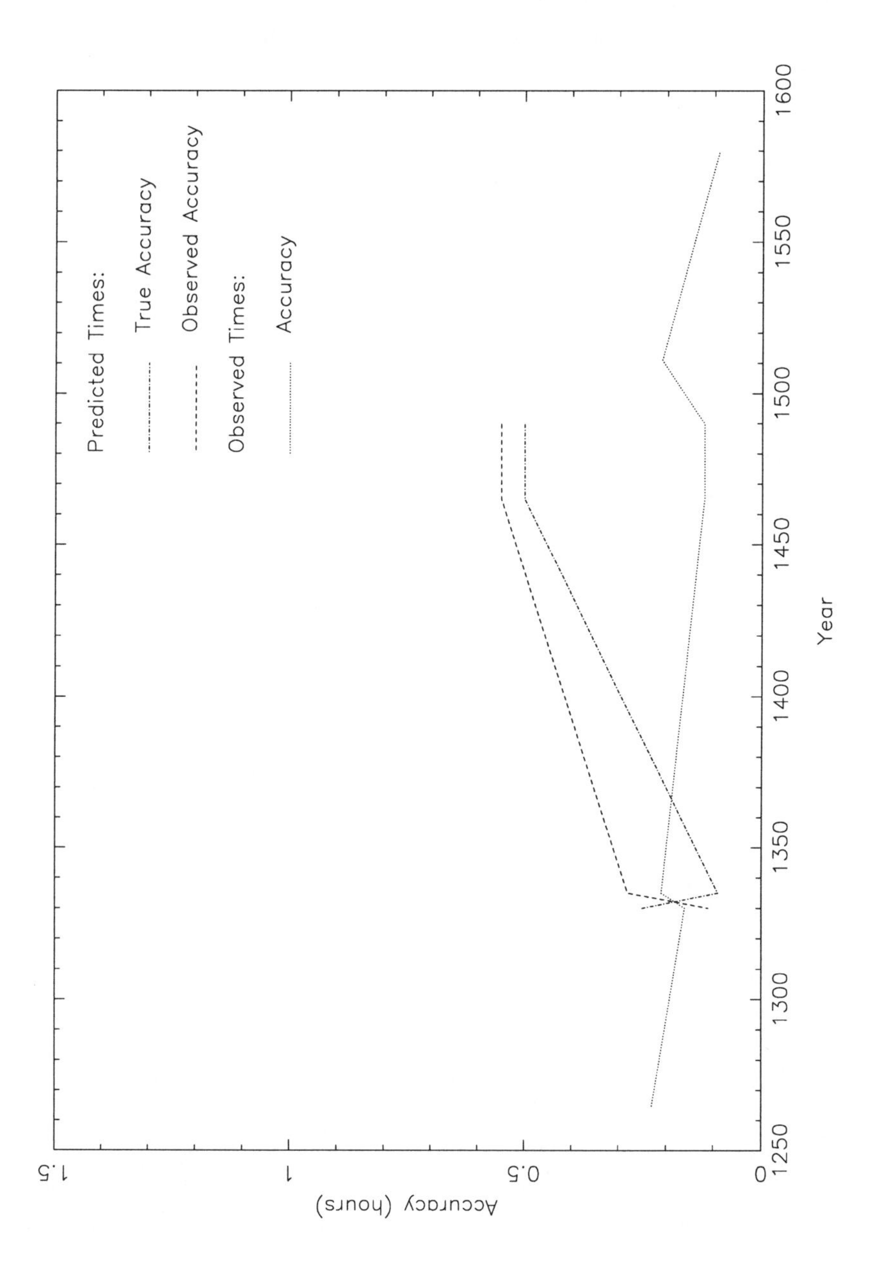

Figure 5.7: Schematic representation of the accuracy of the eclipse times.

(1998) have made a study of a number of Alfonsine texts and found a number of varying coordinates of cities in different sources. It may simply be that Regiomontanus and Walther used longitudes that were inaccurate, but that de Murs' value for Paris was quite good.

Part III

The Eastern Heritage

Chapter 6

China

6.1 Introduction

China has the longest astronomical heritage of any country of the world. Systematic astronomical records began to be kept in the eighth century BC and continue more or less uninterrupted up to the present day. Whilst astronomy in Babylon, the only other great civilization in the ancient world from which a vast array of astronomical records is preserved, had more or less ceased by the first century AD, the traditional astronomy of China that had developed by the Han dynasty (c. 200 BC), continued up until the start of the present century. That is not to say, however, that astronomy in China was not influenced at times by other cultures. For example, during the Yuan dynasty, many Islamic astronomers were invited to the Chinese capital, and an Islamic Observatory was set up.[1] More significantly, when the Jesuits came to China in the seventeenth century AD they brought with them western astronomical knowledge with which they helped to reform the Chinese calendar.[2] However, throughout all of this, Chinese astronomy retained a character all of its own. Furthermore, it was to be at the root of all of the astronomy which was to develop in the two other great ancient and medieval cultures of East Asia, Japan and Korea.

As Needham (1974) has noted, astronomy in China differed from that in the western world in two important respects: it was polar rather than ecliptic, and it was primarily an activity of the bureaucratic state rather than of priests or scholars. The former will not be of great concern to us here; but the second factor, that is, the official nature of astronomy in China, will feature repeatedly throughout this chapter.

In China, the emperor had the ultimate responsibility for the rule of government. But it was believed that he only ruled under the mandate of Heaven, and if Heaven was displeased with his rule, it would send a sign. These signs could take a number

[1] Yabuuti & van Dalen (1997).

[2] For discussions of the activities of the Jesuit astronomers in China, see d'Elia (1960) and Bernard (1973).

of forms such as diseases, floods, earthquakes, or, often, unusual events in the day- or night-sky. Because of this, the Astronomical Bureau, headed by an Astronomer Royal, was one of the most important and secretive governmental institutions. Its purpose was to observe and interpret any celestial events for the emperor. Furthermore, the Astronomical Bureau, or sometimes a separate Calendar Bureau, was responsible for maintaining the calendar, for making predictions of forthcoming astronomical events, and for the maintenance of the clepsydras. The size and structure of the Astronomical Bureau varied from dynasty to dynasty. For example, during the Ming it was comprised of 43 members,[3] whereas during the T'ang, there were more than 80, plus over 300 clerks and 500 students.[4]

6.2 Historical Outline

Before I begin to discuss the astronomical records preserved from ancient and medieval China, it may be useful if I outline some of the main events in Chinese history. But first a word of warning: this outline should not be taken as giving a full account of Chinese history from the earliest times down to the present day. It would be impossible to give a comprehensive history in the three pages I have allowed myself, so I have tried simply to provide what I consider to be a basic historical framework for the ensueing discussions.

The history of China emerges from the legendary period around the middle of the second millennium BC. Legends, written down by Ssu-ma Ch'ien in his *Shih-chi* or *Historical Records*, tell of a Hsia dynasty which dated back to the middle of the third millennium BC. However, there are no contemporary records of any sort from this period. Indeed, it is not until the Shang dynasty, the supposed successor of the Hsia, that there is any direct evidence for the existence of a dynasty. This has been found on the so called "oracle bones" from what is believed to be Yin (the last capital of the Shang dynasty) near An-yang in north-eastern China. These were first recovered at the end of the last century. The oracle bones were used for divination. A hot needle would be placed onto the bone, and the resulting cracks would then be interpreted by the diviner. A record of his divination would then be inscribed onto the bone in a primitive script. These sometimes contain allusions to historical, meteorological, and astronomical events.

During the eleventh century BC, the Shang dynasty was overthrown by the Chou dynasty. This dynasty was to rule China amid growing unrest until the end of the eighth century BC, when a number of feudal semi-independent states grew up which paid little allegiance to the Chou king. This situation continued until 221 BC, first with the various states living in a state of relative harmony during the *Ch'un-ch'iu* or "Spring and Autumn" period, but later, during the *Chan-kuo* or "Warring States" period, with the states constantly at war with one another. However, in 221 BC, the

[3] Ho (1970).

[4] Nakayama (1969: 18–19).

Ch'in ruler succeeded in unifying the whole country, and adopted the title Shih Huang-ti or "First Sovereign Emperor." It was during his rule that the system of bureaucratic government that was to form the basis of all subsequent Chinese history was set up.[5] After Shih Huang-ti's death, his son became Emperor; however, he was a weak leader and was soon overthrown by Liu Pang, the founder of the Han dynasty.

The Han dynasty was to last for nearly four hundred years, with only a brief interregnum between AD 9 and AD 23 when Wang Mang, a former chief minister to the Emperor, seized control of the country. However, following his assassination in AD 23, Liu Hsiu, a relative of the rulers of the former Han dynasty, regained control and started the Later (*Hou-*) Han dynasty. Han society is often characterised as a slightly less harsh version of the bureaucratic system of government developed in the Ch'in dynasty. For our purposes, we may also note that there was a great interest in science during this dynasty. Towards the end of the Later Han period, political machinations became rife, and, after the suppression of a peasant revolt in AD 184, by AD 220 the central government could no longer be sustained.

China then entered another period of division, known as the *San-kuo* or "Three Kingdoms" period. Three states, Wu in the south-east, Shu in the south-west, and Wei in the north, all fought for supremacy over one another. Gradually Wei overcame the other two states, and in AD 265 Ssu-ma Yen, a Wei general, founded the Chin dynasty. However, China was at this time under constant attack from a variety of tribes from the north, and by AD 317 Chin rule had been restricted to the southern part of China. Once more, China became divided: this time into northern and southern states. This period is known as *Nan-pei.* In the south, the Chin dynasty was followed in rapid succession by the Liu-Sung, the Ch'i, the Liang, and the Ch'en. In the north, the Northern Wei ruled for nearly one hundred and fifty years. After the fall of this dynasty there followed in quick succession the Eastern Wei, the Northern Chi, the Western Wei, and the Northern Chou.

After a period of nearly four hundred years of almost constant division, China was once again unified by the Sui dynasty. This dynasty was only to last for 26 years, for in AD 618 it was replaced by the T'ang dynasty. The T'ang dynasty was one of the most successful dynasties in the history of China, lasting for nearly three hundred years. The extent of China's territories and influence were expanded back to the levels of the Han dynasty and then beyond into Tibet and Korea.

From the end of AD 755 to the beginning of AD 763, T'ang government in China was threatened by rebellion. This was led by An Lu-shan, the military governor of the northern P'ing-lu and Fan-yang provinces. Between AD 755 and AD 757, the rebels achieved considerable success, occupying both Lo-yang and the capital Ch'ang-an. The rebels expressed little interest in Ch'ang-an and, sadly, this led to much historical material being lost. Over the next five years the T'ang government began to regain control and by the beginning of AD 763 the rebellion had been quashed.

The T'ang dynasty finally fell in AD 907 and over the next fifty years there followed a period known as the *Wu-tai* or "Five Dynasties" in which five short lived

[5]Needham (1954).

dynasties ruled the country. However, many other small independent states were also active. Most important among these was the semi-nomadic Liao kingdom. In AD 960, the Sung dynasty came to power, setting up its capital at Pien. Initially at peace with its neighbours, in AD 1127 the northern half of China was lost to the Jurchen, who had established the Chin dynasty in AD 1115. The Sung fled south, establishing themselves at Lin-an, their new capital. The Sung dynasty was a period of great scientific and technological achievement in China. In the words of Joseph Needham, "whenever one follows up any specific piece of scientific or technological history in Chinese literature, it is always at the Sung dynasty that one finds the major focal point."[6]

The original capital of the Chin dynasty was Shang-ching in Manchuria, but this was replaced in AD 1153 by Yen, and then sixty-one years later by Pien. The Chin dynasty in northern China was extinguished by the Mongols in AD 1234. The Mongol emperor Kubilai Khan established the Yuan dynasty in AD 1259, and moved his capital from Karakorum in Mongolia to Ta-tu in AD 1264. In AD 1276, the Yuan moved down into southern China and brought to an end the Sung dynasty. For the next hundred years the Yuan dynasty controlled China, but by AD 1369 the indigenous Ming dynasty had overthrown the Mongols and regained control of the country. The Ming dynasty saw a great decline in interest in science in China. This was in part caused by a wish to return to a more traditional way of life, including a return to more traditional types of technology.

By the middle of the seventeenth century, the Ming dynasty had been overthrown by rebels from Manchuria who formed the Ch'ing dynasty. European Jesuits, who had been visiting China since the end of the sixteenth century, now began to bring western astronomical theories and instruments to China, and the traditional astronomical practices began to disappear.

The various dynasties that have ruled China over the last three thousand years are listed in Table 6.1. The imperial capital has been changed many times in the history of China. Often this occurred when a new dynasty came to power, but in some cases, for example during the T'ang dynasty, the capital was moved for other reasons. The capital cities adopted at the various times are also shown in Table 6.1.

6.3 Sources of Astronomical Records

The earliest astronomical references in Chinese history are contained on the Shang dynasty oracle bones.[7] These artifacts, which were first discovered near An-yang, consist of animal bones and turtle shells inscribed with a primitive form of the Chinese characters. Many hundreds of thousands of these bones have so far been uncovered, a small number of which contain allusions to astronomical events such as observations of eclipses, comets, and planets. However, generally it is not possible to date the observations contained on the oracle bones, and so they provide little scope for detailed analysis.

[6] Needham (1954: 134).

[7] Xu, Yau & Stephenson (1989); Xu, Stephenson & Jiang (1995).

Dynasty	Rise	Fall	Capital City	Latitude (°)	Longitude (°)
Shang	c. -1500	c. -1045	Yin	36.06	-108.88
Western Chou	c. -1045	-700	Hao	34.16	-108.71
Eastern Chou	-769	-255	Lo-i	34.75	-112.50
Lu state[1]	-721	-480	Ch'u-fu	35.66	-117.01
Ch'in	-254	-205	Hsien-yang	34.38	-108.85
Former Han	-204	+9	Ch'ang-an	34.35	-108.88
Hsin	+9	+25	Ch'ang-an	34.35	-108.88
Later Han	+26	+190	Lo-yang	34.75	-112.47
	+191	+196	Ch'ang-an	34.35	-108.88
	+197	+220	Hsu-chang	34.05	-113.80
Wu[2]	+221	+280	Chien-k'ang	32.03	-118.78
Shu[2]	+221	+264	Ch'eng-tu	30.61	-104.10
Wei[2]	+221	+265	Lo-yang	34.75	-112.47
Western Chin	+265	+312	Lo-yang	34.75	-112.47
	+312	+317	Ch'ang-an	34.35	-108.88
Eastern Chin	+317	+420	Chien-k'ang	32.03	-118.78
Liu-Sung[3]	+420	+479	Chien-k'ang	32.03	-118.78
Ch'i[3]	+479	+502	Chien-k'ang	32.03	-118.78
Liang[3]	+502	+557	Chien-k'ang	32.03	-118.78
Ch'en[3]	+557	+588	Chien-k'ang	32.03	-118.78
Northern Wei[4]	+398	+493	P'ing-ch'eng	40.20	-113.20
	+493	+534	Lo-yang	34.75	-112.47
Eastern Wei[4]	+534	+550	Yeh	36.31	-114.55
Northern Chi[4]	+550	+557	Yeh	36.31	-114.55
Western Wei[4]	+534	+556	Ch'ang-an	34.35	-108.88
Northern Chou[4]	+556	+581	Ch'ang-an	34.35	-108.88
Sui	+581	+583	Ch'ang-an	34.35	-108.88
	+583	+617	Ta-hsing Ch'eng	34.27	-108.90
	+607	+617	Lo-yang[5]	34.75	-112.47
T'ang	+618	+682	Ch'ang-an	34.27	-108.90
	+682	+701	Lo-yang	34.75	-112.47
	+701	+717	Ch'ang-an	34.27	-108.90
	+717	+718	Lo-yang	34.75	-112.47
	+718	+722	Ch'ang-an	34.27	-108.90
	+722	+723	Lo-yang	34.75	-112.47
	+723	+724	Ch'ang-an	34.27	-108.90
	+724	+727	Lo-yang	34.75	-112.47
	+727	+731	Ch'ang-an	34.27	-108.90
	+731	+732	Lo-yang	34.75	-112.47
	+732	+734	Ch'ang-an	34.27	-108.90
	+734	+736	Lo-yang	34.75	-112.47
	+736	+881	Ch'ang-an	34.27	-108.90
	+881	+885	Ch'eng-tu	30.61	-104.10
	+885	+904	Ch'ang-an	34.27	-108.90
	+904	+907	Lo-yang	34.75	-112.47
Later Liang[6]	+907	+923	Pien	34.78	-114.33
Later T'ang[6]	+924	+936	Lo-yang	34.75	-112.47
Later Chin[6]	+936	+946	Pien	34.78	-114.33
Later Han[6]	+946	+950	Pien	34.78	-114.33
Later Chou[6]	+950	+960	Pien	34.78	-114.33
Sung	+960	+1127	Pien	34.78	-114.33
	+1127	+1276	Lin-an	30.25	-120.17

Table 6.1: Chinese dynasties and their capital cities.

Dynasty	Rise	Fall	Capital City	Latitude (°)	Longitude (°)
Chin	+1115	+1153	Shang-ching	45.50	-127.00
	+1153	+1214	Yen	39.92	-116.42
	+1214	+1234	Pien	34.78	-114.33
Yuan	+1264	+1368	Ta-tu	39.92	-116.42
Ming	+1369	+1420	Ying-t'ien	32.03	-118.78
	+1421	+1644	Pei-ching	39.92	-116.42
Ch'ing	+1644		Pei-ching	39.92	-116.42

1. Ch'un-ch'iu period.
2. San-kuo period.
3. Nan-pei period: Southern dynasties.
4. Nan-pei period: Northern dynasties.
5. Loyang acted as an second eastern capital during this period.
6. Wu-tai period.

Table 6.1 (cont.): Chinese dynasties and their capital cities.

Written history in China effectively starts with an ancient chronicle known as the *Ch'un-ch'iu*, or *Spring and Autumn Annals*. It is the only surviving example of the ancient state chronicles of China, the others having been lost, possibly in the infamous "Burning of the Books" by Emperor Shih Huang-ti in 213 BC. The escape of the *Ch'un-ch'iu* from this fate may be due to the fact that the work was attributed to Confucius, although the compiler of the work is not named in the text. The Chinese characters in use at this period continued virtually unchanged until the middle of the present century when a large number of simplified characters were introduced in the People's Republic of China.

The *Ch'un-ch'iu* contains a detailed record of the state of Lu over the period 722–481 BC. A short supplement to the work continues the records down to 468 BC. Among the records are accounts of a number of celestial phenomena, together with historical events such as the accession or death of a ruler, or details of a war with a neighbouring state, all reported in chronological order. The astronomical records include a series of 37 solar eclipse observations, two reports of meteors, and four references to comets.[8]

There are few other references to astronomical events in Chinese history until the Han dynasty. From this period on, the dynastic histories form our main source of astronomical records. These histories were compiled after the fall of a dynasty, sometimes by officials of the succeeding dynasty and at other times privately. 25 of these histories have been authorized as official histories. In addition, a history of the Ch'ing dynasty, the *Ch'ing-shih-kao* was completed in 1927. However, this work has not yet been approved by the government and so must be regarded only as a draft history. Nevertheless, this work continues the tradition of the earlier dynastic histories, and so is often considered in the same light as the earlier works.

The earliest dynastic history is Ssu-ma Ch'ien's *Shih-chi*. This work details the history of China from earliest times down to the Former Han dynasty. The *Shih-chi* may be split into five divisions: *Pen-chi*, annals of the emperors; *Shih-chia*, details

[8] Stephenson & Yau (1992).

of the noble families; *Piao*, chronological tables; *Lieh-chuan*, biographies; and *Chih*, treatises.[9] The annals of the emperors detail, in chronological order, important affairs of state, with a particular reference to the emperors. The section on the noble families contains a genealogical register of the various important nobles. The chronological tables provide a brief outline of historical events. The biographies are the longest part of the history. As their name suggests, they give detailed biographical information on a number of important people such as scholars and officials. Finally, a number of treatises are included on various aspects of human life. These include: *Li*, the calendar, *Yueh*, Music, and *T'ien-wen*, astronomy/astrology.[10]

The later dynastic histories all followed the general format of the *Shih-chi*, although in many cases the section dealing with the noble families, and on occasions also the chronological tables or the treatises, were omitted. Furthermore, the range of subjects covered by the treatises also varied from dynasty to dynasty. For example, although an astrological treatise was included in most of the histories, a treatise on the civil service was only included in some of the more recent works.

The 26 dynastic histories are listed in Table 6.2, along with their date of composition and name of their editor.[11] In many cases, the Bureau of Historiography was charged with compiling the histories, and so the name of the director of the Bureau is given. Occasionally, later historians would compile new histories of a particular dynasty if they did not consider the present history to be adequate. If this new history was officially adopted it would become known as the *Hsin* or "New" history of the dynasty, and the original version would become the *Chiu* or "Old" history. For example, the *T'ang-shu* was replaced in the middle of the eleventh century AD by the *Hsin T'ang-shu* and became known as the *Chiu T'ang-shu*. Both versions are still available and are of interest to us here. For instance, the *Chiu T'ang-shu* contains some interesting astronomical records that are absent from the *Hsin T'ang-shu*.

The dynastic histories contain many references to celestial events. These are found predominantly in the astrological, calendrical and five-phases treatises, although there are a small number of astronomical observations reported in the annals of the emperors. It is interesting to note the different styles in which astronomical events are described in these various sources. For example, astronomical observations in the annals tend to be very brief, merely noting the occurrence of the event, as is illustrated by an account of a comet seen in 138 BC:

> "(The Emperor) granted to those who moved to Mou-ling two hundred thousand cash to each household and two hundred *mou* of land, and for the first time the Pien Gate Bridge was built. In the autumn, the seventh month, there was a comet in the north-east. The King of Chi-ch'uan, (Liu) Ming, was sentenced for killing his Grand Tutor and Palace Tutor. He was dismissed and exiled to Fang-ling..."
>
> [*Han-shu*, 6; trans. Dubs (1944: 31–32)]

[9] Han (1955).

[10] In modern Chinese usage, *T'ien-wen* means astronomy, but in the context of early Chinese history, it may be more appropriate to translate the term as astrology. See Nakayama (1966: 442).

[11] This Table is based upon the work of Han (1955).

History	Editor	Date of Compilation
Shih-chi	Ssu-ma Ch'ien	104–87 BC
Han-shu	Pan Ku	AD 58–76
Hou-han-shu	Fan Yeh	c. AD 420
San-kuo-chih	Ch'en Shou	AD 285–297
Chin-shu	Fang Hsuan-ling	AD 644
Sung-shu	Shen Yueh	AD 492–493
Nan-Ch'i-shu	Hsiao Tzu-hsien	c. AD 500
Liang-shu	Yao Ssu-lien	AD 628–635
Ch'en-shu	Yao-Ssu-lien	AD 622–629
Wei-shu	Wei Shou	AD 551–554
Pei-Ch'i-shu	Li Po-yao	AD 627–636
Chou-shu	Ling-hu Te-fen	c. AD 620
Nan-shih	Li Yen-shou	AD 630–650
Pei-shih	Li Yen-shou	AD 630–650
Sui-shu	Wei Cheng	AD 629–636
Chiu T'ang-shu	Liu Hsu	AD 940–945
Hsin T'ang-shu	O-yang Hsiu & Sung Ch'i	AD 1043–1060
Chiu Wu-tai-shih	Hsueh Chu-cheng	AD 973–974
Hsin Wu-tai-shih	O-yang Hsiu	AD 1044-1060
Sung-shih	T'o T'o	AD 1343–1345
Liao-shih	T'o T'o	AD 1343–1344
Chin-shih	T'o T'o	AD 1343–1344
Yuan-shih	Sung-Lien	c. AD 1350
Hsin Yuan-shih	K'o Shao-min	AD 1890–1920
Ming-shih	Chang T'ing-yu	AD 1678–1739
Ch'ing-shih-kao	K'o Shao-min	AD 1914–1927

Table 6.2: The 26 Dynastic Histories.

In comparison, the astrological treatises often give both a detailed account of the observation and an astrological interpretation of the event. For example, a record of a comet in AD 349 gives details of its appearance and then connects the observation with war and sickness in the land:

> "On a *chi-mao* day in the eleventh month of the 5th year of the *Yung-Ho* reign-period of (Emperor) Mu Ti a white *Hui* comet measuring 1 *chang* appeared within *K'ang*, with its rays pointing towards the west. On a *ting-ch'ou* day in the first month of the 6th year a *Hui* comet again appeared at *K'ang*. According to the standard prognostication this presaged war and death and also sicknesses and epidemics. During the eighth month of the 5th year Ch'u P'ou embarked upon the northern campaign, but suffered a great setback. During the eleventh month Jan Min executed Shih Tsun and also put to death more than a hundred thousand men of the Hu border tribes, while the central region (of the empire) was in a state of emergency. During the twelfth month Ch'u P'ou died. A wide-spread epidemic also prevailed during the year."
>
> [*Chin-shu*, 13; trans. Ho (1966: 241)]

A wide variety of astronomical phenomena are reported in the dynastic histories. These range from predictable events such as the positions of planets and the occurrence of eclipses, to observations of less frequent events such as sunspots, aurorae, "new stars," and comets.

A relatively small number of astronomical records may also be found in sources other than the dynastic histories. For example, the *Wen-hsien T'ung-k'ao* (*Comprehensive History of Civilization*), which was compiled in AD 1307 by Ma Tuan-lin, contains a section devoted to astronomical observations, including a number of solar and lunar eclipses. Other histories, both local and national, contain astronomical observations. Beijing Observatory (1988) have made a search through both these sources and the dynastic histories for astronomical observations; however, they do not always reproduce the record in full. Accordingly, I have only used this work as a secondary source in searching for observations not included in the dynastic histories or the *Wen-hsien T'ung-k'ao*. Undoubtably, there are many other works containing astronomical records which have not yet been studied, for instance because they are contained only in library collections that have not yet been catalogued. I have been unable to take into account any such material in this work.

6.4 Calendrical Astronomy in China

The Chinese term *Li* is usually translated into English as "calendar" or "astronomical system," but is perhaps best thought of as including both of these meanings. A *Li* was a complete system for calculating the motion of the heavenly bodies and as a direct result this produced the reckoning of days, months and years which we term a calendar. As "calendar" has become the most widely used translation of *Li*, I shall adopt it with the proviso that one remembers it also applies to the other aspects of astronomical calculation that form part of a *Li*.

As I have mentioned in the introduction to this chapter, the emperor received his mandate to govern from Heaven. In return he was expected to supervise the heavenly rituals, for only he could perform this service on behalf of his subjects.[12] Heavenly phenomena fell into two classes: those that could be predicted, and those that could not. Regular, predictable events held no astrological importance, but unexpected events, such as a comet or an unpredicted eclipse, indicated that Heaven was displeased with the emperor's rule. Therefore, the calendar was seen as a way of regulating the sky and so reducing the number of ominous events. If a calendar was failing to predict events such as eclipses, this would provide a motive for reforming the calendar. In addition, at times of political upheaval, such as at the start of a new dynasty, it was necessary for the Emperor to reform the calendar to establish his right to govern:

> "When a new dynasty rose by accepting the Heavenly Ordinance, at first it had to be prudent. It had to obey the will of Heaven by renewing the basis of all things: the calendar and the colour ... In founding a new dynasty, the Emperor should not depend upon former institutions."
>
> [*Shih-chi*; trans. Yabuuti (1974)]

[12] Yabuuti (1973).

Thus, this principle of "Reformation by the Heavenly Ordinance," whereby a new dynasty received its mandate to govern from Heaven, was responsible for many of the early reformations of the calendar.[13] Furthermore, at least in the early period of Chinese history, even when a calendar was failing to make adequate predictions, reformation could only occur when there was the political will for it to happen.[14]

In all there were more than fifty calendars officially adopted in China from the Han to the Ch'ing dynasties. These are listed in Table 6.3, which is based on the work of Yabuuti (1963a, 1963b).[15] It is worth stressing, however, that only in a small number of cases was there any significant difference between a new calendar and those that had gone before. In addition to the fifty or so calendars that were officially adopted by the Chinese state, nearly as many again were compiled but not accepted. Once more, the majority of these calendars varied little from their predecessors. However in a small number of cases, such as the *Chiu-chih-li* which was based on Indian astronomical methods,[16] they were radically different from the traditional Chinese calendars. Similarly, the *Hui-hui-li* compiled by the astronomers of the Islamic observatory, was based, at least in part, upon Ptolemaic astronomy.[17]

6.4.1 Reckoning of Days

The primary use of the Chinese calendar was to produce a reckoning of days, months, and years. We may think of this as being a calendar in the same sense as our Gregorian calendar, although, of course, its operational rules are completely different. Most importantly, the Chinese calendar is a luni-solar calendar, that is, it is based on the movement of both the moon and the sun. The first day of a month is defined as the day on which the sun and the moon are calculated to be at conjunction. Thus, there are either 29 or 30 days in each month, where a day is defined as lasting from one midnight to the next (or, for some purposes, from sunrise to sunrise). However, a solar (tropical) year of about 365.24 days is not equal to an integer number of lunar months; there is an excess of about 11 days over twelve months. In order to reconcile the months and the years it is necessary to make every second or third year a leap year containing an intercalary month. Initially, this month was added whenever it was judged necessary on an *ad hoc* basis, but, in the *Ssu-fen-li*, a calendar probably developed in the fourth century BC, a mathematical method was devised to determine when to intercalate. The basic method presented in the *Ssu-fen-li* was used to determine intercalation in the Chinese calendar until the arrival of the Jesuits in the sixteenth and seventeenth centuries AD.

The intercalation method of the *Ssu-fen-li* and other early calendars required three pieces of information: a value of the mean length of the solar year, a value of the mean synodic month, and a date (usually in the remote past) when it was assumed that the

[13] Yabuuti (1974).
[14] Cullen (1993).
[15] The gaps in the Table relate to dates where it is not clear which calendar was adopted.
[16] Yabuuti (1979).
[17] An English translation of the *Hui-hui-li* with commentary is currently being prepared by B. van Dalen.

Calendar	Compiler	Date of Use
Ssu-fen-li	Unknown	c. 400–104 BC
San-t'ung-li	Liu Hsin	104 BC– AD 84
Ssu-fen-li	Li Fan	AD 85–263
Ch'ien-hsiang-li	Liu Hung	AD 222–280
Ching-ch'u-li	Yang Wei	AD 237–451
San-chi Chia-tzu-li	Chiang Chi	AD 384–417
Hsuan-shih-li	Chao Fei	AD 412–439 & AD 452–522
Yuan-chia-li	Ho Ch'eng-tien	AD 445–509
Ta-ming-li	Tsu Chung-chih	AD 510–589
Cheng-kuang-li	Chang Lung-hsiang	AD 523–565
Hsing-ho-li	Li Yeh-hsing	AD 540–550
T'ien-pao-li	Sung Ching-yeh	AD 551–557
Chou-li	Ming K'o-jang	c. AD 559
T'ien-ho-li	Chen Luan	AD 566–578
Tah-hsiang-li	Ma Hsien	AD 579–583
K'ai-huang-li	Chang Pin	AD 584–596
Ta-yen-li	Chan Chou-hsuan	AD 597–617
Mao-yin-li	Fu Jen-chun	AD 619–665
Lin-te-li	Li Ch'ung-feng	AD 665–728
Ta-yen-li	I Hsing	AD 729–762
Chih-te-li	Han Ying	AD 758–763
Wu-chi-li	Kuo Hsien-chih	AD 763–783
Cheng-yuang-li	Hsu Ch'eng-ssu	AD 783–806
Kuang-hsiang-li	Hsu Ang	AD 807–821
Hsuan-ming-li	Hsu Ang	AD 822–892
Ch'ung-hsuan-li	Pien Kang	AD 893–907
T'iao-yuan-li	Ma Ch'ung-chi	AD 939–944 & AD 947–994
Ch'in-t'ien-li	Wang P'o	AD 958–963
Ying-t'ien-li	Wang Ch'uno	AD 964–982
Ch'ien-yuan-li	Wu Chao-su	AD 983–1000
Ta-ming-li	Ku Chun	AD 995–1136
I-t'ien-li	Shih Hsu	AD 1001–1023
Ch'ung-t'ien-li	Ch'u Yen & Sung Ts'ung	AD 1024–1064 & AD 1068–1074
Ming-t'ien-li	Chou Ts'ung	AD 1065–1067
Feng-yuan-li	Wei P'o	AD 1075–1093
Kuan-t'ien-li	Huang Chu-ch'ing	AD 1094–1102
Chan-t'ien-li	Yao Shun-pu	AD 1103–1105
Chi-yuan-li	Yao Shung-pu	AD 1106–1130 & AD 1168
Ta-ming-li	Yang Chi	AD 1137–1181
T'ung-yuan-li	Chen Tei	AD 1136–1167
Ch'ien-tao-li	Liu Hsaio-jung	AD 1168–1176
Ch'ung-hsi-li	Liu Hsiao-jung	AD 1177–1190
Revised Ta-ming-li	Chao Chih-wei	AD 1182–1280
Hui-yuan-li	Liu Hsiao-jung	AD 1191–1198
T'ung-t'ien-li	Yang Chung-pu	AD 1199–1207
K'ai-hsi-li	Pao Huan-chih	AD 1208–1251
Ch'un-yu-li	Li Tech'ing	AD 1252
Hui-t'ien-li	T'an Yui	AD 1253–1270
Ch'eng-t'ien-li	Ch'en Ting	AD 1271–1276
Pen-t'ien-li	Teng Kuan-chien	AD 1277
Shou-shih-li	Kuo Shou-ching	AD 1280–1368
Ta-t'ung-li	Yuan Tung	AD 1368–1661
Shih-hsien-li	Jesuits	AD 1665–1912

Table 6.3: Officially adopted Chinese calendars.

Number	Name	Approximate Gregorian Date	Remarks
1	*Tung-chih*	22 December	Winter Solstice
2	*Hsiao-han*	6 January	
3	*Ta-han*	20 January	
4	*Li-ch'un*	4 February	
5	*Yu-shui*	19 February	
6	*Ch'ing-che*	6 March	
7	*Ch'un-fen*	21 March	Vernal Equinox
8	*Ch'ing-ming*	5 April	
9	*Ku-yu*	21 April	
10	*Li-hsia*	6 May	
11	*Hsiao-man*	22 May	
12	*Mang-chung*	6 June	
13	*Hsia-chih*	22 June	Summer Solstice
14	*Hsiao-shu*	8 July	
15	*Ta-shu*	23 July	
16	*Li-ch'iu*	8 August	
17	*Ch'u-shu*	24 August	
18	*Po-lu*	8 September	
19	*Ch'iu-fen*	23 September	Autumnal Equinox
20	*Han-lu*	9 October	
21	*Shuang-hsiang*	24 October	
22	*Li-tung*	8 November	
23	*Hsiao-hsueh*	23 November	
24	*Ta-hsueh*	8 December	

Table 6.4: The 24 *Ch'i* (Based on Yabuuti (1963a)).

moon and sun reached conjunction at the same time that a winter solstice occurred at midnight, and a lunar eclipse took place at the following opposition, and this was the first year in the sexagenary cycle (see below). The early calendars all assume slightly different values for these constants. For example, in the *Ts'ang-t'ung-li*, the length of the solar year was assumed to be $365\frac{385}{1539}$,[18] and the synodic month to be $29\frac{43}{81}$, and in the *Ching-ch'u-li* the solar year is taken as $365\frac{455}{1843}$, and the synodic month $29\frac{2419}{4559}$.[19] Despite their apparent precision, the assumed values of the mean solar year and the mean synodic month are not independent in the early calendars. In the two cases detailed above, for instance, they include the well-known relation that 19 solar years is equal to 235 synodic months. Similar relations are implicit in other early calendars.

The solar year was divided into twenty-four equal parts called *Ch'i*, beginning with the moment of the winter solstice. These twenty-four *Ch'i* are listed in Table 6.4. The odd numbered *Ch'i* in Table 6.4 were called *Chung-ch'i*. By counting on from the ancient epoch of a calendar it was possible to determine the date of every *Chung-ch'i* and the beginning of each lunar month. The time interval between two *Chung-ch'i* is, of course, simply equal to one-twelfth of the mean solar year. For example, in the *Ching-ch'u-li*, this is $365\frac{455}{1843}/12 \approx 30.4$ days. Since the mean synodic month (for the *Ching-ch'u-li* this is ≈ 29.5 days) is always smaller than this value, there will

[18]The apparent precision of this value simply comes from a minor alteration of the value from the preceding *Ssu-fen-li* of $365\frac{1}{4} = 365\frac{385}{1540}$. See Sivin (1969: 12).

[19]A list of the values of the constants in the various calendars is given by Yabuuti (1963a, 1963b).

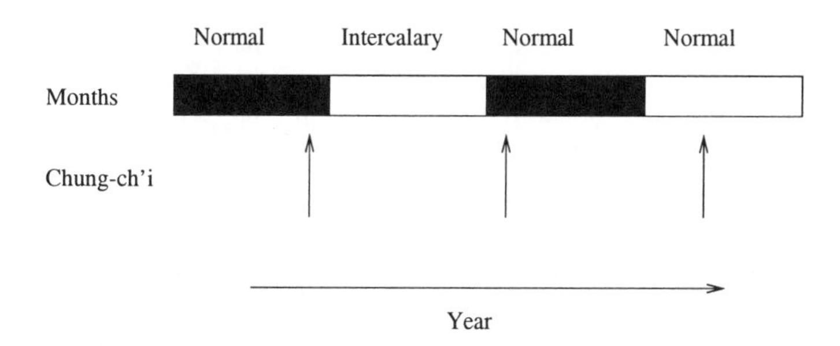

Figure 6.1: The Chinese method for determining intercalary months using the *Chung-ch'i*.

occasionally be a lunar month which does not contain a *Chung-ch'i*. This is shown in Figure 6.1. Here we have four lunar months, illustrated by the four boxes. The positions of the *Chung-ch'i* during the year are indicated by the arrows (note that the excess in length of the intervals between *Chung-ch'i* over the lunar month has been considerably expanded in this diagram). The first, third, and fourth months all contain one of the *Chung-ch'i*, and so are normal months. There is no *Chung-ch'i* during the second month, however. In the Chinese system, such months were taken to be intercalary months.

As we have seen, early Chinese calendar systems used the mean solar year and the mean synodic month to determine the date of the beginning of each month and when to intercalate. However, from as early as the first century BC, Chinese astronomers had discussed the problem of the moon's anomalistic motion, and, at least from the middle of the fourth century AD, the irregulaties in the sun's motion.[20] In about AD 600, Liu Ch'o compiled the *Huang-chi-li* which, for the first time, took these irregularities fully into account. This calendar was not officially accepted, but in the middle of the seventh century Li Ch'ung-feng compiled the *Lin-te-li*, based almost exclusively on the *Huang-chi-li*, and this calendar was used between AD 665 and 728. In AD 727, the Buddhist monk I Hsing compiled the *Ta-yen-li*, which saw some further improvements in coping with the irregular motion of the sun and moon.

In these calendars, the sun's irregular motion was dealt with in a table which gave the number of days it takes the sun to traverse 15° strips of the ecliptic. These were taken to be the same *Ch'i* mentioned earlier. Thus the *Ch'i* had two, irreconcilable, definitions: as 15° lengths on the ecliptic, and as one twenty-fourth of the mean solar year. The former definition (i.e., true *Ch'i*) was used, for example, when calculating eclipses, but the latter (i.e., mean *Ch'i*) continued to be used to determine intercalations for the next thousand years until the Jesuit astronomers compiled the *Shih-hsien-li* in the mid-seventeenth century AD.[21] In the tables of solar motion, the moment of the sun's passage through perigree was customarily (and erroneously) set at winter

[20] Yabuuti (1963a: 460–470).
[21] Liu & Stephenson (1998).

solstice.[22] Interpolation was used to determine the sun's motion at syzygies which did not coincide with the *Ch'i*. Angular positions were given in units called *Tu*, where the number of *Tu* in a circle is equal to the length of the solar year. For example, in the *Ssu-fen-li* there were 365.25 *Tu* in a circle; in the *Ta-ming-li* there were 365.2428. It has become customary to translate *Tu* as "Chinese degree," and to use the symbol ^{c} to distinguish it from $^\circ$.

The moon's irregular motion was also incorporated into tables, this time tabulated day by day. Once more, interpolation was used to determine the motion at syzygies which took place between two entries.[23] In the *Lin-te-li*, Li Ch'ung-feng had advocated changing from calculating the beginning of months using the mean synodic month, to using true month lengths. This change was officially accepted, and all future calendars used true lunar motion to calculate the months. Thus, the calendars used the true lunar and solar motion to calculate month lengths, but mean solar motion to determine intercalation.

Further modifications were made to the calendars over the next one thousand years or so. Towards the end of the thirteenth century AD, Kuo Shou-ching was able to obtain parameters that were significantly more accurate than any determined earlier by building improved instruments, in particular a giant gnomon some 12 metres or so in height. The resulting *Shou-shih-li* was probably the most successful of the traditional Chinese calendars.

As mentioned above, years contained either twelve or thirteen months. The years themselves were given in terms of reign periods. These are similar to the regnal years used, for example, in Babylon. However, they need not necessarily start with an emperor's accession year. Sometimes he might proclaim a new reign period three, four, or, occasionally, even more times during his life. Generally this would be prompted by the occurrence of an unusually auspicious event. In addition to the reign periods, a continuous count of years in a 60 year cycle was stated in the Former Han. Similarly, a 60 day count of days has been in continuous use, perhaps from as early as the Shang period, down to the present day. These counts are known as the sexagenary cycle. Days and years in this cycle are given by a combination of one of the ten *t'ien-kan*, "celestial stems," with one of the twelve *ti-chih*, "earthly branches." For example, the first day or year of a cycle is *chia-tzu*, the second *i-ch'ou*, and so forth. The full sexagenary cycle is shown in Table 6.5. The names of the ten celestial stems are given in the left column, and the twelve earthly branches along the top. The day number in the cycle is given by reading along the row and column containing the stems and branches. Note that because the cycle progresses through the stems and branches in order, only half of the possible combinations are utilized. The celestial stems and the earthly branches are also used in other parts of Chinese life. For example, the stems label each of the twelve double-hours by which time was measured, while both the stems and branches are used as direction indicators.

As dates in the Chinese calendar are generally defined both by the year, month,

[22] Yabuuti (1963a: 462).

[23] Yabuuti (1963a: 469).

	tzu	*ch'ou*	*yin*	*mao*	*ch'en*	*ssu*	*wu*	*wei*	*shen*	*yu*	*hsu*	*hai*
chia	1		51		41		31		21		11	
i		2		52		42		32		22		12
ping	13		3		53		43		33		22	
ting		14		4		54		44		34		24
wu	25		15		5		55		45		35	
chi		26		16		6		56		46		36
keng	37		27		17		7		57		47	
hsin		38		28		18		8		58		48
jen	49		39		29		19		9		59	
kuei		50		40		30		20		10		60

Table 6.5: The sexagenary cycle.

and day, and by the day in the sexagenary cycle, it is usually possible to accurately convert a date to the Julian calendar. In the present study, I have made use of the tables of Hsueh & Ou-yang (1956) and Wong (1902) for this task. In virtually all cases, the dates of the eclipses investigated agreed exactly with those of computed eclipses when converted to the Julian calendar. The exceptions may all be explained as minor scribal errors in the published texts. Such cases will be noted in the translations of the records.

6.4.2 Eclipse Prediction

Besides providing a method for reckoning the passage of days, the Chinese calendar was also used to calculate the positions of the sun, moon, and planets, and resulting phenomena such as passages of the moon through various constellations. However, the most important use for the calendar, after the count of days, was to predict the occurrence of eclipses. As Li Ch'ien wrote in the thirteenth century:

> "The exactitude of an astronomical system stands and falls on its treatment of eclipses. In this computational art exactitude is hard to come by. There is always uncertainty about whether the (predicted) time is early or late, and whether the (forecast) immersion is too shallow or too deep. If exact agreement (with the phenomena) be the goal, there can be no room for chance."
>
> [*Yuan-shih*, 53; trans. Sivin (1997)]

These words not only reflect the importance of eclipse prediction in China, but also an awareness of the inherent difficulties in such a task. Indeed, it may have been the failure of the calendars to be able to accurately predict the circumstances of eclipses that led to early ideas that the physical world is too complex to be fully predictable.[24] As early as AD 175 Ts'ai Yung was to write:

> "The astronomical regularities are demanding in their subtlety, and we are too far removed from the Sages [who founded this art]. Success and failure take their turns, and no technique can be correct forever."
>
> [*Hou-han-shu*, 2; trans. Sivin (1986)]

[24] Sivin (1986).

Various mythological explanations for the causes of eclipses had been given in China from ancient times, the most well known being that the sun or moon was eaten by a celestial dragon. By the first century BC, however, the correct astronomical explanation of eclipses had been proposed,[25] although its acceptance was still being debated by philosophers for at least the next two hundred years. How, it was asked, could the *yang* of the sun be subdued by the *yin* of the moon when at the time of a solar eclipse the light of the moon was very weak?[26] Nevertheless, from the first century BC the Chinese astronomers included methods for predicting eclipses in their calendars which imply at least a partial understanding of the astronomical causes of eclipses. In the early calendars, methods were only given for predicting lunar eclipses.

The first three calendars to deal with eclipse prediction, the *San-t'ung-li*, the *Ssu-fen-li*, and the *Ch'ien-hsiang-li*, all used what was essentially the same method.[27] This made use of the eclipse cycle of 135 months which had presumably been found from inspecting earlier records of observed lunar eclipses, although early lunar eclipse observations, unlike their solar counterparts, are not recorded in surviving works such as the *Ch'un-ch'iu*. The cycle was not simply used to predict eclipses 135 months after a previously observed eclipse, however. The Chinese astronomers instead used the fact that there were 23 eclipse possibilities within the 135 month cycle. Assuming that these were regularly spaced, this meant that eclipse possibilities occur every $5\frac{20}{23}$ months. Adding on $5\frac{20}{23}$ from the epoch of the calendar (when it was assumed an eclipse had occurred), keeping the fractions in the running total, but rounding down to find the month of an eclipse,[28] gives, in theory, the dates of all eclipses throughout eternity.[29] However, since the epoch of the calendar was usually at some date in the distant past (more than 2 million years in the *Ssu-fen-li*), counting on from this epoch would not be practical. To avoid this the Chinese used intermediate cycles in which some, but not all, of the conditions at the epoch would recur.[30] Making use of remainder arithmetic, the count from the epoch to the present year would be reduced to something more manageable, say under one thousand. Combining the eclipse cycle with the rule of 7 intercalations in 19 years, yields that there are $\frac{1081}{513}$ eclipses per year. One could then quickly move on to the present year, knowing how many eclipses had taken place since the epoch of the intermediate cycle. Then it is just a matter of counting eclipse possibilities, moving on by $5\frac{20}{23}$ months each time, and taking into account intercalary months. Finally, the day and time of the eclipse were simply taken

[25] For example, in Liu Hsiang's *Wu-ching T'ung-i*. See Needham (1959: 414).

[26] This is one of several arguments made against the "astronomical" explanation of eclipses by Wang Ch'ung in chapter 32 of his *Lun Heng*. See Forke (1907: 269) and Needham (1959: 411-413).

[27] Translations of the relevant parts of these calendars are given by Sivin (1969). In addition, he gives a detailed discussion of the two (astronomically equivalent) methods described in the *Ssu-fen-li* with worked examples. For additional discussions, see Eberhard & Müller (1936) and, briefly, Aaboe (1972).

[28] Rounding down in this way implies that the initial eclipse was at one extreme of the nodal zone. However, this implication is not explicitly stated in the text, and may not have been considered by the early Chinese astronomers.

[29] In practice, of course, this would break down since the relation 23 eclipse possibilities = 135 months is not without error.

[30] For a general discussion of these cycles, see Sivin (1966, 1969).

as the moment of mean opposition, calculated by adding half a mean synodic month on to the time of mean conjunction which had already been found in the process of constructing the lunar calendar.

Although there are many steps in this process — and many potential places for mistakes — it has the advantage that eclipse prediction is reduced merely to book-keeping, and yet the method still predicts most visible eclipses over the course of a hundred years or so. Furthermore, the calendar tends to predict too many, rather than too few, eclipses. In China, the astrological importance of an event was negated if it was predicted beforehand. If a predicted eclipse failed to appear, then it was assumed that the emperor was sufficiently virtuous to have prevented it. This tendency to err on the side of caution by over-predicting eclipses continued in later calendars. Examples of court officials congratulating the emperor when an anticipated eclipse did not occur are well documented in T'ang sources.[31]

The first mathematical treatment of eclipse calculation without reference to an eclipse cycle is found in the *Ch'ing-ch'u-li* from the third century AD.[32] In this and subsequent calendars, eclipses were predicted by calculating the distance of the moon from a node at the moment of syzygy. However, unlike in Greek astronomy, the "eclipse limits" which were used to determine whether an eclipse would occur were not expressed in terms of degrees of longitude, but in the number of days that it would take the moon to reach the node. This is given by the ratio

$$\frac{\text{distance from node } (^c)}{\text{daily lunar motion } (^c/day)} = x(days)$$

If this ratio was below a certain figure then an eclipse was possible. For example, in the *Ta-yen-li*, a lunar eclipse would occur when x was less than $\frac{3523.9}{3040}$. If x was less than $\frac{779}{3040}$, then the eclipse would be total.[33] These eclipse limits correspond to about 15° and just over 3° elongation from the node respectively.

Magnitudes of partial eclipses were obtained from the difference between x and the eclipse limit for a total eclipse using the formula

$$\text{magnitude } (fen) = 15 - \frac{3040}{183} \times (x - \frac{779}{3040})$$

with 15 being the magnitude of a total eclipse. Durations were calculated as a simple function of the magnitude by adding 2 if the magnitude is less than 5 *fen*, 3 if it is between 5 and 10 *fen*, and 5 if it is between 10 and 15 *fen*, and expressing the result in *k'o*.[34] A small correction may then have been added to this duration to take into account the irregularities in the moon's motion.

Solar eclipses are calculated in essentially the same way as lunar eclipses in the *Ta-yen-li*. However, since the parallax of the moon is relatively large, this factor needs to

[31] Yabuuti (1973).
[32] Yabuuti (1963a: 476).
[33] Yabuuti (1963a: 479).
[34] The *k'o* is a unit of time such that there are 100 *k'o* in one day. See Section 6.5 below.

be taken into account when calculating solar eclipses. The Chinese astronomers never possessed a fully developed theory of parallax and so they used a number of semi-empirical corrections in their eclipse calculations.[35] In the *Ta-yen-li*, this was limited to a correction based upon the *Ch'i* in which a solar eclipse occurred (i.e., upon the declination of the moon). No consideration was made of the hour angle of the moon at the time of the eclipse.[36] However, this was rectified with further empirical corrections in later calendars.

The effect of location on calculating eclipses was first taken into account in the *Ta-yen-li*. In AD 725, its author, I Hsing, organized for the length of the shadow cast by a gnomon at the solstice and equinox to be measured at a number of stations on a roughly north-south line through China.[37] Using these measurements he was able to make crude adjustments to his calculations to take into account differences in latitude. However, the concept of a spherical earth was not generally accepted in China at this time. There could therefore be no concept of latitude and longitude as we think of it today, and so corrections could not be applied to take account of east-west location in the eclipse calculations.

Later calendars followed essentually the same procedure as the *Ta-yen-li* for calculating eclipses. Small changes were made, for example in the eclipse limits used or in the way in which the solar and lunar motions were deduced from interpolation, but it was not until Kuo Shou-ching produced the *Shou-shih-li* in AD 1280 that there were any further significant developments. So successful was the *Shou-shih-li* that it was used from its adoption in AD 1280 until the end of the Ming dynasty with only one minor modification when it became the *Ta-t'ung-li* in AD 1368.[38] Unlike many earlier calendars, the *Shou-shih-li* was able to calculate not only the time of conjunction and the duration of an eclipse, but all of the phases of lunar and solar eclipses.

6.5 Units of Time

From at least the middle of the first millennium BC, three units of time were used simultaneously in China. These were the *shih* or "double hour," the *keng* or "night watch," and the *k'o* or "mark." The day was divided into twelve equal double hours named after the twelve earthly branches. The first double hour, called *tzu*, was centred on midnight (i.e., running from 11 p.m. to 1 a.m.) and was followed by the other earthly branches in order. Sometimes the double hour would be split into two halves: the *ch'u* or "initial" half and the *cheng* or "central" half. Thus the first double hour

---

[35] Nakayama (1969: 144-145).

[36] Yabuuti (1963a: 482–483).

[37] See *Chiu T'ang-shu*, 35 for a description of the work. This has been translated and discussed by Beer et al. (1961), who remark that it is "one of the most remarkable pieces of organized field research in the early middle ages."

[38] Sivin (1997) has translated the *Shou-shih-li I*, a discussion of the calendar contained in the *Yuan-shih*, into English in an unpublished manuscript. A detailed discussion of the *Shou-shih-li* is given by Nakayama (1969: 123–150).

contained an initial half lasting from 11 p.m. to midnight and a central half lasting from midnight to 1 a.m.

The night watch was a seasonal unit of time defined as lasting for one fifth of the length of the night. Strangely, there is no similar seasonal time unit for use during the day. For the purposes of calculating the watches, the night was considered to last from dusk, or the end of evening twilight, to dawn, the start of morning twilight. Twilight was defined as lasting for a set amount of time throughout China with no concession to the fact that the observable length of twilight varies with the observer's latitude. Initially the length of twilight was set as 3 marks (0.72 hours); however, after the start of the Later Han dynasty (AD 25), a length of $2\frac{1}{2}$ marks (0.60) hours was adopted. At the latitude of central China, therefore, the night watch varied in length from about 2.6 hours in winter to about 1.6 hours in summer. Each night watch was further subdivided into five equal parts, known before the Sui dynasty as *ch'ang* or "calls," during the Sui as *ch'ou* or "rods," and later as *tien* or "points."

Finally, the day was usually divided into 100 marks. This meant that there were $8\frac{1}{3}$ marks in a double hour,[39] making it difficult to use the two systems together. To simplify matters, on the advice of Hsia Ho-ling, in 5 BC the emperor Ai decreed that there would be 120 marks in a day, and hence 10 marks in each double hour. However, after only a few months he reversed this decision:

> "In the eighth month, an imperial edict said, 'The Expectant Appointee Hsia Ho-ling and others gave advice that (We) should change the year-period, alter (Our) title, and increase the (number of) graduations on the clepsydra (i.e., the number of marks in a day), whereby (We) could (secure) permanent place for the clan (ruling) the state. We mistakenly listened to the advice of (Hsia) Ho-liang and the others, hoping to obtain blessings for (all) within the (four) seas. (But) in the end there was no happy verification (of their promises); they have all gone contrary to the Classics, turned their backs on ancient (practices), and are not in accordance with the needs of the times. The decree of the sixth month and (the day) *chia-tzu*, except for the order of an amnesty, is all expunged. (Hsia) Ho-liang and the others have gone contrary to the (right) Way and misled the crowd; they are to be committed (to the charge of) the high officials.' They all suffered death for their crimes."
>
> [*Han-shu*, 11; trans. Dubs (1955: 31–32)]

Not for the last time would politics impede scientific progress in China. Indeed, although the experimental increase to 120 marks was repeated by Wang Mang when he overthrew the Han dynasty in AD 9, it was once again rejected upon the restoration of Han power, and it was to be another five centuries before changes to the number of marks in a day were proposed again. This was at the order of the emperor Wu of Liang in AD 507. He decreed that there would be 96 marks in a day, making 8 marks in a double hour. In AD 554 the number was increased to 108; however, neither of these numbers proved popular and so in AD 560 the Ch'en emperor Wen returned to

[39] Or, alternatively, $4\frac{1}{6}$ marks in each of the initial and central halves of a double hour if this system was being used.

the classical definition of 100 marks in a day.[40] This remained unaltered until AD 1628, when, on the recommendation of Jesuit astronomers, the last Ming emperor T'ai-tsung adopted emperor Wu's system of 96 marks in the day. This was used until modern times.

As one mark lasts for $\frac{1}{100}$th of a day (14.4 minutes), it was necessary to split the mark into smaller divisions named *fen*. The number of *fen* in a mark varied from dynasty to dynasty, and indeed from calendar to calendar within a dynasty. At some times there could be as few as 24 *fen* in a mark, at others as many as 150 *fen*.

As I have noted, at most periods of Chinese history there were $8\frac{1}{3}$ marks in each double hour. Thus each double hour contained eight standard marks, plus a period that was only one-third of the length of a standard mark. This short period was called the *hsiao-k'o* or "small mark." Stephenson (1997b: 276), Cohen & Newton (1983), and others have assumed that the small mark was always at the end of each double hour. However, none of these authors give a justification for this assumption. For the period of the Sui dynasty at least, the situation is explained in chapter 19 of the *Sui-shu*. Here it is written that a single mark was split over three double hours, each double hour ending with a third of this mark (i.e., 20 *fen* as there were 60 *fen* in one mark at this period). For the first of the three double hours, the final small mark contained 20 *fen*, numbered 1 to 20. The second of the three double hours ended with twenty *fen* numbered 21 to 40, whilst the third double hour concluded with twenty *fen* numbered 41 to 60.[41] This system was repeated every three double hours with the result that the small mark was always at the end of the double hour.

Maspero (1939) discusses the *Lou-k'o ching* or *Book of the Clepsydra*, written in AD 938 during the *Wu-tai* period. This work explains that at this period the double hours were split into two halves each containing 4 marks followed by 10 *fen*. Once more there were 60 *fen* in each double hour at this time. These are only isolated cases, however, and I am not aware of any other similar written evidence for the rest of Chinese history. Fortunately, it is possible to use some of the astronomical data contained in the dynastic histories to provide information for the other periods.

Contained within the calendar treatise of a number of the dynastic histories are details of the time of sunrise and sunset for the 24 *Ch'i* throughout the year. These are the canonical values for Yang-cheng which were calculated by the various calendars, given in terms of double hours, marks and *fen*. By comparing these times with modern computations, it is possible to establish whether or not the small mark was at the end of the hour. In the following, I will assume that the small mark was at the end of the hour. If this is the case, then the recorded sunrise and sunset times should be in close agreement with the computed values, but if not there should be a constant systematic error between the two sets of values.

Sunrise and sunset times are given in chapter 17 of the *Sui-shu*, chapter 32 of the *Chiu T'ang-shu*, and chapters 70 and 76 of the *Sung-shih*. These are transcribed in Appendix B. Maspero (1939) has translated and discussed the sunrise and sunset

[40] Maspero (1939).
[41] Maspero (1939).

timings from the *Sui-shu*, together with values from the Wei dynasty whose source is unknown to me.[42] In all cases, as the values given in the calendar treatise are for the 24 *Ch'i*, they are symmetrical about the solstices — save for numerous scribal errors. Therefore, only values for the first 13 *Ch'i* will be considered.

Table 6.6 shows the times of sunrise and sunset recorded in the Wei dynasty, together with the times given by modern computation. In the Table, the original times have been converted from double hours, marks and *fen* assuming that the small mark was at the end of the double hour. Figure 6.2 shows the difference between the recorded and the computed times of sunrise and sunset from the Wei dynasty. Clearly there is no evidence for any constant systematic error in the recorded times.[43] Alternatively, if the small mark had been assumed to have been, say, at the beginning of the double hour, then a constant of +0.12 hours would have been added to all of the points on the graph, thus moving them significantly away from the zero error line. Therefore, the recorded sunrise and sunset times throughout the year have their best agreement with the computed values when we adopt the first assumption. This is sufficient evidence to prove that, during the Wei dynasty, the small mark was always at the end of each double hour.

Similarly, Tables 6.7 and 6.8 show the sunrise and sunset times for the Sui and T'ang dynasties respectively. Once more, it is initially assumed that the small mark is at the end of every double hour. The errors in the times are shown in Figures 6.3 and 6.4. As with the Wei times, there is no evidence for any systematic error in the records when the small mark is assumed to be at the end of the double hour. Thus we may safely assume that during these dynasties, this assumption is also correct.

Chapters 70 and 76 of the *Sung-shih* give the canonical times of sunrise and sunset for the *I-t'ien-li* and the *Ch'ung-t'ien-li* calendars respectively. Both sets of times are given in marks and *fen* measured from the central half of a double hour.[44] The normal practice when using the two halves of the double hours was to measure up to $4\frac{1}{6}$ marks in each half separately. In these two cases, however, the full $8\frac{1}{3}$ marks were measured consecutively from the central half of the double hour. Thus the final 4 marks of the double hour were actually during the first half of the following double hour. Despite the unusual terminology used in the sunrise and sunset records from the *Sung-shih*, it is still possible to use them to test our assumption regarding the location of the small mark within the double hour. Tables 6.9 and 6.10 show the sunrise and sunset times from chapters 70 and 76 respectively. The errors in the times are shown in Figures 6.5 and 6.6. Once more, it is clear that there is no systematic error in the times, and so again it would seem that my assumption concerning the placing of the small mark

[42] It should be noted that Maspero analyses the sunrise and sunset timings to try to determine their accuracy, having already assumed that the small mark was at the end of the double hour on the basis of the limited written evidence discussed above.

[43] Although, of course, the distribution of the errors is certainly not random. At least in part, this is probably a factor of the interpolation used in calculating the length of day and night throughout the year. This and the following graphs seem to suggest that the Chinese customarily used a ratio for the longest to the shortest night which was slightly too small.

[44] This is explicitly stated in the second set of times and may be inferred for the first set from the times themselves which would otherwise have sunrise and sunset an hour earlier than the true value.

	Sunrise			Sunset		
Ch'i	Rec. (h)	Comp. (h)	Comp. - Rec. (h)	Rec. (h)	Comp. (h)	Comp. - Rec. (h)
1	7.18	7.07	-0.11	16.82	16.93	+0.11
2	7.08	7.02	-0.06	16.90	16.98	+0.08
3	6.97	6.89	-0.10	17.02	17.11	+0.09
4	6.75	6.69	-0.06	17.23	17.31	+0.08
5	6.48	6.45	-0.03	17.50	17.55	+0.05
6	6.18	6.19	+0.01	17.80	17.81	+0.01
7	5.88	5.92	+0.04	18.10	18.08	-0.02
8	5.58	5.66	+0.08	18.40	18.35	-0.05
9	5.30	5.38	+0.08	18.65	18.62	-0.03
10	5.08	5.15	+0.07	18.90	18.86	-0.04
11	4.93	4.94	+0.01	19.07	19.06	-0.01
12	4.80	4.81	+0.01	19.18	19.19	+0.01
13	4.78	4.77	-0.01	19.22	19.23	+0.01

Table 6.6: Sunrise and sunset times from the Wei dynasty.

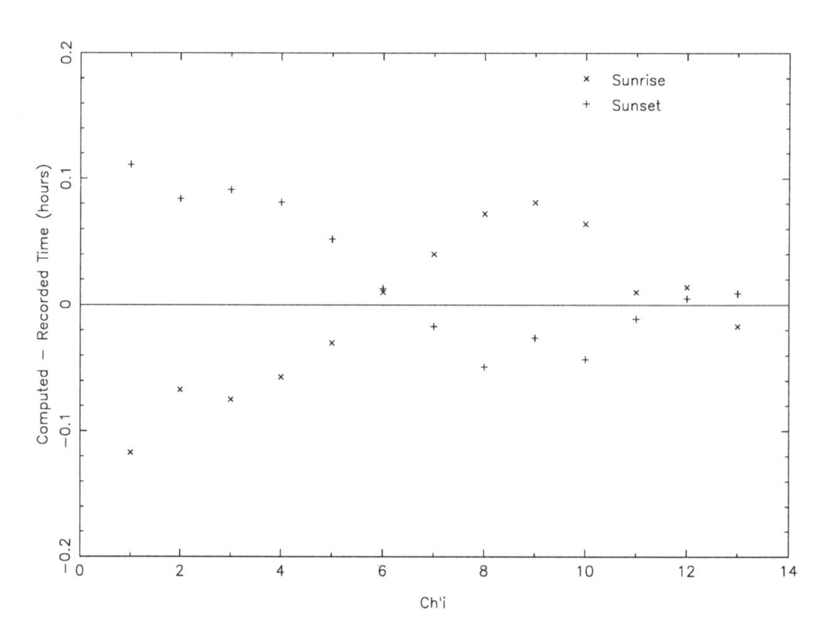

Figure 6.2: The error in the times of sunrise and sunset recorded during the Wei dynasty.

Ch'i	Sunrise Rec. (h)	Comp. (h)	Comp. - Rec. (h)	Sunset Rec. (h)	Comp. (h)	Comp. - Rec. (h)
1	7.20	7.07	-0.13	16.84	16.93	+0.09
2	7.13	7.02	-0.11	16.87	16.98	+0.11
3	6.96	6.89	-0.07	17.00	17.11	+0.11
4	6.80	6.69	-0.11	17.21	17.31	+0.10
5	6.54	6.45	-0.09	17.44	17.55	+0.11
6	6.25	6.19	-0.06	17.75	17.81	+0.06
7	5.94	5.92	-0.02	18.02	18.08	+0.06
8	5.67	5.66	-0.01	18.37	18.35	-0.02
9	5.35	5.38	+0.03	18.64	18.62	-0.02
10	5.11	5.15	+0.04	18.89	18.86	-0.03
11	4.93	4.94	+0.01	19.07	19.06	-0.01
12	4.82	4.81	-0.01	19.18	19.19	+0.01
13	4.84	4.77	-0.06	19.20	19.23	+0.03

Table 6.7: Sunrise and sunset times recorded in chapter 17 of the *Sui-shu*.

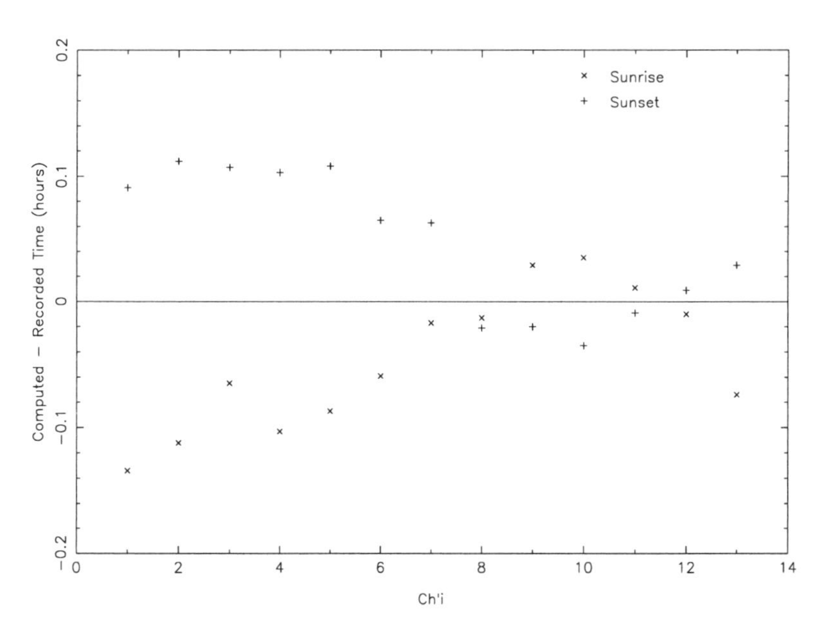

Figure 6.3: The error in the times of sunrise and sunset recorded in chapter 17 of the *Sui-shu*.

		Sunrise			Sunset	
Ch'i	Rec. (h)	Comp. (h)	Comp. - Rec. (h)	Rec. (h)	Comp. (h)	Comp. - Rec. (h)
1	7.20	7.07	-0.13	16.80	16.93	+0.13
2	7.13	7.02	-0.11	16.83	16.98	+0.15
3	6.99	6.89	-0.10	17.01	17.11	+0.10
4	6.79	6.69	-0.10	17.21	17.31	+0.10
5	6.54	6.45	-0.09	17.46	17.55	+0.09
6	6.25	6.19	-0.06	17.75	17.81	+0.06
7	5.94	5.92	-0.02	18.06	18.08	+0.02
8	5.63	5.66	+0.03	18.37	18.35	-0.02
9	5.35	5.38	+0.03	18.65	18.62	-0.03
10	5.12	5.15	+0.03	18.88	18.86	-0.02
11	4.93	4.94	+0.01	18.99	19.06	+0.07
12	4.82	4.81	-0.01	19.18	19.19	+0.01
13	4.80	4.77	-0.03	19.20	19.23	+0.03

Table 6.8: Sunrise and sunset times recorded in chapter 32 of the *Chiu T'ang-shu*.

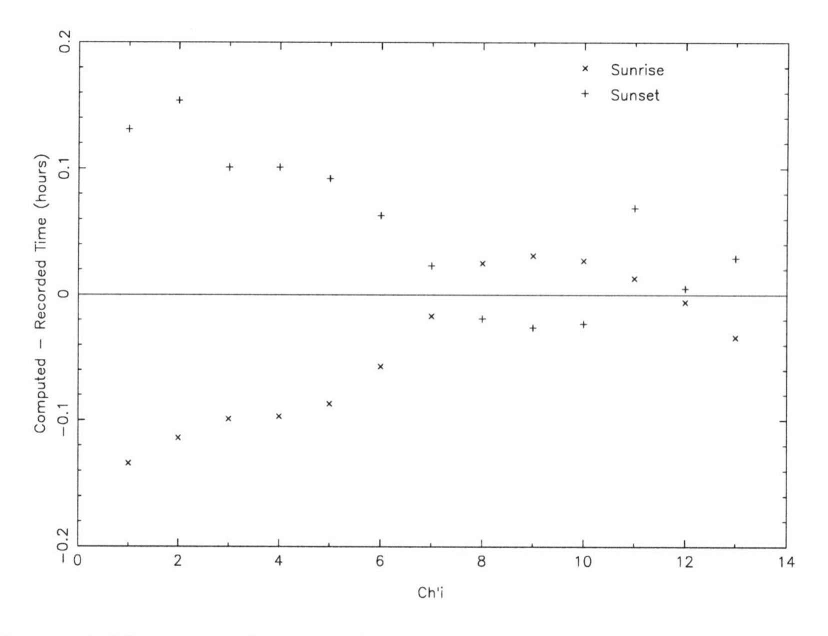

Figure 6.4: The error in the times of sunrise and sunset recorded in chapter 32 of the *Chiu T'ang-shu*.

	Sunrise			Sunset		
Ch'i	Rec. (h)	Comp. (h)	Comp. - Rec. (h)	Rec. (h)	Comp. (h)	Comp. - Rec. (h)
1	7.19	7.07	-0.12	16.80	16.93	+0.13
2	7.15	7.02	-0.13	16.84	16.98	+0.14
3	7.02	6.89	-0.13	16.98	17.11	+0.13
4	6.81	6.69	-0.11	17.18	17.31	+0.13
5	6.57	6.45	-0.12	17.42	17.55	+0.13
6	6.30	6.19	-0.11	17.69	17.81	+0.11
7	6.00	5.92	-0.08	18.00	18.08	+0.08
8	5.69	5.66	-0.03	18.30	18.35	+0.05
9	5.40	5.38	-0.02	18.59	18.62	+0.03
10	5.20	5.15	-0.05	18.84	18.86	+0.04
11	4.95	4.94	-0.01	19.04	19.06	+0.02
12	4.83	4.81	-0.02	19.16	19.19	+0.03
13	4.80	4.77	-0.03	19.19	19.23	+0.04

Table 6.9: Sunrise and sunset times given in chapter 70 of the *Sung-shih*.

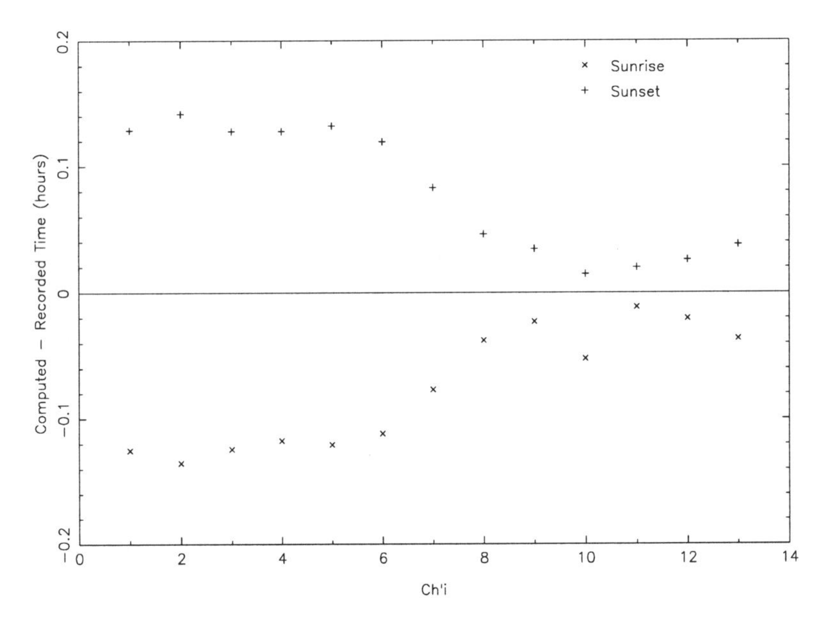

Figure 6.5: The error in the times of sunrise and sunset recorded in chapter 70 of the *Sung-shih*.

		Sunrise			Sunset	
Ch'i	Rec. (h)	Comp. (h)	Comp. - Rec. (h)	Rec. (h)	Comp. (h)	Comp. - Rec. (h)
1	7.20	7.07	-0.13	16.80	16.93	+0.13
2	7.16	7.02	-0.14	16.84	16.98	+0.14
3	7.04	6.89	-0.15	16.96	17.11	+0.15
4	6.81	6.69	-0.11	17.15	17.31	+0.16
5	6.60	6.45	-0.15	17.40	17.55	+0.15
6	6.31	6.19	-0.11	17.69	17.81	+0.11
7	6.00	5.92	-0.08	18.00	18.08	+0.08
8	5.69	5.66	-0.03	18.31	18.35	+0.04
9	5.40	5.38	-0.02	18.56	18.62	+0.06
10	5.15	5.15	+0.00	18.85	18.86	+0.01
11	4.96	4.94	-0.02	19.04	19.06	+0.02
12	4.84	4.81	-0.03	19.16	19.19	+0.03
13	4.80	4.77	-0.03	19.20	19.23	+0.03

Table 6.10: Sunrise and sunset times given in chapter 76 of the *Sung-shih*.

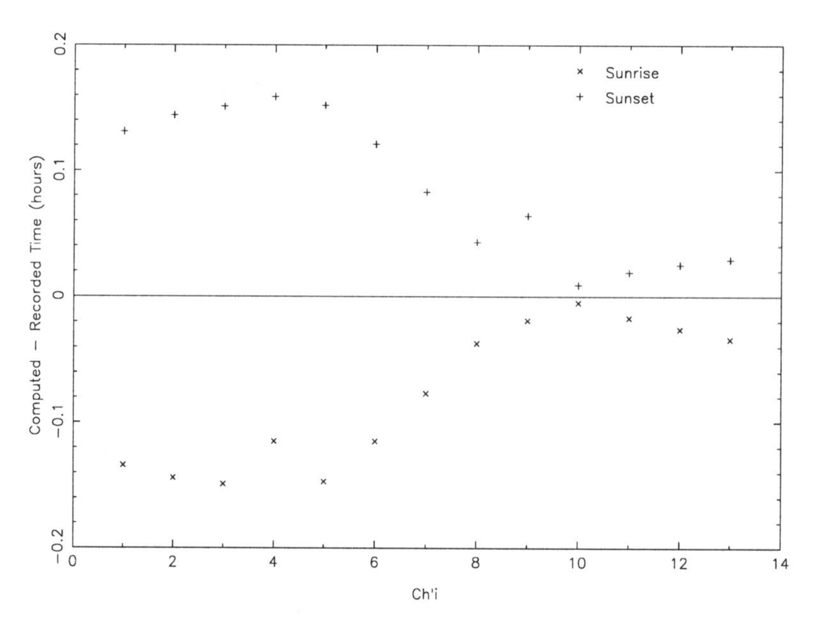

Figure 6.6: The error in the times of sunrise and sunset recorded in chapter 76 of the *Sung-shih*.

Shih	0 *k'o*	1 *k'o*	2 *k'o*	3 *k'o*	4 *k'o*	5 *k'o*	6 *k'o*	7 *k'o*	8 *k'o*
tzu	23.00	23.24	23.48	23.72	23.96	0.20	0.44	0.68	0.92
ch'ou	1.00	1.24	1.48	1.72	1.96	2.20	2.44	2.68	2.92
yin	3.00	3.24	3.48	3.72	3.96	4.20	4.44	4.68	4.92
mao	5.00	5.24	5.48	5.72	5.96	6.20	6.44	6.68	6.92
ch'en	7.00	7.24	7.48	7.72	7.96	8.20	8.44	8.68	8.92
ssu	9.00	9.24	9.48	9.72	9.96	10.20	10.44	10.68	10.92
wu	11.00	11.24	11.48	11.72	11.96	12.20	12.44	12.68	12.92
wei	13.00	13.24	13.48	13.72	13.96	14.20	14.44	14.68	14.92
shen	15.00	15.24	15.48	15.72	15.96	16.20	16.44	16.68	16.92
yu	17.00	17.24	17.48	17.72	17.96	18.20	18.44	18.68	18.92
hsu	19.00	19.24	19.48	19.72	19.96	20.20	20.44	20.68	20.92
hai	21.00	21.24	21.48	21.72	21.96	22.20	22.44	22.68	22.92

Table 6.11: The start of each mark of the double hours.

	chu (initial) half					*cheng* (central) half				
Shih	0 *k'o*	1 *k'o*	2 *k'o*	3 *k'o*	4 *k'o*	0 *k'o*	1 *k'o*	2 *k'o*	3 *k'o*	4 *k'o*
tzu	23.00	23.24	23.48	23.72	23.96	0.00	0.24	0.48	0.72	0.96
ch'ou	1.00	1.24	1.48	1.72	1.96	2.00	2.24	2.48	2.72	2.96
yin	3.00	3.24	3.48	3.72	3.96	4.00	4.24	4.48	4.72	4.96
mao	5.00	5.24	5.48	5.72	5.96	6.00	6.24	6.48	6.72	6.96
ch'en	7.00	7.24	7.48	7.72	7.96	8.00	8.24	8.48	8.72	8.96
ssu	9.00	9.24	9.48	9.72	9.96	10.00	10.24	10.48	10.72	10.96
wu	11.00	11.24	11.48	11.72	11.96	12.00	12.24	12.48	12.72	12.96
wei	13.00	13.24	13.48	13.72	13.96	14.00	14.24	14.48	14.72	14.96
shen	15.00	15.24	15.48	15.72	15.96	16.00	16.24	16.48	16.72	16.96
yu	17.00	17.24	17.48	17.72	17.96	18.00	18.24	18.48	18.72	18.96
hsu	19.00	19.24	19.48	19.72	19.96	20.00	20.24	20.48	20.72	20.96
hai	21.00	21.24	21.48	21.72	21.96	22.00	22.24	22.48	22.72	22.96

Table 6.12: The start of each mark of the single hours.

within the hours is justified.

Whilst there are no sunrise or sunset times recorded in the *Yuan-shih*, the values accepted in the *Shou-shih-li* have fortunately been preserved in the Korean *Chiljŏngsan Naepiŏn* calendar. This calendar from the Chosŏn dynasty was based upon the *Shou-shih-li* and its modified successor, the *Ta-t'ung-li*. Lee (1997) has analysed the sunrise times in this calendar and found them to be in good agreement with modern computation. This implies that the sunrise and sunset times from the *Shou-shih-li* and the *Ta-t'ung-li* must also be close to those given by modern computations. Consequently, during the Yuan and Ming dynasties, the small mark must also have been the final mark in every double or half-double hour.

From the above tests at various periods of Chinese history, it is now possible to safely conclude that the small mark was always at the end of each double hour. By analogy, we may also assume that when the double hour was split into an initial and a central half, each containing $4\frac{1}{6}$ marks, the small mark (i.e., $\frac{1}{6}$ mark) was at the end of each half. Tables 6.11 and 6.12 show the divisions of each double and half double hour into the marks. The double hour commences with the "initial" mark (labelled "0 *k'o*" in the tables), followed by the first mark, and so on up to the eighth "small" mark.

It should be noted that the double hours and marks were time *intervals* rather than discrete moments.[45] Thus, a time merely stated as being at a given mark within a given double hour could be at any moment between the beginning of that mark, as shown in Tables 6.11 and 6.12, and the beginning of the following mark. Therefore, unless the record specifically refers to the start or end of a given double hour or mark, the midpoint of the appropriate interval will be taken in the subsequent analysis.

6.6 Methods of Time Measurement

The early Chinese astronomers almost exclusively used clepsydras to measure the passage of time. These ranged from simple outflow devices to complicated mechanical clocks using water as the motive power source. In its most primitive form the clepsydra used in China was the simple outflow water clock. This consisted of a container out of which water was allowed to flow and the passage of time was determined by the drop in the water level. However, by the first century BC, the inflow clepsydra was in general use.[46] Instead of allowing the water to flow out of a container, the inflow clepsydra works by collecting water in a container and using an indicator rod to determine the amount collected, which is related to the elapsed time. A number of indicator rods would be used, each calibrated to give the time in double hours or in night watches.

The effect of temperature and humidity on the rate of flow of water in a clepsydra was known in China by at least the start of the Later Han dynasty. Needham (1959: 321–322) quotes a certain Huan T'an as saying that when he was Secretary at the Court it was his job to regulate the rate of flow of the clepsydras by comparing them with sundials. Maspero (1939) has asserted that the sundials could not be read with sufficient accuracy to be used for any other purpose than as regulators.

By at least the second century AD it was realised that if the water being collected was flowing from an unregulated container then the fall in the water level would reduce the water pressure and the rate of flow would slow down. Two principal methods were used to maintain a constant water level. The first method used one or more compensating tanks from which water would flow before eventually being collected at the lowest level. This had the effect of keeping the water level in the final outflow tank almost constant. In some clocks as many as six compensating tanks were used to achieve an almost uniform rate of water flow. The second method used to achieve a constant level of water was to place an overflow tank at the head of the series of compensating tanks. This keeps all of the lower tanks at an almost constant level, thus producing a highly uniform rate of flow.

By the 7th century AD devices called "steelyard clepsydras" were being used.[47] These clepsydras measured time not by the use of an indicator rod but by weighing

[45] Needham, Lu, Combridge, & Major (1986: 9).
[46] Needham (1959: 315–317).
[47] Needham (1959: 327).

the water-receiving vessel. The length of time the clepsydra had been running would thus be proportional to the change in weight of the receiving vessel.

Also around the 7th century AD, the Chinese began to use water to power mechanical clocks.[48] This work culminated at the end of the 11th century with the building of a giant clock tower by Su Sung. His clock tower contained an armillary sphere, an automatically rotated celestial globe, and a number of bells and gongs to mark the time of the day in both double hours and marks and night watches, all driven by water dipping from a clepsydra into rotating scoops. It is described in Su Sung's *Hsin I Hsiang Fa Yuo*.[49]

A prototype of Su Sung's clock tower was built of wood in AD 1088 and was soon replaced by a permanent bronze construction in the Sung capital of Pien. Despite being threatened by a change of government in AD 1094, the clock tower remained in full working order until AD 1126 when the Chin Tartars captured Pien. The Chin took all of the captured astronomical instruments to their new capital Yen where they remained in use for some years before gradually wearing out. The last working part of the clock tower, the armillary sphere, continued in use until the Chin fled from the Mongols in about AD 1215 leaving it behind in Yen.[50]

On being driven out of their capital of Pien by the Chin Tartars in AD 1126, the Sung fled south. At their new capital of Lin-an, they soon set about trying to rebuild their lost astronomical instruments. Su Sung's son was asked to assist in the making of a clock to his father's designs, but his father's book containing a description of his clock was lost and family papers did not contain enough details to enable an identical one to be built. Many people investigated the water-drive devices, but none claimed to have been as successful as Su Sung. However, in AD 1172 Su Sung's book was recovered and printed in the south, once more allowing others to use his designs.

Interest in clock design continued throughout the following centuries with astronomers such as Kuo Shou-ching designing several clepsydras.[51] Indeed the last emperor of the Yuan was himself building clocks in the middle of the fourteenth century. However, during the Ming dynasty clock development had virtually ceased and by the time of the coming of the Jesuits in the seventeenth century the mechanical clock had disappeared and more basic clepsydras were again in use.[52] The reason for this decline in clepsydra technology during the Ming was purely political. On the occasion of being presented with a fine crystal clepsydra, the first Ming emperor immediately threw it to the ground where it smashed.[53] Subsequently, he issued an Imperial Edict proclaiming that traditional clepsydras were to be used throughout the empire.

[48] Forte (1988).

[49] The *Hsin I Hsiang Fa Yuo* has been translated into English by Needham, Wang & de Solla Price (1986).

[50] Needham (1965: 494–498).

[51] Bo (1997).

[52] Needham 1965: 508).

[53] Hua (1997).

6.7 Timed Eclipse Records in Chinese History

Throughout Chinese history eclipses have been regarded as among the most astrologically significant of astronomical events. Solar eclipses in particular were seen as major portents and, as such, are one of the most frequently reported types of heavenly observation from the *Ch'un-ch'iu* period onwards. Lunar eclipses held less value as omens, and it is not until the start of the fifth century AD that we begin to find systematic records of their observation.

Preserved eclipse records are almost exclusively contained in the dynastic histories, although, after the tenth century AD, a small number of eclipse observations are also found in other works such as the *Wen-hsien T'ung-k'ao* and the *Sung-hui-yao Chi-k'ao*. Within the dynastic histories, most of the eclipse records are found in the annals, the astrological treatises, and the five-phases treatises. Many of the descriptions are very brief, noting no more than that on a certain day, "the sun was eclipsed." However, if the eclipse was very large then this would also often be noted, for the astrological importance of the eclipse was in direct relation to its magnitude, as is mentioned in the report of the eclipse on 28 April 360 AD:

> "A solar eclipse occurred on a *hsin-ch'ou* day, the first day of the eighth month in the 4th year of the *Sheng-P'ing* reign-period. It was almost a total eclipse observed (in the position of) *chio*. Whenever an eclipse covers a small portion of the sun the calamity will be relatively small, but when it covers a large portion of the sun the consequences will be much more serious."
>
> [*Chin-shu*, 12; trans. Ho (1966: 159)]

In this example, the eclipse occurred when the sun was within *chio*, one of the 28 "lunar lodges" which correspond to zones of right ascension. Sometimes the position of the sun would be given to the nearest *Tu* (^{c}) within a lunar lodge.[54] The astrological interpretation of an eclipse depended upon which lunar lodge the sun was in, as is shown by the record of the eclipse on 22 June 103 AD in the five-phases treatise of the *Hou-han-shu*:

> "The sun was eclipsed in the 22nd degree of *tung-ching*. *Tung-ching* is the mansion (lodge) in charge of wine and food, the duty of a wife: 'It will be theirs neither to do wrong nor to do good, only about the spirits and the food will they have to think.' In the winter of the previous year, the (Lady) Deng had become empress. She had the nature of a man, she participated in and had knowledge of affairs outside of the palace, therefore Heaven sent a symbol. During that year floods and rain damaged the crops."
>
> [*Hou-han-shu*, 27; trans. Beck (1990: 162)]

The above example illustrates two of the most important aspects of early Chinese eclipse records: astrological interpretation and political manipulation. This is particularly evident during the Former and Later Han dynasties when many eclipse observations were interpreted in ways that were intended to criticise either the emperor or the

[54] See Stephenson (1994) for a list of the lunar lodges.

government. Indeed, the level of political manipulation of the astronomical records in the dynastic histories of the Former and Later Han has led to serious questions being asked of their reliability as a genuine record of the observations being made at the time. As Dubs (1938, 1944, 1955), Eberhard (1957), and Bielenstein (1950, 1984) have noted, only a small proportion of the solar eclipses visible in China during the Former and Later Han dynasties are recorded in the *Han-shu* and the *Hou-han-shu*. Even allowing for the effects of adverse weather, the observational record is far from complete. Furthermore, Bielenstein (1950) found that there is significant evidence for a correlation between the amount of missing "observations" during an emperor's rule, and his popularity. Evidently, the record of observations, which it should be remembered is really a record of astrological interpretations, was being manipulated, although it is not clear by whom. Bielenstein (1950, 1984) is of the opinion that this manipulation was being perpetrated by the officials of the time, whilst Eberhard (1957) prefers to think that while the officials may have played some role in this, it was mainly due to the compiler of the dynastic history himself.

In addition to the fact that the observational record in China is obviously far from complete, Foley (1989) has shown that at various periods in Chinese history, a number of the solar eclipses that were recorded could not in fact have been observed. Whilst in the later periods it would seem that many of these records relate to unsuccessful attempts at prediction, failing either because the path of the eclipse passed completely to the north or to the south of China, or because the sun was below the horizon at the time of the eclipse, many of the records during the Han dynasties occur on dates when the sun was far away from one of the lunar nodes, and so do not appear to represent even crude attempts at prediction. Thus, they must have been inserted into the historical records for political purposes.

The eclipse records in the annals, astrological treatises and five-phases treatises very rarely give a precise measurement of the time of the observation. The only exceptions to this are two solar eclipses observed in AD 761 and AD 768. The observations of these eclipses are reported in chapter 36 of the *Chiu T'ang-shu*, as part of the astrological treatise.

In complete contrast to the general style of the eclipse records discussed above are a small number reported in the calendar treatises of the dynastic histories. These are given within discussions of the various systems used for predicting eclipses as examples of the reliability of the calendars. Generally, the descriptions of the eclipses given in the calendar treatises are very detailed, giving the time of the eclipse to the nearest mark or fifth of a night watch. Unlike the vast wealth of eclipse records from the astronomical treatises, only four of the dynastic histories, the *Sung-shu*, the *Sui-shu*, the *Sung-shih*, and the *Yuan-shih*, contain this type of detailed eclipse record in their calendar treatise.

Before discussing the eclipse records in the calendar treatises and other sources in detail, it is necessary to note that very few eclipse records in Chinese history state a place of observation. However, we may assume that this was usually the appropriate capital of the time for the following reasons. Firstly, a small number of early eclipse observations note that the eclipse was *not* seen at the capital, but was instead reported from one of the provinces. This implies that the general practice *was* to observe from the capital, unless the observation was hindered by bad weather. Secondly, the Astronomical Bureau, which was based in the capital as it was necessary to be located near to the emperor so that it could report to him on short notice of any unusual observations that would act as portents, would probably be the only place that had the instruments required to make the detailed observations reported in the calendar treatises. Finally, it is likely that only observations made at the capital would as a rule be sufficiently important to include in an official history.

Finally, it may be helpful at this point to briefly outline the terminology generally used in the eclipse records. Solar eclipses are denoted by *Jih-shih* literally meaning "the sun was eaten." This refers to the ancient myth that an eclipse was caused by a dragon eating the sun. It is more conveniently translated "the sun was eclipsed." For lunar obscurations, *Jih*, "the sun," is replaced by *Yueh*, "the moon." The records may include timings of up to five phases for the eclipse: first and last contact, maximum phase, and, for total eclipses, the start and end of totality. The moment of first contact may be denoted by *Shih-shih*, "beginning of eclipse," *Ch'i*, "it began," or *K'uei-ch'u*, "loss began." Eclipse maximum is usually called *Shen*, and last contact *Fu-man*, "return to fullness." Total eclipses are denoted by *Chi*, with beginning of totality usually called *Shih-chi*, "eclipse total," or, more conveniently, "totality," and end of totality *Sheng-kuang*, "reappearance of light." Of course, other variants upon these terms also exist, but these forms are found most frequently.

6.7.1 The *Sung-shu*, the *Sui-shu*, the *Chiu T'ang-shu*, and the *Sung-shih*

Chapter 12 of the *Sung-shu*, chapter 17 of the *Sui-shu*, chapter 36 of the *Chiu T'ang-shu* and chapter 82 of the *Sung-shih*, contain five, eleven, two, and five detailed records of eclipses respectively. Generally, these are compared with the details of the eclipses as predicted using the appropriate calendar of the time. I give below an example of a record from each of these sources. Full translations of all of these records are given in Appendix C.

- 26 October 440 AD

  > "Yuan Chia reign period, 17th year, 9th month, 16th day, night of the full moon. A lunar eclipse was calculated for the start of the hour *tzu*. The eclipse actually began at the end of the 15th day at the 1st call of the 2nd watch. At the 3rd call it was $\frac{12}{15}$ eclipsed. This was at $1\frac{1}{2}$ degrees in *mao*."
  > [*Sung-shu*, 12]

- 1 August 585 AD

  > "K'ai-Huang reign period, 5th year, 6th month, 30th day. According to the solar eclipse calculations the sun should have been 6 degrees in *ch'i hsing*. At the calculated time of the start of the hour of *wu*, the Sun should have

been $\frac{1}{15}$ eclipsed, the loss beginning from the south-west side. Now when observed, the sun began to be eclipsed after the 6th mark of the hour *wu*. The loss came from the north-east side and the sun was $\frac{6}{15}$ eclipsed. After the 1st mark of the hour *wei* it began to return. At the 5th mark it was returned to fullness."[55]

[*Sui-shu*, 17]

- 5 August 761 AD

"[Shang Yuan reign period], 2nd year, 7th month, *kuei-wei*. On the first day of the month the sun was eclipsed. All of the great stars were visible. Ch'u T'an, the Head of the Astronomy Bureau, proclaimed to the Emperor that on (the day) *kuei-wei*, the sun was dimmed. It began after the 6th mark of *ch'en*. After the 1st mark of *ssu* it was total. It was returned to fullness at the start of the 1st mark of *wu*. This was at 4 degrees in *Chang*."

[*Chiu T'ang-shu*, 36]

- 26 March 1168 AD

"Ch'ien Tao reign period, 4th year, 2nd month, the night of the full moon. At the 5th point of the 2nd watch the moon was eclipsed 9 divisions. It rose above the ground and returned to fullness. I ... said to the prime minister that the moon should have been totally eclipsed when it rose above the ground. The *Chi-yuan-li* also gave the eclipse as total when it rose above the ground. The light should have reappeared at the 2nd mark of the initial half of the hour of *hsu*, and it should have been returned to fullness at the 3rd mark of the central half of the hour of *hsu*. That evening, the moon was concealed by cloud at the time of moonrise. By the time of dusk, it was seen that the moon was already totally eclipsed. By the 3rd mark of the initial half of the hour of *hsu*, the shine had reappeared, and so we may know that the eclipse was total when it rose above the ground. It returned to fullness at the 3rd mark of the central half of the hour of *hsu*. This was at the 2nd point of the 2nd watch."

[*Sung-shih*, 82]

The predicted and observed times of the various phases of the eclipses found in these sources, together with the times as deduced using modern computations, are given in Table 6.13. The errors in the observed times are shown in Figure 6.7. Clearly, there is no evidence for any systematic error in the timing of the eclipses from the Liu-Sung (c. AD 450), the T'ang (c. AD 760) or the Sung (c. AD 1200) dynasties. However, the eclipses observed during the Sui dynasty appear at first to be systematically early.

[55] Some commentators, for example Stephenson (1997a), have interpreted phrases such as *Chih wei hou 1 k'o* as meaning "the 1st mark in the central half of the hour of *wei*." However, the practice of splitting the double hour into an initial and a central half did not come into general use until later times and a more likely reading is "after the 1st mark of the hour of *wei*," as I have given above. I am grateful to Dr. Liu Ciyuan of Shaanxi Observatory, China, for a helpful discussion of this issue.

Source	Date	Type	Contact	Local Time (h) Predicted	Observed	Computed
Sung-shu	434 Sep 4	Lunar	1	6.00	1.61	0.32
	434 Sep 4	Lunar	2	-	2.41	1.42
	437 Jan 8	Lunar	1	18.00	-	17.83
	437 Jan 8	Lunar	2	-	18.95	19.26
	437 Dec 28	Lunar	1	20.00	22.04	21.87
	437 Dec 28	Lunar	2	-	22.98	23.04
	438 Jun 23	Lunar	1	20.00	-	16.65
	440 Oct 26	Lunar	1	23.00	20.70	20.66
	440 Oct 26	Lunar	M	-	21.64	22.14
Sui-shu	585 Jan 21	Lunar	1	-	18.00	18.95
	585 Jan 21	Lunar	M	18.00	19.50	20.36
	585 Jan 21	Lunar	4	-	20.50	21.78
	585 Aug 1	Solar	1	-	12.68	14.51
	585 Aug 1	Solar	M	11.00	13.48	15.54
	585 Aug 1	Solar	4	-	14.32	16.47
	586 Jul 6	Lunar	M	18.00	-	21.63
	586 Dec 16	Solar	M	7.00	7.72	8.40
	590 Apr 25	Lunar	M	20.00	-	19.59
	590 Oct 19	Lunar	M	2.00	-	1.69
	592 Aug 28	Lunar	1	20.00	20.06	21.71
	593 Aug 17	Lunar	1	-	2.84	3.00
	594 Jul 23	Solar	1	-	13.84	14.54
	594 Jul 23	Solar	M	10.00	-	15.18
	595 Dec 22	Lunar	1	-	19.34	19.77
	595 Dec 22	Lunar	M	22.00	21.41	21.18
	595 Dec 22	Lunar	4	-	22.96	22.60
	596 Dec 11	Lunar	M	2.00	-	22.71
	596 Dec 11	Lunar	4	-	2.58	2.59
Chiu T'ang-shu	761 Aug 6	Solar	1	-	8.68	8.32
	761 Aug 6	Solar	M	-	9.48	9.60
	761 Aug 6	Solar	4	-	11.24	10.97
	768 Mar 23	Solar	M	-	11.48	12.81
Sung-shih	1168 Mar 26	Lunar	3	19.60	19.84	19.62
	1168 Mar 26	Lunar	4	20.84	20.84	20.65
	1173 Jun 12	Solar	1	-	12.32	12.36
	1173 Jun 12	Solar	M	-	13.60	14.12
	1173 Jun 12	Solar	4	-	15.36	15.59
	1185 Apr 18	Lunar	M	21.69	23.61	22.85
	1202 May 23	Solar	1	-	11.36	11.35
	1202 May 23	Solar	4	-	13.12	13.13
	1245 Jul 25	Solar	M	13.84	14.98	16.50

Table 6.13: Timed eclipse records in the *Sung-shu*, the *Sui-shu*, the *Chiu T'ang-shu*, and the *Sung-shih*.

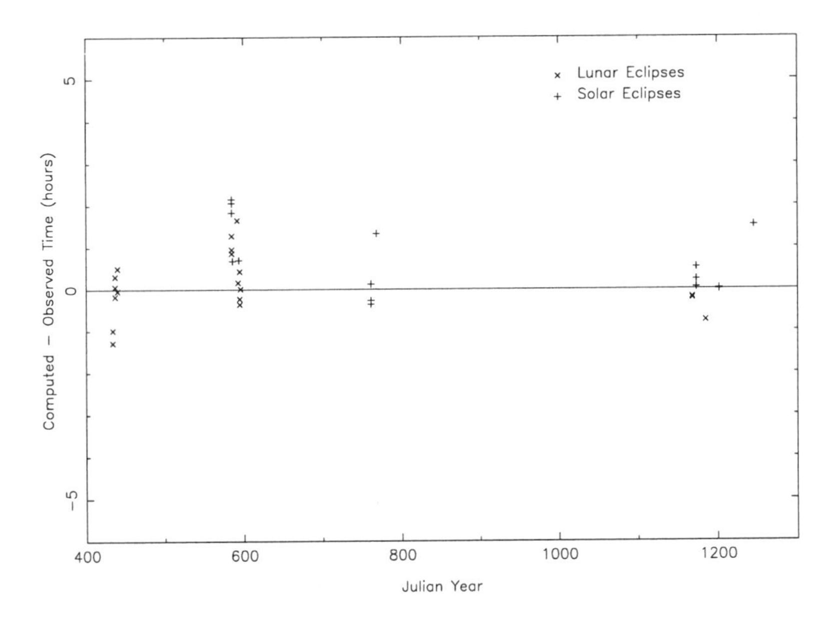

Figure 6.7: The error in the measured times of the eclipses in the *Sung-shu*, the *Sui-shu*, the *Chiu T'ang-shu*, and the *Sung-shih*.

This is mainly caused by the timings of the start, maximum, and end, of the eclipse on 1 August 585 AD. Presumably a poorly calibrated clepsydra was used during this observation.

Figure 6.8 shows the errors in the predicted times of the eclipses given in these sources (there are no predicted times in the T'ang records). These are shown both as the difference between the computed and the predicted times, which I have called the "true" error as we evaluate it, and as the difference between the observed and the predicted times, which I have called the "observed" error as the contemporary Chinese astronomers would perceive it.

The five eclipse predictions from the *Sung-shu* were all made using the *Ching-ch'u-li* calendar. The mean observed accuracy of these predictions is about 2.9 hours, and the mean true accuracy is about 2.7 hours. The eclipses from the *Sui-shu* were predicted using the *K'ai-huang-li* calendar. They have a mean observed accuracy of about 1.8 hours, but a mean true accuracy of only about 2.3 hours. The first of the three predictions from the *Sung-shih* was made with the *Chi-yuan-li* and the second with the *Chung-hsi-li*; however, the name of the calendar used to make the third prediction is not given. This prediction is of noticebly poorer accuracy than the other two eclipses. As there are so few records from this period, it is not possible to determine the accuracies of the individual calendars. However, the first and second predictions

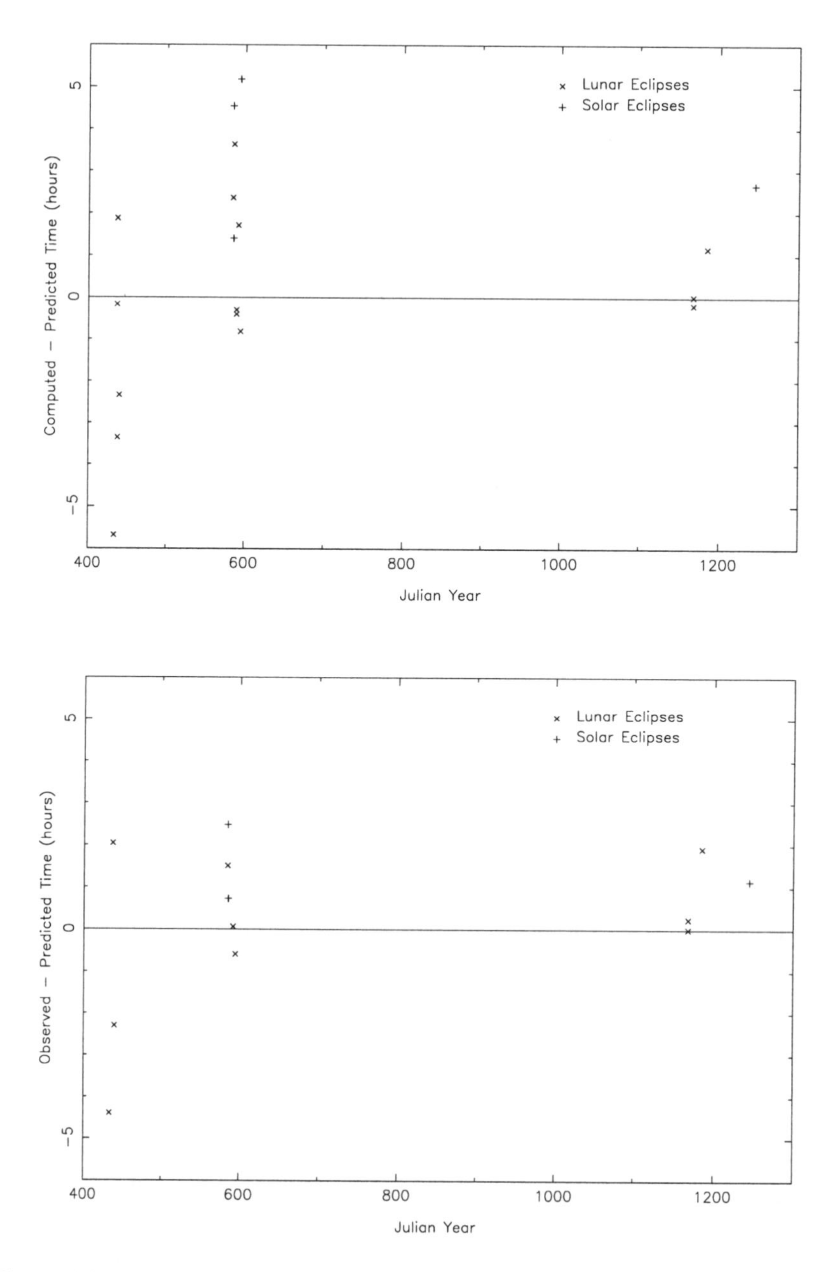

Figure 6.8: The true (above) and observed (below) errors in the predicted times of the eclipses in the *Sung-shu*, the *Sui-shu*, and the *Sung-shih*.

seem to be of similar quality, and so it would seem that the Sung astronomers were making predictions to an observed accuracy of about 0.7 hours. The true accuracy of these Sung predictions is about 0.5 hours.

6.7.2 The *Yuan-shih*

The calendar treatise of the *Yuan-shih* contains more than 60 eclipse timings — only slightly less than all of the other sources in pre-Jesuit Chinese history put together. These are contained within a discussion of Kuo Shou-ching's *Shou-shih-li* calendar by Li Ch'ien. This discussion, entitled the *Shou-shih-li I*, forms chapters 52 and 53 of the *Yuan-shih*. An identical copy of the *Shou-shih-li I* is reprinted as chapters 38 and 39 of the *Hsin Yuan-shih*.

After comparing the calculation of the *Ch'i* by the *Shou-shih-li* with both observation and earlier calendars in chapter 52, Li Ch'ien devotes most of chapter 53 to discussing the reliability of the *Shou-shih-li* in calculating eclipses. To this end he compares the circumstances of eclipses observed over the preceding two thousand years of Chinese history, with those calculated by the *Shou-shih-li*. Li Ch'ien also calculated the circumstances of the eclipses using the *Ta-ming-li* calendar, which had first been used about a century earlier, to show the superiority of the *Shou-shih-li* methods. The earliest records used by Li Ch'ien are the untimed solar eclipse observations recorded in the *Ch'un-ch'iu*. From about AD 400 onwards, however, he made use of detailed timed observations of both solar and lunar eclipses.

The records in the *Shou-shih-li I*, which are split into separate sections for solar and lunar eclipses, always give a detailed account of the date (including the dynasty and the cyclical year and day), the measured time of whichever of the eclipse contacts were observed, the corresponding times as calculated using the *Shou-shih-li* and the *Ta-ming-li*, and an assessment of the accuracy of the two systems. This is on a scale of five, ranging from *mi*, "exact," when there is no difference between the calculated and the observed time, to *shu yuan*, "far off," when the two times differ by 4 or more marks.[56]

The eclipse records in the *Shou-shih-li I* are all set out in the same style. Full translations of all of the timed eclipse records in the *Shou-shih-li I* are given in Appendix C. As examples, I give below translations of a solar and a lunar eclipse record:

- 27 November 680 AD

> "T'ang dynasty, Yung Lung reign period, 1st year, *keng-ch'en*, 11th month, *jen-shen*, first day of the month, (solar) eclipse maximum at the 4th mark of the hour of *ssu*. The *Shou-shih-li* gives the eclipse maximum at 7th mark of *ssu*. The *Ta-ming-li* gives the eclipse maximum at 5th mark of *ssu*. The *Shou-shih-li* is off. The *Ta-ming-li* is close."
> [*Yuan-shih*, 53]

[56] I follow Sivin (1997) in the translation of these terms.

Date	Contact	Local Time (h) Shou-shih-li	Ta-ming-li	Observed	Computed
547 Feb 6	M	15.36	15.84	16.00	16.82
576 Jul 11	M	5.60	6.08	5.00	6.13
680 Nov 27	M	10.80	10.32	10.08	9.49
681 Nov 16	M	8.84	8.36	9.12	7.96
691 May 4	M	4.96	5.12	5.60	5.09
700 May 23	M	15.60	16.12	16.00	15.55
702 Sep 26	M	15.36	16.08	15.84	15.33
707 Jul 4	M	12.60	13.12	12.50	11.93
721 Sep 26	M	12.36	12.60	12.84	11.42
1046 Apr 9[2]	4	16.84	16.36	16.84	16.10
1049 Feb 5	M	11.84	12.12	12.50	12.14
1053 Nov 13[2]	M	13.84	13.12	13.36	13.88
1054 May 10[1]	M	16.36	16.60	16.36	16.55
1059 Feb 15[1]	4	13.60	13.60	13.84	14.16
1061 Jun 20[2]	1	13.12	13.36	13.50	13.20
1066 Sep 22	M	13.84	14.08	13.60	13.76
1069 Jul 21[2]	M	8.32	8.08	7.84	7.58
1080 Dec 14	M	10.32	9.60	10.56	10.01
1094 Mar 19[1]	M	14.32	14.32	14.56	15.09
1107 Dec 16	1	13.84	13.12	13.60	13.61
1107 Dec 16	M	15.12	14.80	14.96	15.23
1107 Dec 16	4	16.56	16.32	16.56	16.61
1162 Jan 17	1	15.36	14.80	15.50	15.54
1183 Nov 17	M	10.60	10.36	10.60	10.63
1195 Apr 12	1	11.36	11.60	11.60	11.67
1202 May 23	1	10.84	11.84	11.36	11.37
1216 Feb 19	M	16.84	16.60	16.98	16.91
1243 Mar 22	M	9.36	9.12	9.60	9.53
1260 Apr 12	M	16.36	15.84	16.60	16.56
1277 Oct 28	1	12.12	12.84	12.12	12.02
1277 Oct 28	M	13.33	14.36	13.36	13.34
1277 Oct 28	4	14.36	15.60	14.60	14.63

1. Eclipses with an identical observational record in another source.
2. Eclipses with an observational record that contradict another source.

Table 6.14: Timed solar eclipse records in the *Yuan-shih*.

- 28 January 948 AD

  "Later Han dynasty (of the 5 dynasties), T'ien Fu reign period, 12th year, *ting-wei*, 12th month, *i-wei*, full moon, (lunar) eclipse. Beginning of loss at the 4th point of the 4th watch. The *Shou-shih-li* gives the beginning of loss at the 5th point of the 4th watch. The *Ta-ming-li* gives the beginning of loss at the 1st point of the 4th watch. The *Shou-shih-li* is close. The *Ta-ming-li* is fairly close."

  [*Yuan-shih*, 53]

The solar and lunar eclipse times recorded in the *Shou-shih-li I* are summarized in Tables 6.14 and 6.15 respectively. Before making any conclusions about the accuracy of the eclipse predictions in the *Shou-shih-li I*, it is necessary to consider the reliability of the observational reports. A number of the observed eclipse times are recorded

		Local Time (h)			
Date	Contact	*Shou-shih-li*	*Ta-ming-li*	Observed	Computed
434 Sep 4[1]	1	2.01	1.61	1.61	0.32
434 Sep 4[1]	2	2.41	2.81	2.41	1.42
437 Jan 8[1]	2	18.95	19.45	18.95	19.26
437 Dec 28[1]	1	22.47	21.96	21.96	21.87
437 Dec 28[1]	2	23.49	23.49	22.98	23.04
543 May 4	1	23.26	0.00	0.00	23.81
592 Aug 28[1]	1	20.46	20.85	20.06	20.35
595 Dec 22[1]	1	18.82	19.85	19.34	19.77
595 Dec 22[1]	M	20.89	21.41	21.41	21.18
595 Dec 22[1]	4	22.44	22.44	22.96	22.60
596 Dec 11[1]	4	3.09	3.60	2.58	2.59
948 Jan 28	1	3.43	1.47	2.94	3.64
1052 Dec 8	1	3.60	3.36	4.08	4.29
1063 Nov 8[1]	M	7.12	7.12	6.80	7.32
1069 Dec 30[2]	1	22.56	23.12	22.56	22.95
1069 Dec 30[2]	M	0.32	0.56	0.32	0.45
1069 Dec 30[2]	4	1.84	2.08	2.08	1.95
1071 Dec 9[1]	1	5.12	6.08	5.60	5.34
1071 Dec 9[1]	M	6.32	6.80	6.56	6.47
1073 Apr 24[1]	1	20.80	21.60	21.36	21.26
1073 Apr 24[1]	M	22.32	22.80	22.56	22.60
1073 Apr 24[1]	4	23.84	0.08	0.08	23.95
1074 Oct 7[2]	1	2.88	2.06	2.88	3.29
1074 Oct 7[2]	2	4.12	3.70	4.12	4.36
1106 Jan 21	M	17.36	17.84	17.84	17.61
1106 Jan 21	4	18.80	19.60	19.12	19.09
1270 Apr 7	1	1.60	2.08	1.84	1.32
1270 Apr 7	M	3.12	3.36	3.12	2.83
1270 Apr 7	4	4.56	4.80	4.56	4.35
1272 Aug 10	1	0.80	1.60	1.12	0.91
1272 Aug 10	M	2.08	2.56	2.56	2.18
1272 Aug 10	4	3.60	3.60	3.84	3.47
1277 May 18	1	0.56	1.12	0.56	0.61
1277 May 18	2	2.08	2.80	1.84	1.70
1277 May 18	M	2.32	2.80	-	2.31
1277 May 18	3	2.56	2.96	2.80	2.92
1277 May 18	4	4.08	4.56	4.08	4.01
1279 Mar 29	1	0.32	0.80	0.32	0.31
1279 Mar 29	M	1.60	1.84	1.60	1.43
1279 Mar 29	4	2.80	2.80	2.80	2.56
1279 Oct 21	1	1.84	2.80	2.32	2.38
1279 Oct 21	M	3.12	3.60	3.12	3.24
1279 Oct 21	4	4.08	4.08	4.08	4.10
1280 Oct 10	4	19.36	20.08	19.36	19.53

1. Eclipses with an identical observational record in another source.
2. Eclipses with an observational record that contradict another source.

Table 6.15: Timed lunar eclipse records in the *Yuan-shih*.

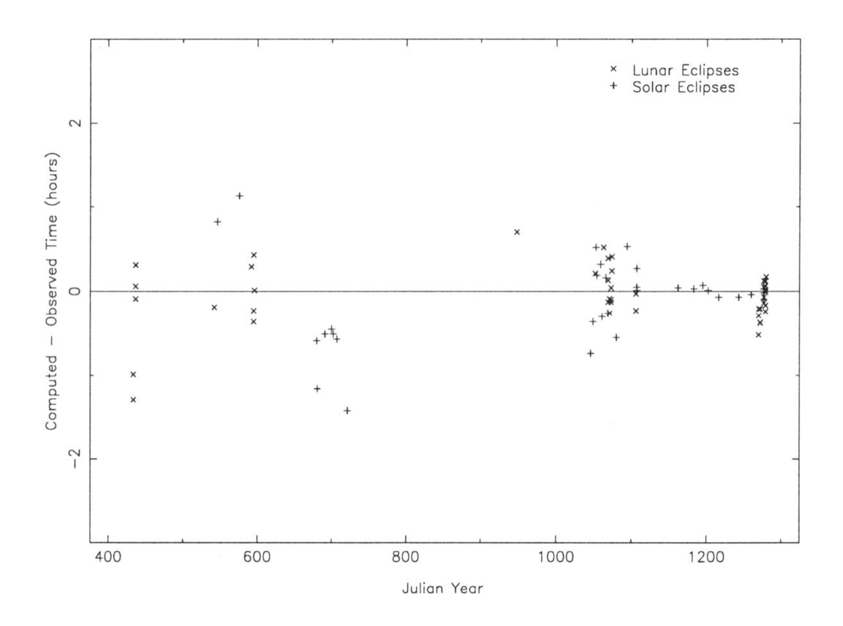

Figure 6.9: The error in the measured times of the eclipses in the *Yuan-shih*.

identically in other sources. These are noted by the superscript 1 to the dates in the Tables. However, a small number of the eclipses, indicated by the superscript 2, contradict the reports given in other sources. The solar eclipses in AD 1061 and AD 1069 are said in the astrological treatise of the *Sung-shih* to have not been seen. However, in the light of the remarks made above about the general reliability of the astrological treatise as an observational record, I choose to accept the detailed reports in the *Shou-shih-li I* as being the more genuine accounts of the observations. In addition, the AD 1061 eclipse is recorded as having been observed in the *Wen-hsing T'ung-k'ao*.

The times of the solar eclipses in AD 1046 and AD 1053, and the lunar eclipses in AD 1069 and AD 1074, contradict those given in the *Wen-hsien T'ung-k'ao*. However, the errors in the times given in the *Shou-shih-li I* are much smaller (and are of comparable size to other contemporary records) than those given in the *Wen-hsien T'ung-k'ao*. Therefore, I shall consider the *Shou-shih-li I* reports as genuine, and the contradictory times in the *Wen-hsien T'ung-k'ao* as anomalous. These anomalous timings will be discussed further in Section 6.7.3 below.

There are no records of the times of the other eclipses in any other available source. This has led Cohen & Newton (1983) to suggest that they may not be observed times, but rather calculated times given by an unknown calendar system. However, this view clearly comes from a misunderstanding of the nature of the *Shou-shih-li I*; it would have been meaningless for Li Ch'ien to have compared two calendar systems with a third to try and show that one was more accurate than the other. It is much more likely that these observed times were recorded in sources that were available to the Yuan astronomers, but which have now been lost.[57]

The error in the observed times of the eclipses recorded in the *Shou-shih-li I* is shown in Figure 6.9. It is noticable that around AD 1200 there are a number of solar eclipse timings which appear to be more accurate than the eclipses from other periods.

Figure 6.10 shows the error in the times calculated using the *Shou-shih-li* calendar. From this figure, there is only a slight suggestion that there is any change in the accuracy with which the calendar could predict the times of eclipse contacts with year. It might have been expected that the earlier predictions would be noticably less accurate than those made around the epoch of the calendar as the small inaccuracies in the parameters used by the calendar would accumulate over time. Evidently, these inaccuracies were sufficiently small to be more or less negligible over about one thousand years. This is a considerable achievement. The mean true accuracy of the *Shou-shih-li* in calculating the times of eclipses over the whole of this period is about 0.34 hours. The mean observable accuracy is about 0.27 hours.

The errors in the times calculated using the *Ta-ming-li* are shown in Figure 6.11. Once more, there is no evidence for any decrease in accuracy of the calendar as it was used further back in time. The mean true accuracy of the *Ta-ming-li* in calculating the times of eclipses is about 0.50 hours over the whole of the period from AD 200 to AD 1300. The mean observable accuracy is about 0.41 hours.

6.7.3 The *Wen-hsien T'ung-k'ao*

The *Wen-hsien T'ung-k'ao*, an encyclopedia compiled by Ma Tuan-lin in AD 1307, contains a number of timed observations of both solar and lunar eclipses. However, in no cases are any predicted times of the eclipses reported. The solar eclipses are contained in chapter 283, whilst the lunar eclipses are in chapter 285. They mainly date from the latter half of the eleventh century AD, although there are two solar eclipses from the AD 1040s. I give below in translation an example of a lunar and a solar eclipse reported in this work. Full translations of all of the eclipse records in the *Wen-hsien T'ung-k'ao* are given in Appendix C.

- 19 March 1094 AD

> "[Yuan Yu reign period,] 9th year, 3rd month, *jen-shen*. On the 1st day of the month, according to the Astronomer Royal, the sun should have been eclipsed, but on account of thick clouds it was not (fully) seen. The loss began (to be seen) at the 3rd mark of *wei*. It was seen through the clouds that the sun was eclipsed on the south-western side in excess of 1 division. At the 6th mark it reached a maximum of 7 divisions. On account of the clouds, its recovery was not seen."
>
> [*Wen-hsien T'ung-k'ao*, 283]

[57] Or, at least, are not among the limited collection of work that is readily available to modern scholars.

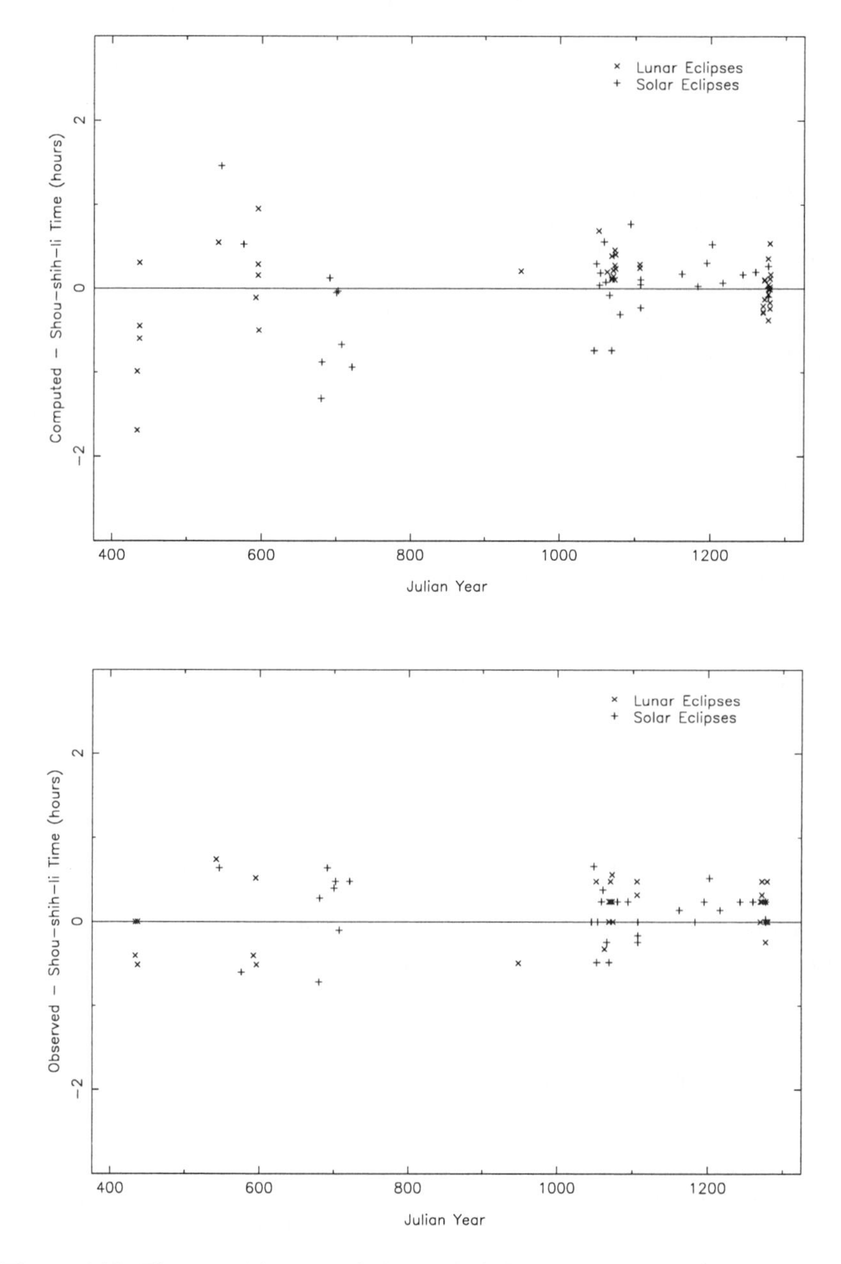

Figure 6.10: The true (above) and observed (below) errors in times predicted using the *Shou-shih-li*.

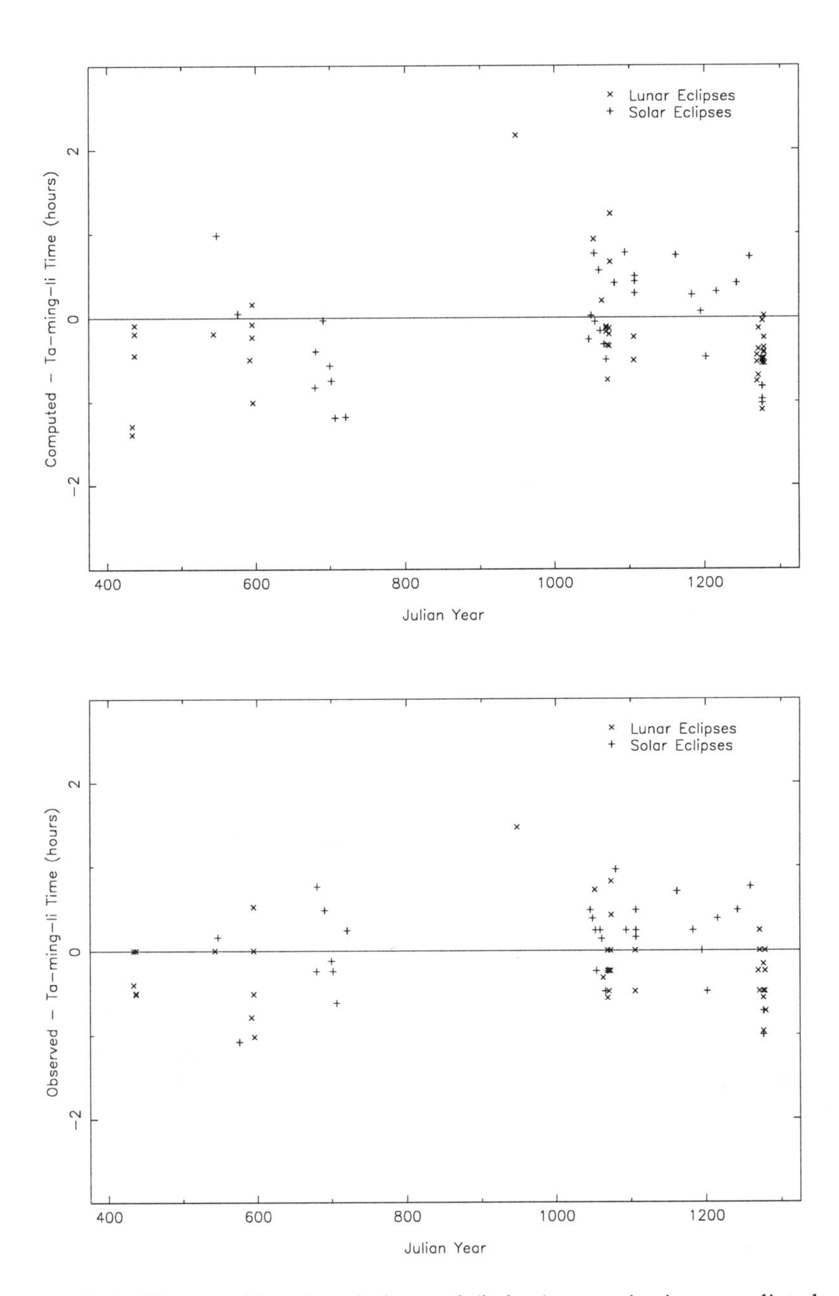

Figure 6.11: The true (above) and observed (below) errors in times predicted using the *Ta-ming-li*.

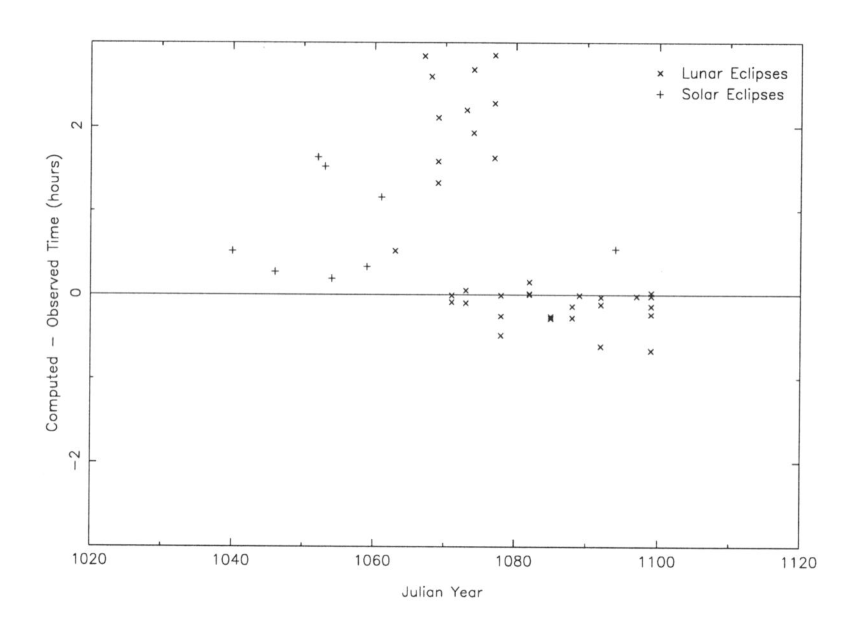

Figure 6.12: The error in the measured times of the eclipses in the *Wen-hsien T'ung-k'ao*.

- 9 December 1071 AD

> "[Hsu Ning reign period, 4th year,] 11th month, *ping-shen*. The moon was eclipsed at the 2nd mark of *mao*. The loss passed from the south-east side to the west side. At the 6th mark, the eclipse reached its maximum of a little less than $4\frac{1}{2}$ divisions. This was at 1 degree in *tung-ching*. As dawn broke, the moon set, and the end of the eclipse could not be discerned."
> [*Wen-hsien T'ung-k'ao*, 285]

The timed eclipse records from the *Wen-hsien T'ung-k'ao* are given in Table 6.16. As noted above, a small number of these observations are also reported in the *Yuan-shih* calendar treatise, in some cases with contradictory details. The records that give the same details are denoted by the superscript 1 to the date. Those that contradict the *Yuan-shih* report have the superscript 2. The error in the measured times of these eclipses is shown in Figure 6.12.

The fact that a number of the reports in the *Wen-hsien T'ung-k'ao* contradict those in the calendar treatise of the *Yuan-shih* suggests that Li Ch'ien and Ma Tuan-lin obtained their eclipse reports from different sources. This is not as surprising as it may at first seem, for there were two astronomical observatories operating at the capital during the eleventh century AD. In addition to the Bureau of Astronomy and Calendar, the Astronomical Department of the Imperial Academy was established within the

Date	Type	Contact	Local Time (h) Observed	Computed
1040 Feb 15	Solar	4	15.36	15.88
1046 Apr 9[2]	Solar	4	15.84	16.11
1052 Nov 24	Solar	4	13.36	15.00
1053 Nov 13[2]	Solar	M	12.36	13.89
1054 May 10[1]	Solar	M	16.36	16.55
1059 Feb 15[1]	Solar	4	13.84	14.17
1061 Jun 20	Solar	M	13.36	14.52
1063 Nov 8[1]	Lunar	M	6.80	7.32
1067 Mar 3	Lunar	1	2.08	4.92
1067 Mar 3	Lunar	M	2.56	6.79
1068 Aug 15	Lunar	1	2.32	4.92
1069 Dec 30[2]	Lunar	1	21.36	22.95
1069 Dec 30[2]	Lunar	M	23.12	0.45
1069 Dec 30[2]	Lunar	4	23.84	1.95
1071 Dec 9[1]	Lunar	1	5.36	5.35
1071 Dec 9[1]	Lunar	M	6.56	6.47
1073 Apr 24[1]	Lunar	1	21.36	21.26
1073 Apr 24[1]	Lunar	M	22.56	22.61
1073 Apr 24[1]	Lunar	4	0.08	23.95
1073 Oct 18	Lunar	1	2.08	4.28
1073 Oct 18	Lunar	M	2.56	7.12
1074 Oct 7[2]	Lunar	1	1.36	3.29
1074 Oct 7[2]	Lunar	M	2.56	5.24
1077 Feb 10	Lunar	1	23.84	1.47
1077 Feb 10	Lunar	M	0.80	3.08
1077 Feb 10	Lunar	4	1.84	4.69
1078 Jan 30	Lunar	1	2.32	1.83
1078 Jul 27	Lunar	2	19.60	19.34
1078 Jul 27	Lunar	4	21.84	21.83
1081 May 25	Lunar	4	20.56	17.51
1082 Nov 8	Lunar	1	17.60	17.75
1082 Nov 8	Lunar	M	18.80	18.80
1082 Nov 8	Lunar	4	19.84	19.85
1085 Sep 6	Lunar	1	19.84	19.56
1085 Sep 6	Lunar	2	20.80	20.51
1085 Sep 6	Lunar	4	23.36	23.10
1088 Jul 6	Lunar	1	22.32	22.18
1088 Jul 6	Lunar	2	0.58	23.49
1088 Jul 6	Lunar	4	2.08	1.80
1089 Jun 25	Lunar	4	2.56	2.55
1092 Apr 24	Lunar	1	21.36	21.24
1092 Apr 24	Lunar	2	22.80	22.18
1092 Apr 24	Lunar	4	0.80	0.77
1094 Mar 19[1]	Solar	M	14.56	15.10
1097 Jan 30	Lunar	M	19.36	19.34
1099 Jun 5	Lunar	1	23.84	23.82
1099 Jun 5	Lunar	2	1.60	0.93
1099 Jun 5	Lunar	4	3.60	3.46
1099 Nov 30	Lunar	1	22.08	22.10
1099 Nov 30	Lunar	2	0.08	23.14
1099 Nov 30	Lunar	4	2.08	1.84

1. Eclipses with an identical observational record in the *Yuan-shih*.
2. Eclipses with a contradictory observational record in the *Yuan-shih*.

Table 6.16: Timed eclipse records in the *Wen-hsien T'ung-k'ao*.

Imperial Palace. These two institutions were instructed to observe independently and then to report to the Palace every morning to compare their results. This was supposed to avoid false reports. However, as P'eng Ch'eng wrote in his *Mo-k'o Hui-hsu*, by about AD 1050, "the officials of the two observatories secretly copied from each other before reporting, and this went on for years."[58] Between AD 1068 and AD 1077, P'eng Ch'eng held the position of Astronomer Royal. One of his main acts was to expose the officials who had perpetrated this deception. This may account for the fact that of the five eclipses reported in the *Wen-hsing T'ung-k'ao* that were observed during his tenure, in two cases the reports contradict those in the *Yuan-shih*. These cannot all simply be put down to scribal errors — the eclipse in AD 1074 was timed in double hours and marks in the report contained in the *Wen-hsing T'ung-k'ao*, but in fifths of a night watch in the *Yuan-shih* report. Thus, it seems very likely that the eclipse records in the *Yuan-shih* and in the *Wen-hsien T'ung-kao* come from two different sources, probably the two observatories.[59]

Now that we have established a plausible reason why contradictory reports of eclipses might be found in different sources, we must ask how it is possible that the two sets of observers apparently made measurements of the times of the eclipses that were so widely different. In many cases the recorded times can vary by an hour or more. However, two observers, at most a few miles apart and presumably using similar clepsydras, would not measure the time of an event to be different by, say, more than a couple of marks, and so there must be another explanation. I will therefore examine each of the contradictory reports in turn to try to find one.

- 2 April 1046 AD

  > "It was returned (to fullness) at the 3rd mark of *shen*." [*Wen-hsien T'ung-k'ao*, 283]
  >
  > "Return to fullness at the 3rd mark of the central half of *shen*." [*Yuan-shih*, 53]

  The simplest explanation in this case is that the character *cheng*, indicating the central half of the hour, has been missed by a careless scribe. However, recalling the sunrise and sunset times from chapter 70 of the *Sung-shih*, it is possible that the convention of counting the marks within a double hour from the middle, rather than the start, of each double hour was being used. Either explanation is quite plausible.

---

[58] *Mo-k'o Hui-hsu*, 7; trans. Needham, Wang, & de Solla Price (1996: 16).

[59] We might speculate that the records in the *Yuan-shih*, which were compiled by an author involved in calendar reform, are more likely to have come from the Bureau of Astronomy and Calendar where he presumably worked, albeit many years later, and the records in the *Wen-hsien T'ung-k'ao* to have come from the Astronomical Department of the Imperial Academy. However, there is no firm evidence either way.

- 13 November 1053 AD

> "... at the 1st mark of the central half of *wu*, (the sun) was eclipsed by $4\frac{1}{2}$ divisions." [*Wen-hsien T'ung-k'ao*, 283]
>
> "Eclipse maximum at the 1st mark of the hour of *wei*." [*Yuan-shih*, 53]

In this case it is possible that either of the records could contain a scribal error and that the time of the middle of the eclipse could be either at the 1st mark of *wei*, or at the 1st mark of the central half of *wu*.

- 30 December 1069 AD

> "At the 1st mark of *hai*, the loss was seen on the north-eastern side. At the initial mark of *tzu*, the eclipse reached its maximum ... At the 3rd mark, it was returned to fullness." [*Wen-hsien T'ung-k'ao*, 285]
>
> "Beginning of loss at the 6th mark of *hai*. Maximum at the 5th mark of *tzu*. Return to fullness at the 4th mark of *ch'ou*." [*Yuan-shih*, 53]

If we interpret the 1st mark of *hai* as the 1st mark in the central half of *hai*, then the time of the start of the eclipse is 22.36 hours. This is close to the time given in the *Yuan-shih* of 22.56 hours. A difference of 0.20 hours (12 minutes) is quite possible when two observers are measuring the time to the nearest mark (14.4 minutes). Similarly, if for the time of the middle of the eclipse, the initial mark of *tzu* is interpreted as being the initial mark in the central half of *tzu* (i.e. 0.12 hours), it is close to the time of 0.32 hours given in the *Yuan-shih*. The time of the end of the eclipse recorded in the *Wen-hsien T'ung-k'ao* cannot be reconciled with the time given in the *Yuan-shih* by this method. However, it is also inconsistent with the other times reported in the *Wen-hsien T'ung-k'ao* — the time from the beginning of the eclipse to the middle is 7 marks, but the time from the middle to the end is only 3 marks.

- 7 October 1074 AD

> "At the 1st mark of *ch'ou* the loss began on the eastern side. At the 6th mark, the eclipse reached its maximum ... Dawn broke and its return to fullness was not seen." [*Wen-hsien T'ung-k'ao*, 285]
>
> "Beginning of loss at the 5th point of the 4th watch. Eclipse total at the 3rd point of the 5th watch." [*Yuan-shih*, 53]

According to the report in the *Wen-hsien T'ung-k'ao*, the end of the eclipse was not seen on account of dawn breaking. If the times given in the report are accepted at face value then the eclipse started at 1.36 and reached its maximum at 2.56. Thus the end of the eclipse cannot have been much after 4 am. However, the moon did not set until about 5.45, and the sun did not rise until 6.28. Therefore we cannot accept the recorded times as they are. If we once more assume that the marks were measured from the middle rather than the start of the double hour, however, then the eclipse would have started at 2.36, reached it maximum at 3.56, and so ended sometime around 5 am. Thus it is just conceivable that the moon might have set, or at least been very close to the horizon, when the eclipse ended. Furthermore, within this assumption, the times are quite close to those given in the *Yuan-shih* (2.88 hours for the beginning of the eclipse and 4.12 hours for the beginning of totality). Discrepancies of this magnitude are consistent with our hypothesis of two different observers.

Three of the four cases outlined above seem to indicate that it is at least plausible, and indeed probable, that the marks were measured from the middle rather than the start of the double hour, in the same fashion as in the sunrise and sunset times in the *Sung-shih* discussed earlier. This may account for the fact that some of the other eclipses in the period from about AD 1065 to AD 1080 also appear to be systematically early. However, in some of these cases the times are early by as much as four or five hours. Clearly these cases must be due to scribal errors.[60] Of course, not all of the eclipses times at this period use this system. For example, the eclipses in AD 1071 and AD 1073, which have an identical record in the *Yuan-shih*, obviously do not. This suggest two possibilities: either the original records contained a mix of timings in this unusual system and in the normal system, or all of the timings were in the unusual system and Ma Tuan-lin converted them to the normal system, but inadvertently missed a small number.

6.7.4 Other Sources

A further nine timed eclipse observations in other pre-Jesuit sources have been uncovered by Beijing Observatory (1988). Two of these, recorded in the *Sung-hui-yao Chi-k'ao*, date from the Sung dynasty; the others are all from the Ming dynasty. They are listed in Table 6.17. The errors in the observed times are shown in Figure 6.13. Full translations of these records are given in Appendix C.

6.8 Accuracy of the Observed Times

Having outlined all of the available measurements of eclipse times, it is now possible to study them as a whole. I have accepted the corrections to the anomalous timings in the *Wen-hsien T'ung-k'ao* that I have considered plausible in Section 6.7.3 above. However, as some of the other measurements in the *Wen-hsien T'ung-k'ao* for which there is no firm evidence for making such corrections are in error by large amounts,

[60]Despite P'eng Ch'eng's comment that the astronomers in the observatories "contented themselves with copying out the positions of the sun, moon, and planets according to very rough ephemerides ... never using the (astronomical equipment in the) observatory" [*Mo-k'o Hua-hsi*, 7; trans. Needham, Wang & de Solla Price (1986: 16)], these very inaccurate times cannot have been calculated. The Sung calendars were able to predict the time of an eclipse to at least within an hour. Furthermore, calculating an eclipse time using these methods is not an easy task — certainly requiring more effort than actually observing the eclipse.

Date	Type	Contact	Local Time (h) Observed	Computed
1068 Feb 6	Solar	1	10.96	11.11
1068 Feb 6	Solar	M	12.44	12.41
1068 Feb 6	Solar	4	13.84	13.66
1100 May 11	Solar	4	10.84	10.11
1572 Jul 10	Solar	1	6.84	6.68
1572 Jul 10	Solar	4	9.84	9.41
1577 Apr 2	Lunar	1	2.84	2.09
1578 Mar 23	Lunar	1	20.84	19.32
1587 Mar 24	Lunar	4	22.84	21.55
1596 Oct 22	Solar	1	10.60	9.89
1596 Oct 22	Solar	M	12.98	11.19
1610 Dec 15	Solar	1	14.36	14.61
1610 Dec 15	Solar	M	15.60	16.07
1610 Dec 15	Solar	4	17.60	17.32
1617 Feb 20	Lunar	1	19.60	18.56
1617 Feb 20	Lunar	M	19.84	20.47

Table 6.17: Timed eclipse records in other sources.

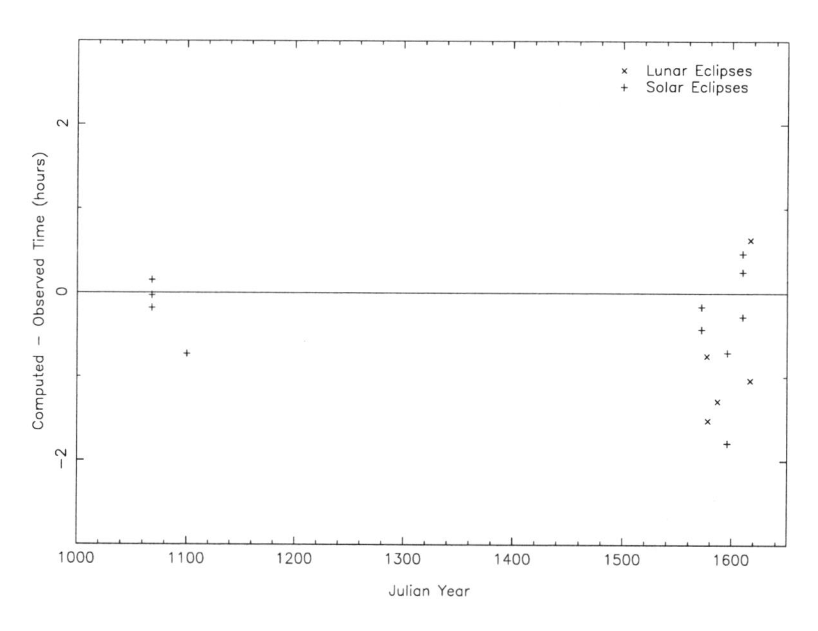

Figure 6.13: The error in the measured times of the eclipses mentioned other sources.

I have discarded any observations with an error of greater than two hours from the following analysis.

Looking first at the solar eclipse records, Figure 6.14 shows the error in all of the measured times. There is no evidence for any systematic error over the whole of the period from about AD 500 to AD 1625; this provides further proof that my assumption that the small mark is at the end of the double or half double hour, made in Section 6.5 above, is correct. The mean accuracy of the solar eclipse timings is about 0.41 hours over the whole of this period. Between about AD 1090 and AD 1300 there appears to be some improvement in the accuracy of the timings; the mean accuracy of the records at this period is about 0.16 hours.

Figure 6.15 shows the error in all of the measured times of lunar eclipses. The mean accuracy of the lunar eclipse timings over the whole of the period from about AD 400 to AD 1625 is about 0.52 hours. Once more, there appears to be an improvement in accuracy in the period from about AD 1090 to AD 1350; the mean accuracy of these timings is about 0.21 hours. Surprisingly, there is little difference in the accuracy of the times measured in hours and marks, and those measured in fifths of a night watch, despite the fact that a fifth of a night watch is on average approximately twice as long as a mark.

In both the solar and lunar eclipse times, there appears to be an improvement in accuracy in about AD 1090. This date coincides with the construction of Su Sung's clock tower in Pien. It would be naïve to attribute this improvement in timing accuracy directly to the construction of Su Sung's clock — indeed when the Sung were driven out of Pien in AD 1127 they clearly could not have used it in their observations — nevertheless it does suggest that there was a significant improvement in the methods of time keeping, of which Su Sung's designs are at least an example, around this period. Towards the end of the Ming dynasty, however, the accuracy of the eclipse timings had decreased back to the levels seen five hundred years earlier. The level of accuracy achieved by the Sung and Yuan astronomers was not bettered until the middle of the eighteenth century when the Jesuit astronomers constructed new instruments for timing eclipses.[61]

In Figure 6.16, the error in the measured times of the eclipse contacts is shown as a function of the interval between the eclipse and either sunrise (for solar eclipses) or sunset (for lunar eclipses). The data are split into two subsets — eclipses before AD 1090 and eclipses after AD 1090. This is the date at which an improvement in the accuracy of the timings was noted. It appears that in the first case there may be a slight increase in the dispersion of the error as the interval from sunrise or sunset increases. This suggests that the clocks used by the early Chinese astronomers were regulated every morning and evening. However, after AD 1090 this is not the case, suggesting that regulation every sunrise and sunset, whether it was performed or not, was not needed.

Finally, Figure 6.17 shows the error in the measured eclipse times as a function of the month of the year. It might be expected that there would be greater errors

[61] Stephenson & Fatoohi (1995); Fatoohi & Stephenson (1996). .

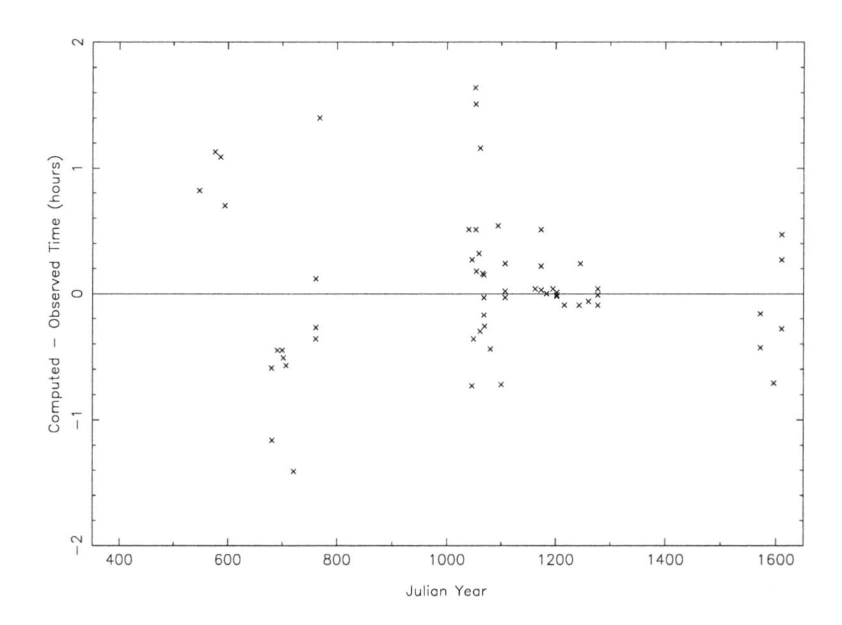

Figure 6.14: The error in the measured times of solar eclipses.

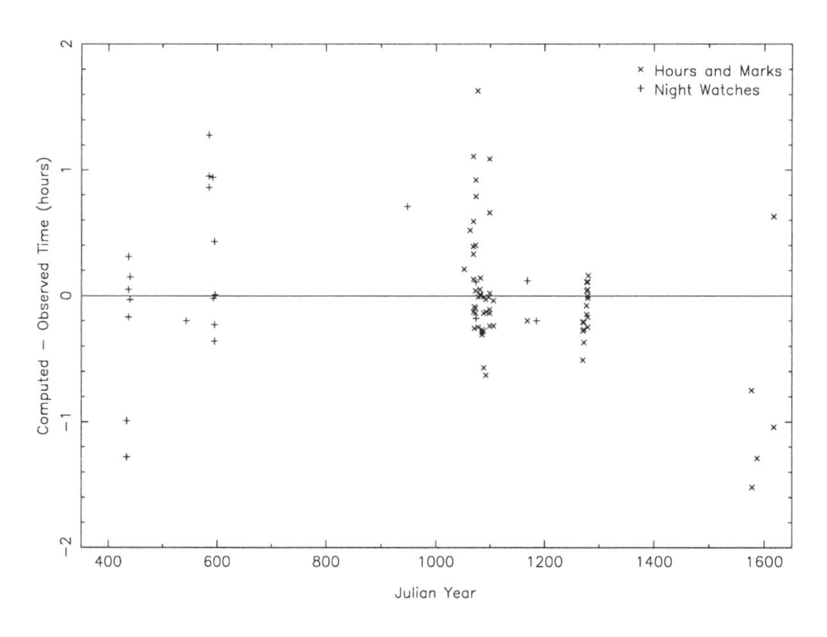

Figure 6.15: The error in the measured times of lunar eclipses.

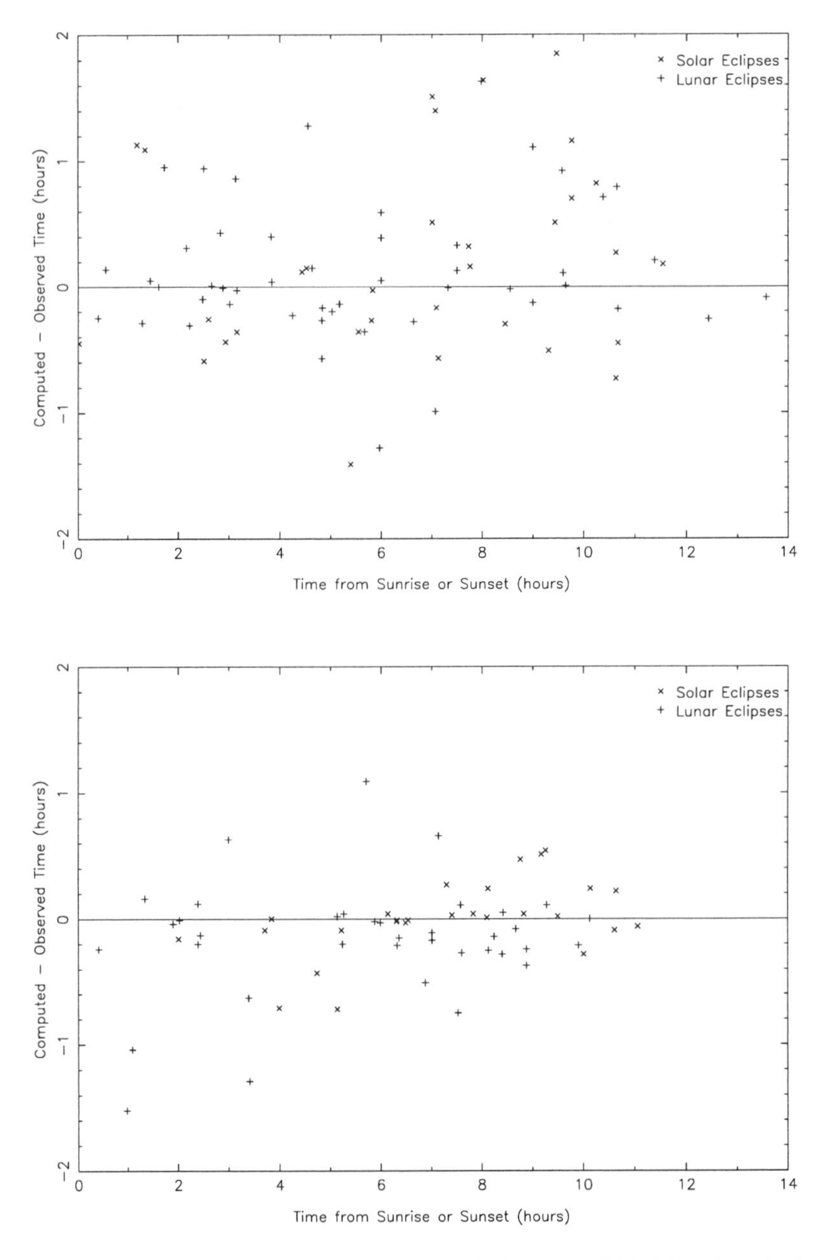

Figure 6.16: The error in the eclipse timings before AD 1090 (above) and after AD 1090 (below).

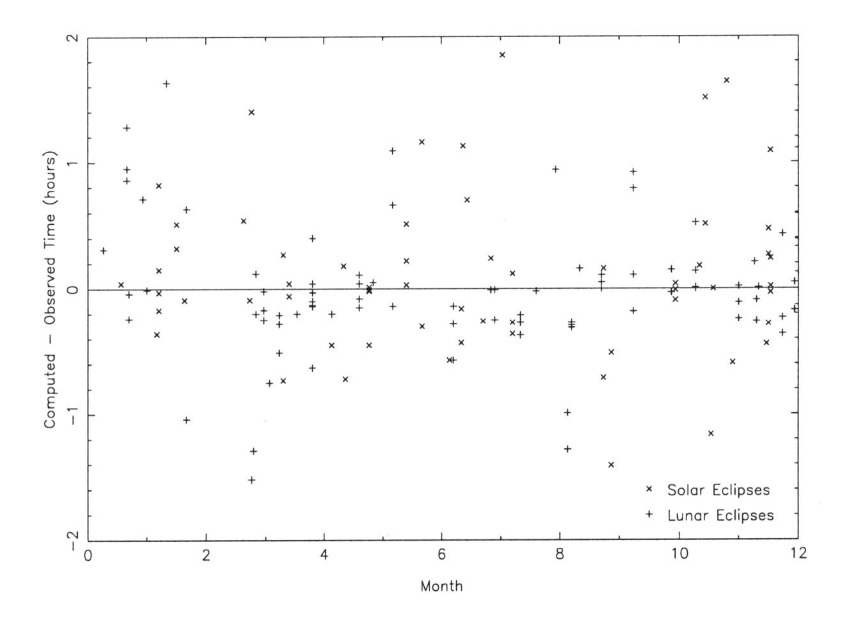

Figure 6.17: The error in the measured eclipse times throughout the year.

in the timing of eclipses in the winter months than in the summer, due, perhaps, to the effects of the change in temperature upon the viscosity of the water in the water clocks.[62] However, as is clear from Figure 6.17, this is not the case. Evidently the Chinese astronomers were successful in their attempts to regulate the flow of water throughout the year.

6.9 Accuracy of the Predicted Times

It is possible to trace a great improvement in the ability to predict the time of an eclipse down through Chinese history. The *Ching-ch'u-li* system used by the Liu-Sung astronomers could only crudely predict the time of an eclipse to an observed accuracy of about 2.9 hours and a true accuracy of about 2.7 hours. By the Sui dynasty, the *K'ai-huang-li* was being used to make predictions to an observed accuracy of about 1.8 hours and a true accuracy of about 2.3 hours. Five hundred years later, the Sung astronomers of the middle of the twelfth century AD were predicting eclipses with an observed accuracy of about 0.7 hours and a true accuracy of about 0.5 hours. By the end of that century, the *Ta-ming-li* was able to predict eclipses with an observed accuracy of about 0.4 hours, and a true accuracy of about 0.5 hours. The Yuan dynasty

[62] And, for the lunar eclipses, to the greater length of the night watches in winter than in summer.

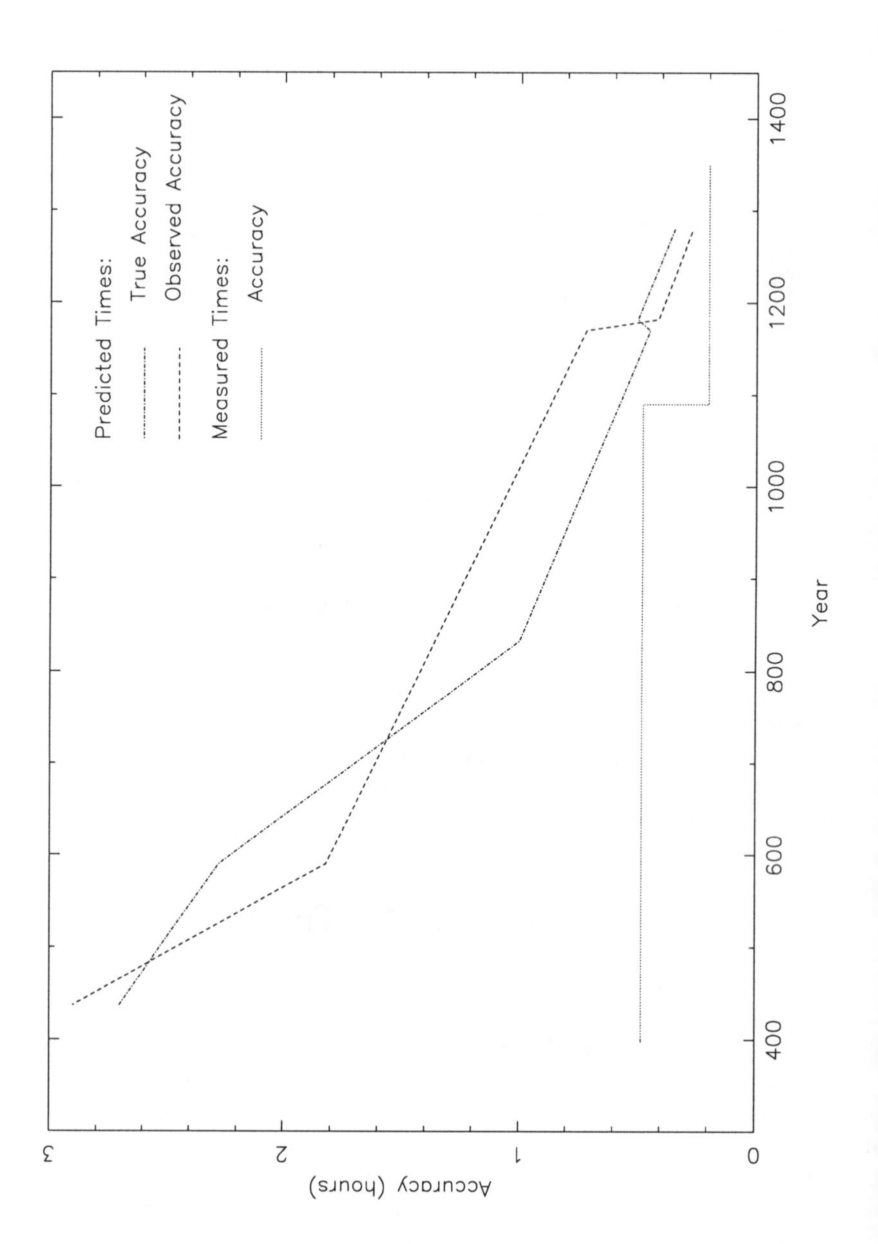

Figure 6.18: Schematic representation of the accuracy of the prediction of eclipse times.

Shou-shih-li calendar achieved even higher levels of accuracy than this, with both an observed and a true accuracy of about 0.3 hours.

It is unfortunate that there are no records of eclipse predictions made with the many calendar systems of the T'ang dynasty. One of the most important calendars of this period was the *Ta-yen-li*, compiled by I Hsing in AD 727. This calendar is discussed in chapter 27b of the *Hsin T'ang-shu*, where it is stated that it has been tested against a number of solar and lunar eclipses. However, no details of these tests are given, and so it is not possible to assess the accuracy of any eclipse predictions made with this calendar. Another important calendar from the T'ang dynasty is the *Hsuan-ming-li*. Again, no eclipse predictions made with this calendar are preserved in Chinese history. However, this calendar was used in Japan from the middle of the ninth until the end of the seventeenth century AD, and over one hundred eclipse predictions made with it are recorded in various Japanese sources. As I shall show in Chapter 7, the true accuracy of this calendar in predicting eclipses can be deduced to be about 1.0 hours.

Figure 6.18 shows schematically the improvement in the accuracy of the Chinese calendar systems to predict the time of eclipses. For each set of predictions the mean accuracy has been plotted against the approximate year that they were made. The points have been connected with straight lines to illustrate the improvement in accuracy over the period. However, these lines must not be taken to imply a linear improvement in the techniques used between two points, but are simply a way of indicating the general development down the centuries. Also shown in Figure 6.18 is the typical accuracy with which the times of eclipses were observed over the same period.

It is interesting to contrast the gradual, but steady, improvement in the ability of the Chinese astronomers to predict the time of an eclipse, with the complete lack of progress in their ability to measure the time of an eclipse before the end of the eleventh century AD, when there was a sudden improvement by a factor of more than 2 in the accuracy of the measured times. Furthermore, this occurred at around the time when the accuracy of the predictions was reaching about the same level as the observations. Indeed without this improvement in the ability to accurately measure the time of an eclipse, it is doubtful whether the improvements in the *Shou-shih-li* could have been formulated. Only when the accuracy of observation was higher than that of prediction could uncertainties in the predictions be reliably identified.

Chapter 7

Japan

7.1 Introduction

The history of Japan emerges from the legendary period around the sixth century AD. The *Nihongi*, a chronicle written in about AD 720, recounts the history of Japan from the time of the gods down to AD 679. However, at least before the fifth century AD, it is clear that much of this history is fictitious, albeit a fiction that is probably based partly upon fact.[1] From archaeological evidence it seems that even as late as the third century BC, Japan was still in the stone age.[2] Chinese culture began to enter Japan from Korea around this time, and by the third century AD, local rule had been established in parts of the country.[3] One of these small village states, Yamato, began to grow in power and by around AD 350 had subdued the others to form Japan's first unified state.[4]

Over the next three centuries, Japan began to form links with its Korean and Chinese neighbours. Envoys and tributes were sent to the Chinese court and, more often, the Paekche state of Korea. During this period, the Japanese were profoundly influenced by Chinese culture, perhaps most significantly by the adoption at the start of the fifth century AD of the Chinese characters for writing. In AD 646, the Japanese clans were federated under an emperor, and a bureaucratic state, based upon that of T'ang China, was established. One result of this was that the Japanese began to imitate the Chinese practices in astronomy and other fields. In AD 628, the Japanese adopted the Chinese system of timekeeping and a clepsydra was constructed. By AD 675, the first astronomical observatory had been set up, and in AD 702, the Taiho civil

[1]As Aston (1972: xv-xvi), who has translated the *Nihongi*, has written, "The earlier part furnishes a very complete assortment of all the forms of Untrue of which the human mind is capable, whether myth, legend, fable, romance, gossip, mere blundering, or downright fiction ... Then we have a series of legendary stories full of miraculous incidents, but in which grains of truth may here and there be discerned ... The narrative becomes more and more real as it goes on, until about the 5th century (AD) we find ourselves in what, without too violent a departure from truth, may be called genuine history..."

[2]Sugimoto & Swain (1978: 2).

[3]Tuge (1968).

[4]Sugimoto & Swain (1978: 3).

code was promulgated. For the first time, this code set up regulations governing the administration and teaching of astrology and calendar making.[5]

During the eighth century AD, the imperial family and the court nobles began to accumulate vast wealth, and consequently significant power. Because of this they began to monopolize the governmental positions, with the result that these positions effectively became hereditary. The nobility began to turn away from Chinese culture, and with the abandonment of the practice of sending missions to the T'ang court in AD 894, official Chinese-Japanese relations were effectively cut.[6] It was not until the fifteenth century AD that Japan again began to take in significant foreign influence on its society.

In this chapter I shall, after giving some necessary background on calendrical astronomy in Japan, examine the numerous reports of eclipse predictions made using the Chinese *Hsuan-ming-li* calendar system recorded in Japanese history. Not only does this allow the accuracy with which the Japanese astronomers were able to predict eclipses to be assessed, but it also provides a convenient method of determining the accuracy of the Chinese *Hsuan-ming-li* in predicting eclipses. As I have noted in Chapter 6 above, there are no eclipse predictions made with this calendar preserved in Chinese history, and so the Japanese records are of great value in tracing the development of the Chinese calendar at this period.

7.2 Calendrical Astronomy in Japan

The earliest reference to the calendar in Japanese history is in the 19th book of the *Nihongi*. The Emperor of Japan sent a mission to Paekche in AD 553 requesting, among other things, that "the men learned in medicine, divination, and in the calendar, have to take it in turn to come up (to the Japanese Court)."[7] The following year, Paekche responded to this request by sending, among others, Wang Po-son, an expert on the Chinese calendar. It is now generally believed that before this time there was no native calendrical system in Japan, and that earlier dates given in the *Nihongi* were calculated retrospectively, and in some cases inaccurately, using Chinese systems.[8]

Over the next two centuries, Korea continued to supply Japan with the Chinese calendar. In AD 602, Paekche sent a priest to the Japanese Court to present books on calendar-making, astrology and geography (including geomancy), and to teach in these disciplines:

> "A Paekche priest named Kwal-leuk arrived and presented by way of tribute books of Calendar-making, of Astronomy, and of Geography, and also books of the art of invisibility and magic. At this time three or four pupils were selected, and made to study under Kwal-leuk. Ochin, the ancestor of the Yako no Fumibito, studied the art of Calendar-making. Koso, Otomo no Suguri, studied Astronomy

[5]Nakayama (1969: 10).

[6]Sugimoto & Swain (1978: 28).

[7]*Nihongi*, 19, 38; trans. Aston (1972: II, 68)

[8]Nakayama (1969: 7–8).

> and the art of invisibility. Hinamitatsu, Yamashiro no Omi, studied magic. They all studied so far as to perfect themselves in these arts."
>
> [*Nihongi*, 22, 7; trans. Aston (1972: II, 126)]

Following Ochin's study of the calendar, two years later the old calendar was repealed, and the *Yuan-chia-li* calendar was adopted.[9] Little more is known of the adoption of the Calendar in Japan at this early period. The situation is made confusing by a statement in the *Nihongi* that in AD 690 "in compliance with an Imperial order, the use of the *Yuan-chia* and the *I-feng* calendars was begun."[10] At the start of the last century, Kozawa Masakata concluded that the *Yuan-chih-li* was used from AD 604 to AD 697, and the *I-feng-li* was used from AD 692 to AD 763.[11] Presumably both calendars were used simultaneously in the overlapping years.

In AD 735, the *Ta-yen-li* was brought to Japan. Unlike the earlier calendars which had come to Japan via Korea, the *Ta-yen-li* was brought directly from China by Kibi no Makibi. However, the calendar was not actually adopted in Japan until AD 764, by which time it had already been replaced in China. Similarly, the *Wu-chi-li* calendar was not adopted by the Japanese until AD 764, nearly eighty years after it had been brought from China.[12] Finally, in AD 862, the *Hsuan-ming-li* was adopted. This calendar continued in use until AD 1685 when it was replaced by Japan's first indigenous calendar.

The long period without calendar reform in Japan may be attributed to a number of factors. In China, the calendar was often reformed when a new government came to power. However, in Japan there was an unbroken imperial line, and so there was no need for new calendars to be adopted to keep establishing the heavenly mandate to govern.[13] Furthermore, Japan's close links with China were cut towards the end of the ninth century when the country entered a state of semi-seclusion. During this period, which lasted until the start of the fifteenth century, the Japanese were either unable or unwilling to adopt more recent Chinese calendar systems.[14] Finally, in Japan the practices of calendar-making and of divination were all part of the same *Yin-yang* office. Members of this office, like all of the Japanese government, were selected on a hereditary rather than a merit basis. By the tenth century AD, the *Yin-yang* office was divided into two factions — the calendrical officials of the Kamo clan, and the more important divinatory officials of the Abe clan.[15] Even when new calendars were proposed, because the Abe clan controlled the *Yin-yang* office, they were not put into use.

It was not until AD 1685 that the *Hsuan-ming-li* was finally replaced by the *Jokyo* calendar. This was to be the first calendar to be compiled by a native astronomer,

---

[9] Wang (1988).

[10] *Nihongi*, 30, 19; trans. Aston (1972: II, 400). Note that the *I-feng-li* was another name for the *Lin-te-li*, used in China between AD 665 and AD 728.

[11] Quoted by Nakayama (1969: 70).

[12] Nakayama (1969: 70).

[13] Sugimoto & Swain (1978: 72).

[14] Sugimoto & Swain (1978: 128).

[15] Wang (1988).

Calendar	Date of Use in China	Date of Use in Japan
Yuan-chia-li	AD 445–509	c. AD 604–697
I-feng-li	AD 665–728	c. AD 692–763
Ta-yen-li	AD 726–762	AD 764–858
Wu-chih-li	AD 763–783	AD 858–862
Hsuan-ming-li	AD 822–892	AD 862–1685
Jokyo calendar	-	AD 1685–

Table 7.1: Calendars adopted in ancient and medieval Japan.

namely Yasui Harumi. It was based upon the methods of Kuo Shou-ching's *Shou-shih-li* as applied to Yasui Harumi's own observations.[16]

Table 7.1 lists the calendars that were used in Japan up until AD 1685, together with the date of their use, both in Japan and in China. As we have seen, in many cases the calendar had already ceased to be used in China before it was adopted in Japan.

7.3 Timed Eclipse Records in Japanese History

The earliest astronomical records in Japanese history come from the seventh century AD. These are contained in the *Nihongi* and include observations of the eclipsed sun or moon, occultations of stars and planets, and the appearance of comets or meteors.[17] The astronomical observations in the *Nihongi* are of a similar style to those contained in the annals of the emperors found in the Chinese dynastic histories, that is, they are reported in amongst other affairs of state in chronological order. For example, on 10 April 628 AD the sun was seen to be totally eclipsed:

> "Summer, 5th month. Flies gathered together in great numbers. They clustered together for ten rods, and floated away in the air across the Shinano pass with a sound like thunder. They reached as far east as the province of Kamitsuke, and then spontaneously dispersed. 36th year, Spring, 2nd month, 27th day. The Empress took to her sick bed. 3rd month, 2nd day. There was a total eclipse of the sun. 6th day. The Empress's illness became very grave, and (death) was unmistakably near..."
>
> [*Nihongi*, 22, 40-41; trans. Aston (1972: II, 155)]

In common with all of the other observations in the *Nihongi*, this record is not very detailed. No times are ever recorded for any of the eclipse records in this work.

From the eighth century AD, many hundreds of astronomical records are preserved. Unlike China, where the dynastic histories contain the vast majority of astronomical reports, Japanese astronomical records are found in many works, ranging from privately and officially compiled histories, to diaries and temple records. Kanda (1935) has made a detailed search through these sources up to AD 1600 and collected

[16]For a detailed discussion of the *Jokyo* calendar, see Nakayama (1969: 116–152).

[17]Knobel (1905); Stephenson (1968).

all of the of the available astronomical records together, preserving the original language (Chinese) of the texts. From around the end of the ninth century AD, a large number of the records of solar and lunar eclipses are preserved. These often contain details of the eclipses as calculated using the *Hsuan-ming-li*, including the expected times of the beginning, middle, and end of the eclipses, given in double hours, marks and *fen*. I should note, however, that many of these records do not explicitly mention that these times are calculated, rather than observed.

Most of the records are extremely formulaic, giving the date of an eclipse possibility, the (predicted) magnitude of the eclipse, and the (predicted) times of the eclipse phases. The record of the eclipse on 30 January 1097 AD is typical:

> "1st month, 15th day, *keng-tzu*, full moon. At dusk there was an eclipse of the moon. It was a little more than $6\frac{1}{2}$ fifteenths (eclipsed). The loss (should have) began at 2 marks and 10 *fen* in the hour of *hsu*, as the calculated time (of the middle of the eclipse) was at 4 marks and 10 *fen* in the hour of *hsu*, and (the moon should have) returned to fullness at 2 marks and 5 *fen* in the hour of *hai*. The moon (should have) moved within *hsing*."
>
> [*Shen-yu-chi*]

In these cases, it is not possible to tell that these times are predicted and not observed by simply reading the record. However, as I will discuss below, some of the records relate to dates of eclipses that were not visible in Japan, and so clearly must be predictions.

In a small number of cases, however, it is made explicit that the times relate to the predicted times of the eclipse. The solar eclipse on 21 February 1319 AD provides a good example of such a case:

> "2nd month, new moon, *ting-hai*. The previous night heavy rain fell and a great wind blew. At dawn the wind and rain stopped and the sky cleared. That day the sun was $\frac{14}{15}$ eclipsed. The loss (should have) began at 4 marks and 17 *fen* in the hour of *mao*, as the calculated time (of the middle of the eclipse) was at 4 marks and 22 *fen* in the hour of *ch'en*, and (the sun should have) returned to fullness at 4 marks and 27 *fen* in the hour of *ssu*, according to the honourable Doctors of Astronomy. Investigating at that time (it was seen that) the loss began during the hour of *ch'en* and it returned to fullness during the hour of *wu*. The calendar give the eclipse as $\frac{14}{15}$, but there was only $\frac{7}{15}$ eclipsed."
>
> [*Hua-yuan Yuan-ch'en-chi*]

It should be noted at this point that in none of the records is the time when the eclipse was actually observed ever reported to better than the nearest double-hour. In this example, the observed time is considerably later than the predicted time of the eclipse. Using the *Hsuan-ming-li*, it was calculated that the eclipse should begin at a local time of 6.00 hours, reach its maximum at 8.01 hours and end at 10.02 hours. It was observed, however, to begin sometime between 7.00 and 9.00 hours, and end between 11.00 and 13.00 hours. It is perhaps surprising that discrepancies such as these do not seem to have concerned the Japanese astronomers.

	Predicted Time (hours)			Computed Time (hours)		
Date	1st	M	4th	1st	M	4th
937 Aug 24	21.24	-	1.72	22.84	-	2.34
938 Feb 17	17.48	-	22.84	19.24	-	2.34
939 Jan 18	21.00	21.24	21.48	21.19	21.37	21.54
982 Mar 13	-	21.72	23.48	-	-	-
1023 Dec 29	-	3.48	5.72	-	5.00	6.89
1027 Oct 18	11.72	-	17.24	14.41	–	18.11
1028 Oct 6	17.72	18.68	19.72	19.92	20.78	21.63
1078 Jan 30	1.50	3.73	5.97	3.25	5.15	7.05
1080 Nov 29	1.96	-	5.48	5.16	-	8.11
1081 Nov 19	11.48	13.72	15.96	15.20	17.07	18.93
1092 Apr 24	21.25	23.50	1.74	22.66	0.43	2.20
1093 Apr 14	13.48	15.24	15.97	14.96	16.22	17.47
1097 Jan 30	19.50	19.96	21.49	19.68	20.77	21.85
1111 Oct 18	23.99	1.75	3.74	2.18	3.56	4.94
1118 Jun 5	19.28	21.04	21.98	20.83	22.47	0.11
1127 May 27	19.24	19.96	21.24	21.60	22.52	23.43
1133 Aug 17	21.74	-	1.01	0.59	-	2.51
1155 Jun 17	3.51	3.75	3.99	5.72	6.15	6.58
1161 Aug 7	1.13	3.74	6.28	2.50	4.46	6.41
1167 Apr 6	15.17	16.63	18.35	15.84	17.02	18.19
1169 Mar 15	3.50	4.68	5.72	4.25	5.58	6.91
1203 Apr 28	3.48	5.24	6.68	6.65	7.33	15.03
1203 Oct 22	13.24	14.68	15.24	15.03	16.21	17.39
1210 Dec 2	16.92	-	19.24	18.62	-	20.95
1212 Nov 10	22.71	0.23	1.56	0.53	1.99	3.44
1221 Nov 1	21.43	22.60	23.80	23.60	0.75	1.89
1222 Apr 27	15.09	17.53	20.06	15.78	17.63	19.47
1226 Feb 14	6.46	8.96	11.42	8.81	10.73	12.64
1231 May 17	2.68	-	3.24	-	-	-
1233 Mar 27	18.54	20.73	22.93	19.33	21.90	22.86
1240 May 7	0.41	2.83	5.31	22.91	0.72	2.52
1244 Feb 25	14.99	-	19.98	16.63	-	20.47
1245 Feb 13	17.41	19.03	20.34	18.01	19.34	20.68
1245 Aug 9	17.24	18.44	19.60	18.52	19.88	21.24
1283 Jul 11	15.72	17.72	19.48	17.06	18.86	20.66
1287 Apr 29	17.96	20.20	22.20	18.97	20.70	22.42
1294 Jun 9	19.72	21.72	23.72	20.07	21.68	23.28
1308 Sep 1	16.68	-	-	18.06	-	-
1339 Jan 26	12.89	14.27	15.63	18.86	20.32	21.78
1339 Jul 21	18.87	20.34	21.80	18.86	20.32	21.78
1340 Jun 10	0.10	0.33	0.57	-	-	-
1340 Dec 5	5.30	6.50	7.62	-	-	-
1345 Sep 12	18.11	20.53	22.96	20.27	22.12	23.96
1346 Mar 8	15.31	16.80	18.31	15.68	17.26	18.83
1346 Sep 1	23.48	1.29	3.10	0.28	1.92	3.56
1349 Jul 1	3.42	5.65	7.88	7.25	9.04	10.82
1355 Feb 27	14.46	15.59	16.75	15.31	16.3	17.29
1355 Aug 23	19.95	20.83	22.03	20.72	21.31	21.90
1356 Feb 17	4.20	6.44	8.68	6.49	8.23	9.97
1359 Jun 11	0.20	1.72	3.72	1.10	2.64	4.17
1359 Dec 5	14.20	16.68	19.96	14.63	16.45	18.27
1363 Mar 30	12.68	115.26	17.96	12.33	14.09	15.95
1364 Mar 18	23.12	0.76	2.37	23.77	1.32	2.94
1364 Sep 12	6.20	8.01	9.77	7.37	9.08	10.78
1367 Jan 16	1.74	4.23	6.73	2.81	4.71	6.60
1367 Jul 12	12.20	14.44	16.68	14.68	16.46	18.24

Table 7.2: Lunar eclipse predictions made with the *Hsuan-ming-li*.

	Predicted Time (hours)			Computed Time (hours)		
Date	1st	M	4th	1st	M	4th
1374 Feb 27	14.47	16.62	18.82	14.96	16.70	18.44
1393 Feb 27	15.18	16.68	18.21	14.23	15.81	17.39
1397 Dec 4	5.41	5.62	5.87	-	-	-
1401 Mar 30	8.20	-	8.68	-	-	-
1401 Sep 22	23.27	-	0.38	-	-	-
1404 Jul 22	22.36	23.72	1.04	23.18	0.68	2.18
1407 May 22	5.03	-	-	8.42	-	-
1411 Sep 2	-	2.48	3.06	-	1.85	3.57
1423 Jan 26	5.19	5.49	5.75	-	-	-
1428 Sep 23	1.57	-	5.57	0.93	-	4.42
1432 Jun 13	17.00	19.58	0.00	19.02	20.89	22.76
1438 Mar 12	3.72	-	-	5.58	-	-
1464 Oct 15	16.68	19.08	21.37	16.44	18.10	19.79
1478 Jan 18	0.79	1.72	2.35	2.02	2.34	2.66
1484 Oct 4	21.33	21.87	22.45	23.17	0.01	0.85
1485 Mar 1	22.14	23.17	0.27	-	-	-
1526 Jun 24	17.49	-	21.96	19.88	-	23.41

Table 7.2 (cont.): Lunar eclipse predictions made with the *Hsuan-ming-li*.

Since the majority of the Japanese records follow the style of the first example given above, I have chosen not to give full translations of all of the records, but simply to summarize the predicted times in Tables 7.2 and 7.3. These tables list, respectively, all of the lunar and solar eclipse times predicted by the Japanese astronomers before AD 1600, together with the contact times of the eclipses as deduced from modern computations. In AD 794, the Imperial Residence was transfered to Heian, later to be renamed Kyoto (latitude 35.03°, longitude −135.75°). Kyoto remained the capital until the second half of the nineteenth century AD when Tokyo, the present day capital, was established. Although an earlier observatory had been built at Asuka during the seventh century AD,[18] the Japanese astronomers who made the eclipse predictions and observations were probably based in Kyoto. Most of the other major Japanese settlements, for example Kamakura, were military rather than cultural centres. Accordingly, Kyoto has been taken as the place of observation when making modern computations of the circumstances of the eclipses.

Many of the eclipses predicted by the Japanese astronomers were not actually visible in Japan. Usually this was because the eclipse occurred when the luminary was below the horizon; that is, a lunar eclipse was expected to occur during the hours of daylight, or a solar eclipse during the night. Nine of the lunar eclipses listed in Table 7.2 actually refer to occasions when the moon just missed the Earth's umbral shadow and a large penumbral eclipse occurred instead. In these cases, no computed times are given in the table, and the eclipse has been omitted from further analysis. Five of the solar eclipse predictions listed in Table 7.3 relate to eclipses that would have passed completely to either the north or the south of Kyoto. In these cases, only the time of maximum phase, which corresponds to the moment when the eclipse made

[18] Kuniji (1979).

	Predicted Time (hours)			Computed Time (hours)		
Date	1st	M	4th	1st	M	4th
875 Dec 2	-	1.72	-	-	4.10	-
877 May 17	-	1.24	-	-	3.36	-
975 Aug 10	5.25	7.48	9.00	6.83	7.89	9.55
982 Mar 28	7.72	9.24	9.73	9.29	10.16	11.07
1021 Aug 11	9.96	11.48	-	12.88	14.25	-
1029 Sep 11	5.25	7.24	9.24	6.72	7.71	8.80
1080 Dec 14	9.24	9.98	11.74	10.54	12.43	14.22
1085 Feb 27	3.25	3.99	5.75	4.60	5.06	5.53
1100 May 11	8.02	9.51	11.00	10.60	11.24	11.90
1106 Dec 27	13.56	13.81	14.02	13.48	14.36	15.18
1118 May 22	15.72	17.24	17.72	18.99	19.81	20.58
1143 Jan 18	-	5.24	6.20	-	5.61	-
1147 Oct 26	15.99	17.73	19.48	20.03	20.91	21.73
1148 Apr 20	11.98	13.06	13.77	14.82	15.96	17.00
1149 Apr 10	3.48	5.24	7.24	5.31	6.10	6.94
1187 Sep 4	17.24	17.72	-	20.39	21.12	-
1203 May 13	3.48	4.92	-	4.55	5.13	-
1210 Dec 18	5.24	-	7.96	6.38	-	8.13
1230 May 14	12.92	13.24	13.72	-	3.05	-
1245 Jul 25	16.11	17.38	19.24	16.27	17.43	18.47
1246 Jan 19	14.44	-	16.20	16.91	-	18.57
1267 May 25	15.72	-	-	18.57	-	-
1304 Nov 28	4.92	6.20	7.72	6.63	7.61	8.69
1307 Apr 3	16.34	18.16	20.01	-	20.31	-
1319 Feb 21	6.00	8.01	10.02	8.12	9.52	11.05
1346 Feb 22	11.49	13.02	14.53	13.26	14.44	15.52
1361 May 5	15.41	17.01	18.55	18.65	19.47	20.24
1374 Mar 14	8.01	8.39	8.81	5.54	8.60	9.40
1383 Aug 29	7.72	-	8.68	7.71	-	8.31
1397 May 27	4.87	8.68	8.47	6.70	7.70	8.81
1401 Mar 15	9.84	10.37	10.94	10.40	12.04	1.66
1484 Sep 20	9.03	11.12	13.48	7.67	8.99	10.45
1527 May 30	10.13	11.17	12.35	10.63	10.93	11.24

Table 7.3: Solar eclipse predictions made with the *Hsuan-ming-li* in Japan.

its closest approach to Kyoto, is given in the table.

7.4 Accuracy of the Predicted Times

As I have noted above, when the Japanese records contain both a predicted time of an eclipse, and the rough timings of when it was actually seen, there is often a considerable difference between the two. Figures 7.1 and 7.2 show the difference between the precise times of the eclipse contacts as recorded by the Japanese astronomers and those given by modern computations, for the lunar and solar eclipses respectively. From these figures, it is clear that there is also a systematic error between the computed times and these times. This mean (systematic) error is shown as the dotted line in the two figures. It is equal to +0.97 hours for the lunar eclipses and +1.05 hours for the solar eclipses. However, as there is a considerable scatter in both sets of data, the difference between the two systematic errors is not significant. All that may be said is that in each case there is a systematic error making these times early by about one hour. This immediately indicates that these times were predicted rather than observed by the Japanese astronomers, confirming my earlier supposition. They were presumably made using the Chinese *Hsuan-ming-li*.

It is now natural to ask what would have caused this systematic error in the predicted eclipse times. The most plausible answer to this problem is in a failure of the Japanese astronomers to correct for the differing locations of China and Japan. As noted in Section 6.4 above, by the time of the *Hsuan-ming-li* it was known that the circumstances of eclipses vary depending on the location of the observer. However, the early attempts made to correct for this were very crude, being hindered as they were by the lack of the notion of a spherical Earth in Chinese cosmology. Today, we know that the effect of moving east is to make the local time of a simultaneous event later in the day by 1 hour for every 15 degrees of longitude.[19] Lunar eclipses are such events, and so as the longitude of Yang-cheng, the meridian of the *Hsuan-ming-li*, is about 113° east, and that of Kyoto is about 136° east, the local time of a lunar eclipse in Kyoto is about $1\frac{1}{2}$ hours greater than that at Yang-cheng. Solar eclipses, however, are not simultaneous events because of the effects of lunar parallax. A solar eclipse can appear between about 1.1 and 1.9 hours later in Japan than China. However, as the distribution of solar eclipse times is random, the average difference in time between a solar eclipse at Kyoto and Yang-cheng evens out to about $1\frac{1}{2}$ hours.

It is possible to determine the accuracy of the *Hsuan-ming-li* in predicting eclipses by removing the systematic error from the predicted times, and taking the modulus of the result. The accuracy of the *Hsuan-ming-li* in predicting lunar and solar eclipse times is shown in Figures 7.3 and 7.4 respectively. For the lunar eclipses, the mean accuracy of the predictions is 0.98 hours. In predicting solar eclipse times, however, the mean accuracy, is 1.18 hours.

[19] By a simultaneous event, I mean an event that is observed at the same moment of an absolute time scale, such as is given by an atomic clock, no matter where they are observed from.

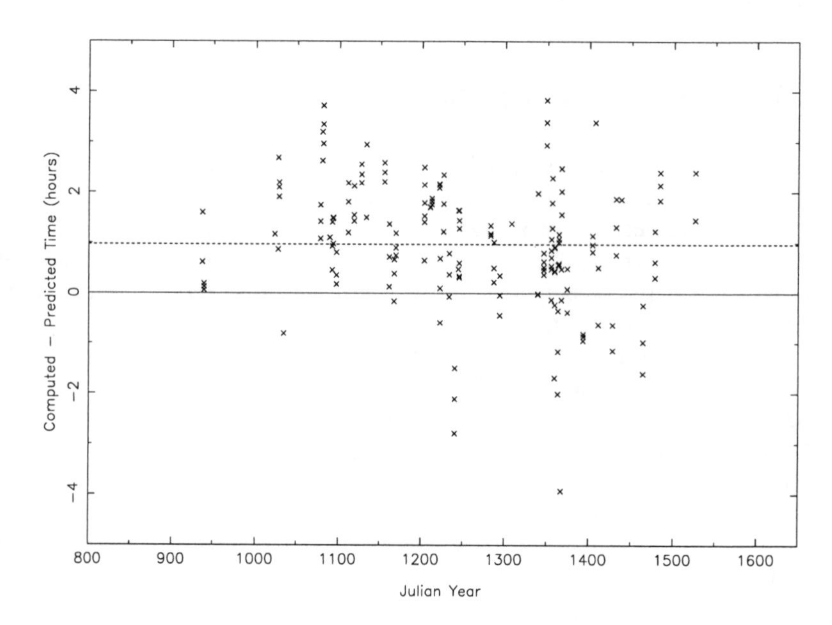

Figure 7.1: The error in the time of the lunar eclipse predictions made using the *Hsuan-ming-li*.

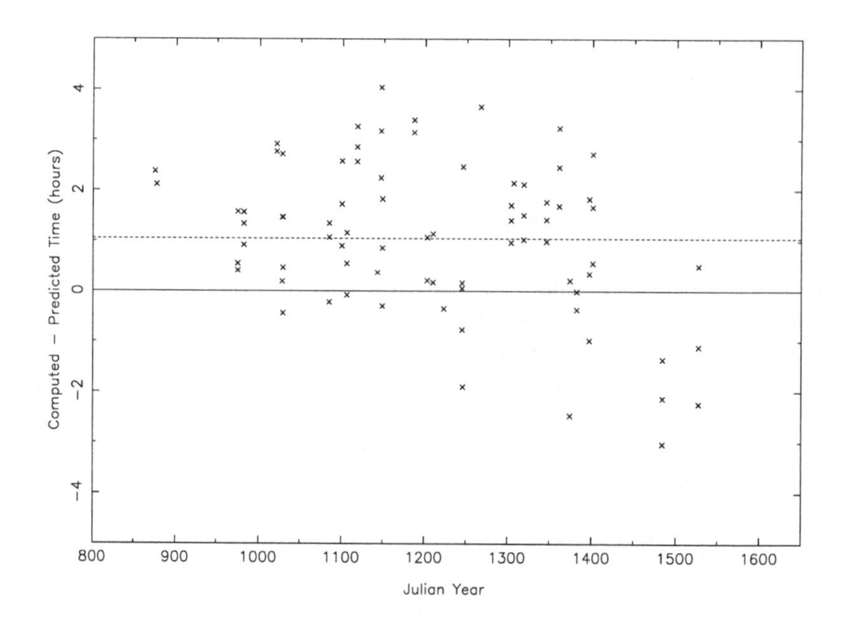

Figure 7.2: The error in the time of the solar eclipse predictions made using the *Hsuan-ming-li*.

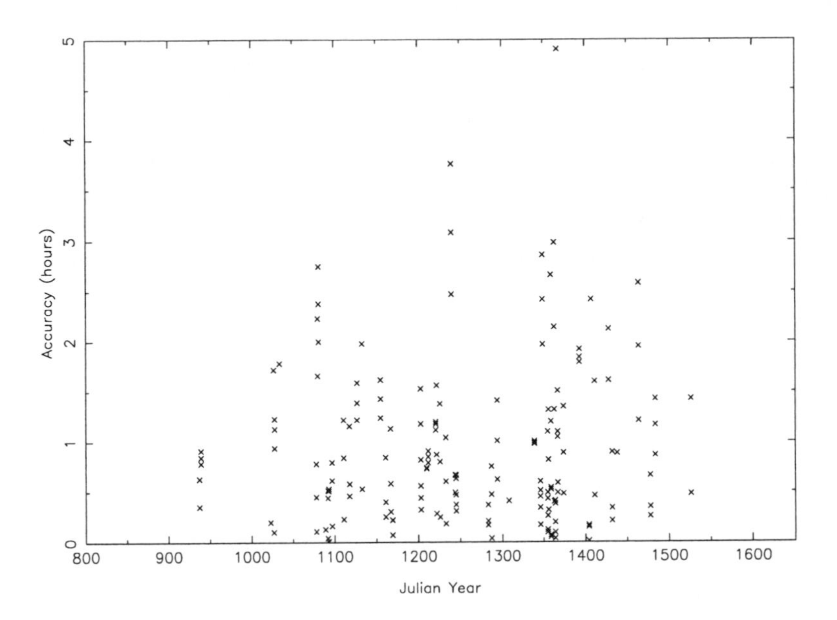

Figure 7.3: The accuracy of the *Hsuan-ming-li* in predicting the time of a lunar eclipse contact.

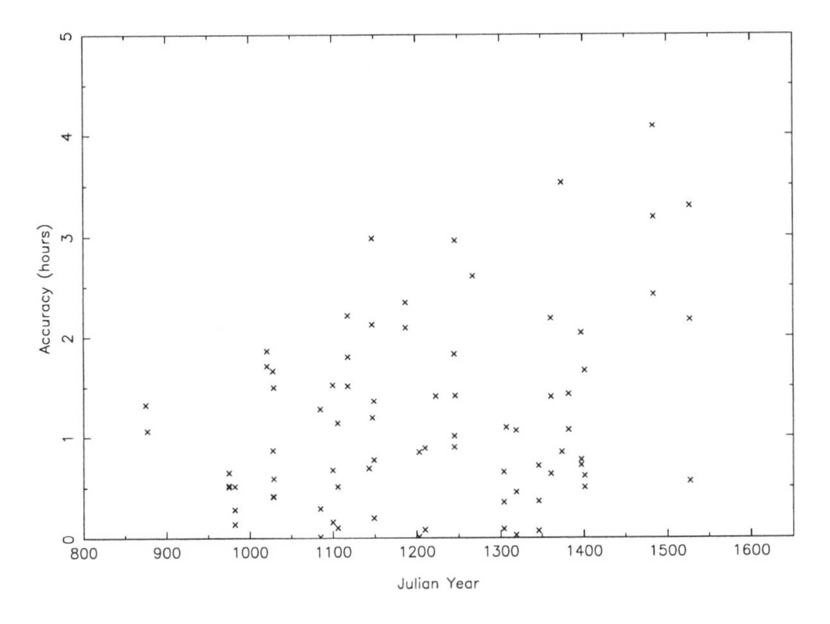

Figure 7.4: The accuracy of the *Hsuan-ming-li* in predicting the time of a solar eclipse contact.

From Figures 7.3 and 7.4, it is evident that the longer the calendar was in use, the less accurate it became. This is in contrast to the later *Ta-ming-li* and *Shou-shih-li* calendars, discussed in Section 6.7.2 above, which do not show any appreciable decrease in accuracy as they were used to make predictions at periods further removed from their epoch. As might be expected, the decrease in accuracy of the *Hsuan-ming-li* occurs at a faster rate for solar eclipses than it does for lunar eclipses. In the former case, this corresponds to about 0.17 hours per century, and in the latter to about 0.04 hours per century. This decrease in accuracy over time is largely caused by small errors in the parameters used in the calendar increasing cumulatively over time at an approximately linear rate.

Part IV

Conclusions

Chapter 8

Discussion

8.1 Historical Implications

In Chapters 2 to 7 of this work, I have discussed the eclipse records preserved in the history of each of the following cultures: Mesopotamia (principally Babylon), Ancient Europe, the Islamic Near East, and Late Medieval and Renaissance Europe, which form what I have termed the Western Heritage; and China and Japan, which form what I have termed the Eastern Heritage. In the present section I shall review the main findings of these chapters and their relevance to the history of astronomy. Some of the results of this work also have implications for present-day studies of the long-term changes in the Earth's rate of rotation. I shall discuss these in Section 8.2 below.

The main purpose of this work has been to collect together all known records of observations and predictions of eclipse times made by astronomers in the pre-telescopic period, and to obtain some understanding of the accuracy of these times. As I have shown in the preceding chapters, there exist marked differences between the methods used to time eclipses by astronomers in various cultures. For example, the Chinese astronomers seem to have used clepsydras almost exclusively in their observations, whereas the Islamic astronomers of the Near East determined the time by measuring the altitude of either the eclipsed luminary or a clock-star. Nevertheless, all the astronomers seem to have been able to measure the time of an eclipse to within about half an hour, which is no small achievement.

Figure 8.1 shows schematically the typical accuracy of each set of timed eclipse observations. Clearly there is a general trend of improvement in accuracy down the centuries. This is shown by both the observations I have grouped into the Western Heritage, and those in the Eastern Heritage. The measurements made by the Babylonian, Greek, and early Chinese astronomers were all made using fairly primitive clepsydras. It is unsurprising, therefore, that these are the least accurate of the observations. Much more accurate are the timings made by the Islamic astronomers of the Near East and the astronomers of Late Medieval and Renaissance Europe. These timings were made by either an altitude measurement or, occasionally during the Renaissance, a mechanical clock. However, it is interesting to note that, around the beginning of the twelfth century AD, the Chinese astronomers had developed water-powered clocks that were of comparable accuracy to these later methods used in the west.

It would seem that the optimum accuracy in timing eclipses that could be achieved by astronomers in the pre-telescopic period was about 0.1 hours or 6 minutes. Indeed, even after the invention of the telescope at the start of the seventeenth century AD, European astronomers were not able to determine the time of eclipses significantly more accurately until after about AD 1650.

Figure 8.2 shows schematically the accuracy of the methods used to predict the time of an eclipse in the different cultures.[1] It is interesting to contrast the lack of improvement in the accuracy of the Babylonian predictions down the centuries with the definite trend of improvement shown by the Chinese predictions. By the end of the first millennium AD, the Chinese predictions had reached a comparable level of accuracy to those made in the Islamic Near East and Europe. Unfortunately, there are no records of Chinese eclipse predictions from after the thirteenth century AD. However, as the calendar system that was used for making these predictions remained essentially unchanged from AD 1280 to the time of the Jesuits (c. AD 1660), it would be expected that the accuracy of the predictions would remain at a more or less stable level over this period.

It is also noticeable that there was not necessarily a straightforward improvement in the accuracy of the predictions made by the astronomers of the Near East and Europe. The predictions made by al-Battānī are significantly less accurate than those made by his contemporaries; however, this is unsurprising since he used Ptolemy's tables for calculating his eclipses, rather than more recently compiled tables as used by the other astronomers. Similarly, the predictions made by Regiomontanus and Walther in the fifteenth century AD are significantly less accurate than those made by earlier European astronomers. All the European predictions, except those made by Levi ben Gerson, were made with the *Alfonsine Tables*. However, it is possible that *Alfonsine Tables* which circulated in the fourteenth century AD were not the same as those that had been available in the thirteenth century AD. Furthermore, it may be that the later astronomers used inaccurate meridians when making their calculations.

8.2 Geophysical Implications

One important byproduct of the present study is that it provides useful information on the reliability of investigations into the long-term variations in the Earth's rate of rotation. Stephenson & Morrison (1995) have recently made an extensive study of these changes in Earth's rotation based largely upon records of observations of historical

[1] Note that the Japanese predictions have been incorporated into the Chinese data since they were made using a Chinese calendar. See Section 6.9 above.

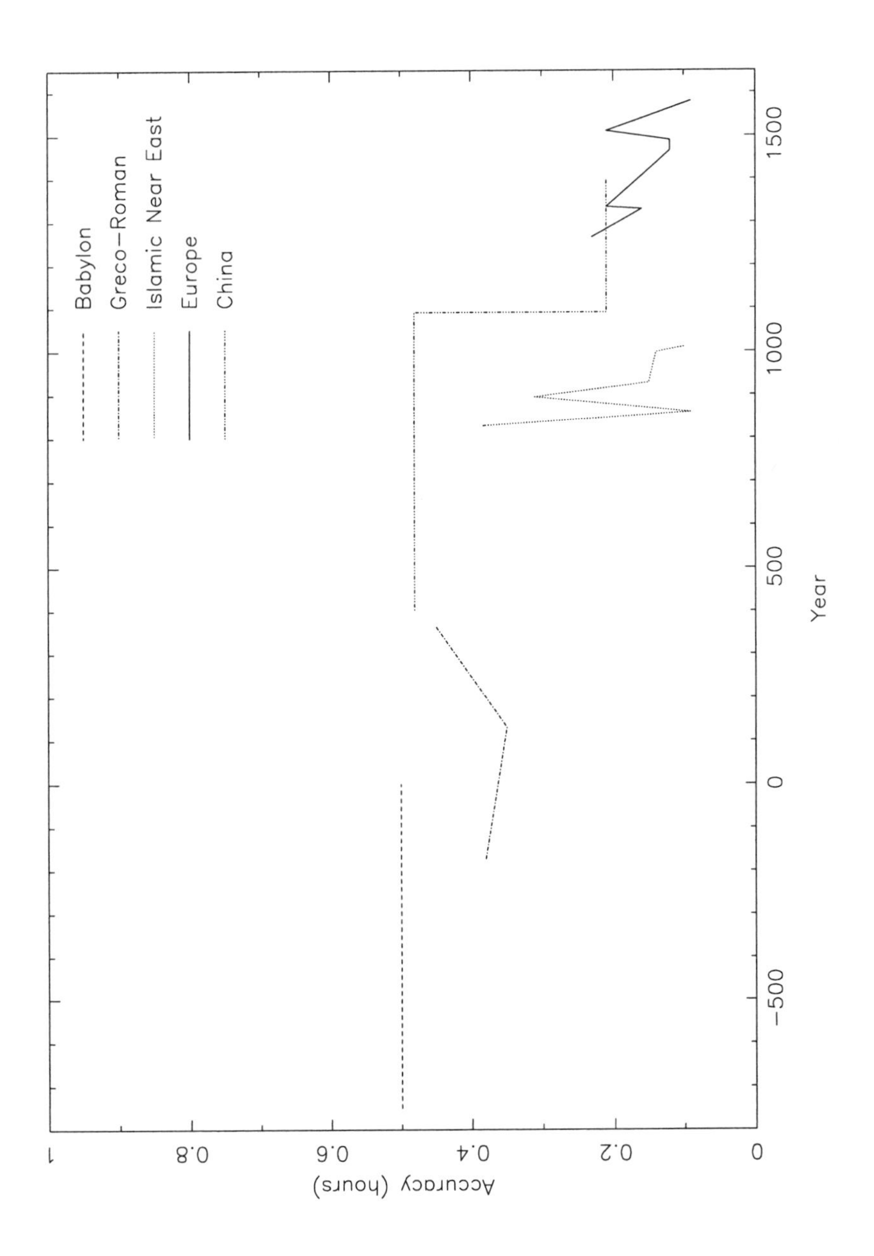

Figure 8.1: Schematic representation of the accuracy of observation of the time of an eclipse in different parts of the world.

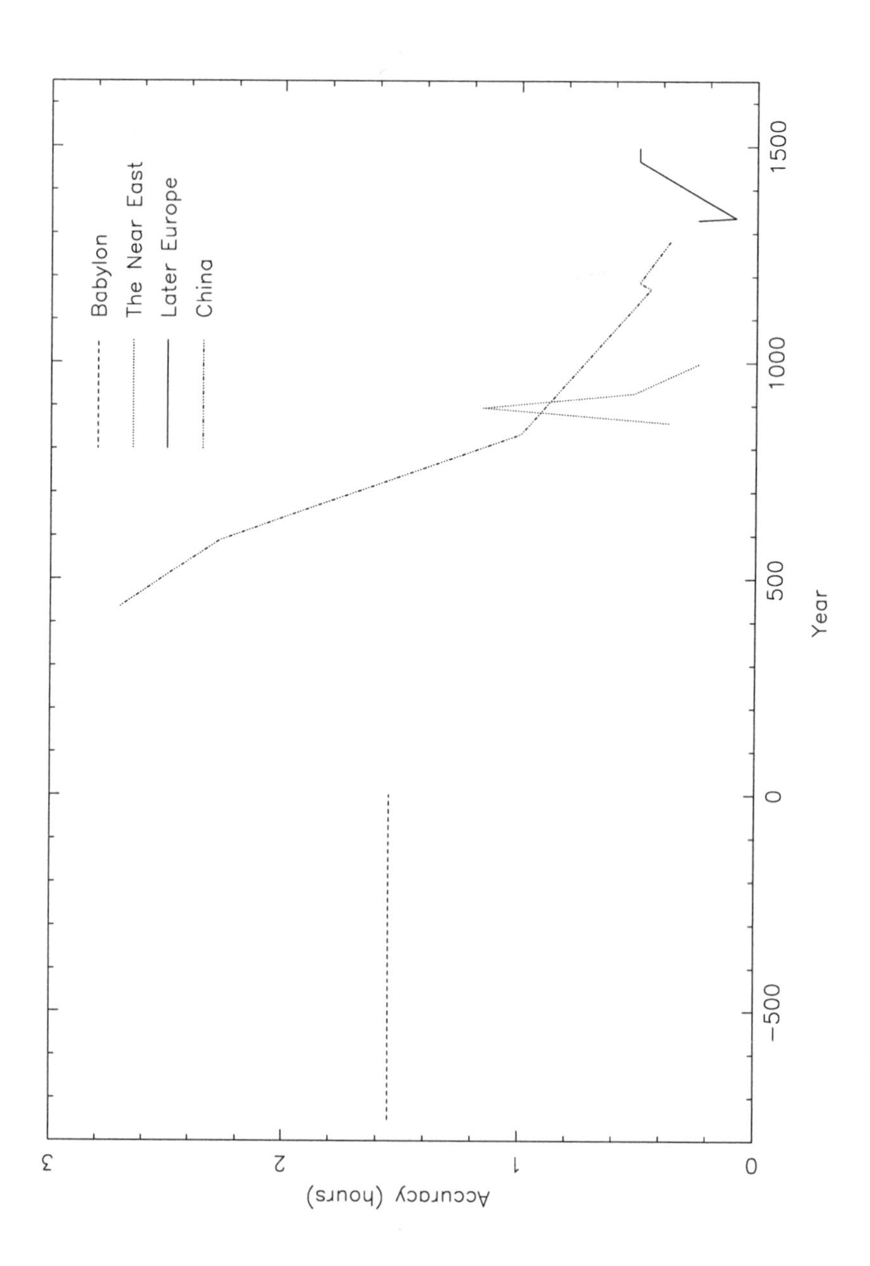

Figure 8.2: Schematic representation of the accuracy of prediction of the time of an eclipse in different parts of the world.

eclipses. In contrast to earlier studies by Fotheringham (1920b) and others, Stephenson & Morrison (1995) have not only used observations of total solar eclipses, but also a large number of eclipse timings made by early astronomers. However, some authors, for example Cohen & Newton (1983) and Rochberg-Halton (1989b), have raised the question: how do we know that these times were measured, and not predicted, by the early astronomers? The present study provides an answer: the times were indeed measured because, in general, the preserved observed times are significantly more accurate than the times that are known to have been predicted. Furthermore, the nature of the errors of some of the observed times — such as the existence of random clock-drifts in the Babylonian timings — are characteristic of times that were indeed measured.

Another source of concern regarding the use of timed eclipse observations by Stephenson & Morrison (1995) is the small number of group of timings from China that give radically different times for the same eclipse in two sources. As I have shown in Section 6.7.3, however, this can be explained by the historical facts. At this time there were two observatories functioning in the Chinese capital, each perhaps using a different system of time measurement, and so it is not surprising that two different accounts of the same eclipse are preserved.

Finally, this study has uncovered a number of eclipse observations that were not available to Stephenson & Morrison (1995). It is hoped that these will prove to be of use in future studies of the Earth's rotation.

8.3 Concluding Remarks

The present study has demonstrated some of the ways in which elements of history and astronomy can be used together for their mutual benefit. By making use of modern computations of the circumstances of eclipses in the past it has been possible to gain a greater understanding of the ways in which eclipses were observed and predicted by early astronomers. Furthermore, by using modern computations of eclipses and planetary and stellar visibilities, a number of astronomical records have been dated. This is not only interesting in providing information regarding the dates of the individual records, but also in allowing a greater overview of the periods in which astronomers were active in various cultures. Finally, the study of early astronomical records is of reciprocal use in modern science by providing long-baseline data for investigations into the variations in the Earth's rate of rotation.

This study also enhances our knowledge of the interaction between observation and theory in early astronomy. Many previous studies made in the history of astronomy have concentrated on the various aspects of astronomical theory reported in antiquity without considering the actual records of observations and predictions made by the astronomers themselves. For example, although many studies have been made of Babylonian mathematical astronomy, it has not been until recently that the actual observational records have been investigated. They have revealed that some of the predictions found in these texts were made using different schemes from those claimed in the mathematical astronomical texts. Similarly, I have shown that the many predictions of eclipses contained in the Babylonian Astronomical Diaries were not made using the schemes found in the ACT type texts. Only by considering both types of astronomy, observational and theoretical, in tandem may we gain a full insight into historical astronomical practice.

Part V

Appendices

Appendix A

Late Babylonian Eclipse Records

The following tables list all of the lunar and solar eclipse recorded in the Late Babylonian Astronomical Diaries (D), the Eclipse Texts (ET), the Normal Star Almanacs (NSA), the Almanacs (A), the Goal-Year Texts (GYT), and the Horoscopes (H). The tables are arranged chronologically by the date of the earliest record on the tablet. For convenience, tablets are listed both by their museum number and, where appropriate, by their position in the LBAT catalogue of Sachs (1955). Observed eclipses are denoted by an "O" in the fifth column; predicted eclipses by a "P." Eclipses with preserved times used in this study are indicated by a "Y" in the sixth column; those without a fully preserved time are indicated by an "N." Finally, the seventh column contains any comments on the record, including cross references to other records of the same eclipse.

In compiling these tables, I have used the edition of the Astronomical Diaries by Sachs & Hunger (1988, 1989, 1996), of the Eclipse Texts by Sachs & Hunger (1999), and of the Horoscopes by Rochberg (1998). For the Goal-Year Texts, I have mainly relied on Huber (1973), but have also searched though the copies published in Sachs (1955). In searching for records in the Almanacs and Normal Star Almanacs I have used a variety of sources including: Sachs (1955), Sachs (1976), Sachs & Walker (1984), Walker & Roughton (1999), Hunger (1999), Epping (1889), Kugler (1900, 1907, 1909, 1912, 1913, 1914, 1924), and Kugler & Schaumberger (1935). Many of these tablets have been read only from copies and have not been collated. Undoubtedly, when these texts are edited properly there will be a small number of records that I have missed.

Tablet	LBAT	Lines	Date of Eclipse	O/P	Times	Comments
BM 32315		II, 1 III, 4'	-651 Jul 2 -651 Dec 27	P P	N N	
VAT 4956	**160	Obv. 17	-567 Jul 4	P	N	
BM 34638	116	Obv. 16'	-382 Jun 29	P	N	
BM 35333	181	Rev. 8'–10'	-370 Nov 11	O	Y	cf. BM 32238 (ET)
BM 40122	*218	Obv. 6'–8'	-366 Mar 6	O/P	Y	
BM 35184	186	III, 5–6	-366 Aug 30	O	N	
BM 36913		Obv. 5'	-356 Feb 14	P	Y	
BM 46229	*189	Obv. 20–23	-345 Jan 14	O	Y	
BM 37231		Obv. 7'–8'	-333 May 29	O	N	
BM 36761 + 36390	*196	Obv. 3'	-330 Sep 20	O	N	
BM 34794 + 34919 + 34990 + 35071 + 35329	201f	Obv. 5'	-324 May 19	P	N	
BM 34093 + 35758	212 + 213	Obv. 9	-321 Apr 17	P	N	
Rm 803	217	Obv. 5'	-307 Jul 9	O	Y	
BM 34616 + 45901	220	Obv. 18' Rev. 17'	-302 Sep 11 -301 Mar 7	P ?	N N	
BM 32272 + 32288 + 32422 + 32501 + 32624	*224	17	-291 Aug 11	P	Y	
BM 55541	*228	Obv. 9'–10'	-286 May 20	O	N	
BM 41660	232	Rev. 5	-280 Jan 16	P	N	
BM 132279	239	Obv. 15	-277 Nov 3	P	N	cf. BM 48090 (ET)
BM 36710 + 92688 + 92689	242f	Rev. 6'–8'	-272 Feb 16	O	Y	
BM 45723	267	Obv. 3'	-248 Apr 19	P	Y	cf. BM 32154 (GYT)
BM 45949	268	Obv. 3–4	-247 Oct 3	O	N	
BM 46006	269	Rev. 2'	-246 Mar 29	O	N	
BM 32889 +	271f	Rev. 11'	-246 Sep 22	P	Y	

Table A.1: Lunar eclipses recorded in the Astronomical Diaries

Tablet	LBAT	Lines	Date of Eclipse	O/P	Times	Comments
32967 + 41614 + 41618						
BM 132276		Rev. 10'	-245 Sep 11	P	N	
BM 55511			-238 Apr 28	O	Y	
BM 33837	**284	Obv. 30	-232 Dec 14	P	Y	
BM 41647 + 34285 + Rm 728	285 + 550 + 287	Obv. 7	-230 Apr 30	P	N	
BM 41007		Rev. 9'	-225 Feb 6	P	Y	
BM 33655		Rev. 4–8	-225 Aug 2	O	Y	
BM 36402 + 36865	*294		-214 Dec 25	O	Y	
BM 40116		7'	-209 Apr 9	P	N	
BM 41477	320	Obv. 10'	-194 Jun 20	P	Y	cf. BM 42053 (ET) cf. BM 34048 (GYT)
BM 35331	324	Obv. 18'–19'	-193 Nov 5	O	Y	cf. BM 34236 (ET)
BM 32349 + 32428		Rev. 2'	-185 Dec 6	P	N	
BM 35330	336	Obv. 7'–9'	-184 Nov 24	O	Y	
BM 45613 + Rm 693 + 734	340 + 341	Rev. 4'–5'	-182 Oct 4	O	N	
BM 34806 + 35610 + 35812 + 55569	361 + 608 + 756 + *358	Rev. 11'	-172 Mar 21	P	Y	
BM 45654 + 45745	365	Rev. 12'–13'	-170 Aug 23	O/P	Y	
BM 40092		Rev. 10'	-169 Feb 16	P	Y	
BM 41462 + 41941	380 + 920	Rev. 20'–21'	-162 Mar 31	O	N	cf. BM 34037 (GYT)
BM 36763 + 36891		Rev. 6'	-161 Aug 14	P	Y	cf. BM 34615 (GYT)
BM 46003	385	Rev. 4'–5'	-159 Jan 26	O	Y	cf. BM 34236 (ET)
BM 45898	*389	Obv. 10	-158 Jul 12	P	Y	
BM 45774	390	Obv. 10	-158 Dec 7	P	N	
BM 45731 +	*393f	Obv. 6'–7'	-156 Nov 15	O	Y	

Table A.1: Lunar eclipses recorded in the Astronomical Diaries

Tablet	LBAT	Lines	Date of Eclipse	O/P	Times	Comments
45802						
BM 34632	400	Obv. 5'–8'	-149 Jul 3	O	Y	
BM 34609 + 34788 + 77617 + 78958	403f	Obv. 23'–24'	-144 Oct 3	O	N	
BM 41486	874	Rev. 6'	-141 Aug 3	P	N	
BM 45703 + 45741 + 45748	416f	Rev. 15'	-140 Jul 22	P	Y	
BM 34050	420	Obv. 12	-140 Dec 18	P	Y	
BM 45709	422	Obv. 7'	-137 May 22	P	Y	
BM 33820	*424	Rev. 3'	-137 Nov 15	P	N	
BM 41841 + 41881	430 + 431	Rev. 2'	-136 Oct 5	P	Y	
BM 45745	*429	Rev. 6'–7'	-135 Apr 1	O	Y	cf. BM 34034 (GYT)
BM 34174	533	Rev. 8'	-134 Mar 21	O	Y	
BM 34669 + 35740	432f	Rev. 16'–17'	-133 Mar 10	O	Y	
BM 32843	*435	Obv. 8'–9'	-133 Sep 3	O	Y	cf. BM 34051 (A)
BM 34175 + 34039	437 + 436	Obv. 7'	-132 Jan 29	P	Y	
BM 35411	728	6'–8'	-130 Jul 2	O	Y	
Rm 701 + BM 41478 + 41646	*296 + 446	Obv. 7'	-129 May 23	P	N	
BM 34112		Obv. 15'	-125 Mar 11	P	N	cf. BM 34949 (A)
BM 45850	*449	Rev 10'–11'	-124 Aug 24	O	N	
BM 33024 + 33045 + 45745	*452	Rev. 14–15	-123 Aug 13	O	Y	
BM 45947A		Rev. 6'	-122 Aug 2	O	N	cf. BM 33044 + ... (D) cf. BM 34033 (NSA)
BM 33044 + 33047		Obv. 8'–10'	-122 Aug 2	O	N	cf. BM 45947A (D) cf. BM 34033 (NSA)
BM 35759 + 45621		Obv. 3'–5'	-119 Jun 2	O	Y	cf. BM 45845 (ET)
BM 45762		Rev. 5'	-118 Oct 16	P	N	

Table A.1: Lunar eclipses recorded in the Astronomical Diaries

Tablet	LBAT	Lines	Date of Eclipse	O/P	Times	Comments
BM 46029 + 46035 + 46084	463	Obv. 14'	-111 Jul 2	P	N	cf. BM 46196 + ... (D) cf. BM 38212 (A)
BM 46196 + 45875	*465 + 464	Obv. 5'	-111 Jul 2	P	N	cf. BM 46029 + ... (D) cf. BM 38212 (A)
BM 35086 + 46149 + 77619	658	Obv. 5'–6'	-109 Nov 5	O	Y	
BM 40622	*469	7–8	-108 May 1	O	Y	
BM 45750 + 45983 + 45984	*472f	Rev. 8'	-106 Mar 11	P	Y	
BM 77670		5'–6'	-105 Feb 28	O	Y	
BM 34937 + 34957 + 35558 + 35662 + 35776 + 45647 + 45700+ 46033	477 + 479 + 480 + *481 + 627 + 777 + 909	Rev. 11'–12'	-105 Aug 24	O	Y	
BM 140677		Obv. 4–6 Obv. 11-12 Rev. 4–7	-98 Apr 11 -98 Oct 6 -97 Mar 31	O O O	N N N	
BM 41689	487	4'–5'	-96 Aug 14	P	N	
BM 45874	492	6'–8'	-95 Aug 4	O	Y	
BM 32884	*494	Rev. 3–4	-93 Jul 13	O	N	
BM 41529 + 41546 + 13227 + Böhl 1332	495f	Obv. 6'–7'	-90 Nov 5	O	N	
Rm 695 + Sp 172 + BM 41921	*504f	Rev. 18'–20	-86 Mar 1	O	Y	cf. BM 34963 + ... (ET)
BM 41018		Rev. 11'	-86 Aug 24	P	Y	
BM 46227		6'–8'	-80 Apr 21	O	Y	cf. BM 42145 (ET) cf. BM 42025 (H)
BM 45659 + 45685	515f	Rev. 5'	-76 Feb 9	P	Y	
BM 41559	881	3'–7'	-75 Jan 28	O	N	
BM 45625		16'–18'	-72 Nov 16	O	Y	
BM 34753	520	Obv. 7	-62 May 3	P	Y	

Table A.1: Lunar eclipses recorded in the Astronomical Diaries

Tablet	LBAT	Lines	Date of Eclipse	O/P	Times	Comments
BM 41985	1413	Obv. 2	-746 Feb 6	O	N	Date Uncertain
		Obv. 3–4	-746 Aug 2	O	N	Date Uncertain
		Obv. 5	-745 Jan 26	O	N	Date Uncertain
		Obv. 6	-745 Jul 22	O	N	Date Uncertain
		Obv. 7	-744 Jan 15	P	N	Date Uncertain
		Obv. 8	-744 Jul 11	P	N	Date Uncertain
BM 32238	*1414	Obv. I', 1'–3'	-730 Apr 9	P	Y	
		Obv. II', 1'–3'	-712 Apr 19	O	Y	Wrongly filed?
		Obv. III', 1'–3'	-694 May 1	O	D	
		Obv. IV', 1'	-676 May 11	P	N	Restoration
		Obv. V', 1'–3'	-658 May 22	P	N	
		Rev. I', 1'–2'	-388 Oct 31	P	N	
		Rev. II', 1'–3'	-370 May 17	O	Y	
		Rev. II', 4'–14'	-370 Nov 11	O	Y	cf. BM 35333 (D)
		Rev. III', 1'	-352 May 28	P	Y	
		Rev. III', 2'–10'	-352 Nov 22	O	Y	
		Rev. IV', 1'–2'	-334 Dec 3	P	Y	
		Rev. V', 1'–4'	-316 Jun 18	O	Y	
		Rev. V', 5'–12'	-316 Dec 13	O	Y	
BM 35115 +	1415 +	Obv. I', 1'	-702 Sep 23	P	N	
35789 +	1416 +	Obv. I', 2'–5'	-701 Mar 20	O	N	
45640	1417	Obv. II', 1–6	-685 Apr 22	O	Y	
		Obv. II', 8–9	-685 Oct 15	P	N	
		Obv. II', 1'–4'	-684 Oct 3	O	Y	
		Obv. II', 5'–6'	-683 Mar 30	P	N	
		Obv. III', 1–4	-667 May 2	P	Y	
		Obv. III', 5–6	-667 Oct 25	P	Y	
		Obv. III', 1'–3'	-666 Oct 15	O	N	
		Obv. III', 4'–6'	-665 Apr 10	O	N	
		Obv. IV', 1–4	-649 May 13	P	Y	
		Obv. IV', 5–8	-649 Nov 6	O	N	
		Obv. V', 1–6	-631 May 24	O	Y	
		Rev. I', 1'–2'	-414 Mar 26	P	Y	
		Rev. I', 3'–8'	-414 Sep 19	O	N	Wrongly filed?
		Rev. I', 9'–12'	-413 Mar 16	P	N	
		Rev. II', 1'	-397 Oct 12	P	N	
		Rev. II', 2'–8'	-396 Apr 5	O	Y	
		Rev. II', 9'–11'	-396 Sep 30	O	Y	Wrongly filed?
		Rev. II', 12'–13'	-395 Mar 26	P	Y	
		Rev. II', 14'–15'	-395 Sep 19	?	N	
		Rev. III', 1'–2'	-379 Oct 22	P	Y	
		Rev. III', 3'–9'	-378 Apr 17	O	Y	Wrongly filed?
		Rev. III', 10'–11'	-378 Oct 11	P	Y	
		Rev. III', 12'–20'	-377 Apr 6	O	Y	
		Rev. IV', 2'–5'	-359 Apr 17	O	N	
BM 38357		Obv. I, 4'–10'	-611 Nov 5	O	N	
		Obv. II, 7'–15'	-609 Sep 16	O	N	
		Rev. I, 1–9	-607 Aug 24	O	N	Date uncertain
BM 32234	*1419	Obv. I', 1'–8'	-608 Sep 4	O	N	
		Obv. II', 1'–2'	-590 Mar 22	O/P	Y	cf. BM 38462(ET)
		Obv. II', 3'–5'	-590 Sep 15	P	Y	cf. BM 38462 (ET)
		Obv. II', 6'–7'	-589 Mar 12	P	N	cf. BM 38462 (ET)
		Obv. III', 1'–8'	-572 Apr 2	O	Y	
		Obv. III', 9'–11'	-572 Sep 25	P	Y	
		Obv. III', 12'–13'	-571 Mar 22	?	N	

Table A.2: Lunar eclipses recorded in the Eclipse Texts

Tablet	LBAT	Lines	Date of Eclipse	O/P	Times	Comments
		Obv. IV', 2'–8'	-554 Oct 6	O	Y	
		Obv. IV', 9'	-553 Apr 3	?	N	
		Obv. V', 1'–3'	-536 Apr 23	O	N	
		Obv. V', 4'–10'	-536 Oct 17	O	Y	
		Rev. I', 1'–4'	-518 Oct 28	P	N	
		Rev. II', 1'–6'	-500 Nov 7	O	Y	
		Rev. III', 1'–6'	-482 Nov 19	O	Y	
		Rev. IV', 1'–3'	-464 Jun 5	O	Y	
		Rev. IV', 5'–9'	-464 Nov 29	O	N	
		Rev. V', 1'–2'	-446 Dec 11	?	N	
BM 38462	*1420	Obv. I, 1	-603 Jun 13	P	N	
		Obv. I, 2	-603 Dec 6	P	N	
		Obv. I, 3	-602 May 3	P	N	
		Obv. I, 4–6	-602 Oct 27	O	Y	
		Obv. I, 7	-601 Apr 22	P	N	
		Obv. I, 7	-601 Oct 17	P	N	
		Obv. I, 8–10	-600 Apr 11	O	Y	
		Obv. I, 11	-600 Oct 5	P	N	
		Obv. I, 12	-599 Mar 31	P	N	Restoration
		Obv. I, 12	-599 Sep 24	P	N	
		Obv. I, 13–15	-598 Feb 20	O	Y	
		Obv. II, 1–2	-594 Nov 27	O	N	
		Obv. II, 3–4	-593 May 23	O	Y	
		Obv. II, 4	-593 Nov 17	P	N	
		Obv. II, 5	-592 May 12	P	N	
		Obv. II, 5	-592 Nov 5	P	N	
		Obv. II, 6–7	-591 Apr 2	O	Y	
		Obv. II, 8	-591 Sep 26	P	N	
		Obv. II, 9	-590 Mar 22	O/P	N	cf. BM 32234 (ET)
		Obv. II, 10	-590 Sep 15	P	N	cf. BM 32234 (ET)
		Obv. II, 11	-589 Mar 12	P	N	Restoration. cf. LBAT *1419 (ET)
		Obv. II, 12	-589 Sep 4	P	N	
		Obv. II, 12	-588 Feb 29	P	N	
		Obv. II, 13	-588 Jul 25	P	N	
		Obv. II, 14–15	-587 Jan 19	O	Y	
		Obv. II, 16	-587 Jul 15	P	N	
		Obv. II, 17–18	-586 Jan 8	O	Y	
		Rev. III, 1'–2'	-579 Feb 19	O	N	
		Rev. III, 3'	-579 Aug 15	O	Y	
		Rev. III, 4'	-578 Feb 8	?	N	
		Rev. III, 5'	-578 Aug 4	P	N	
		Rev. III, 5'	-577 Jan 28	P	N	
		Rev. III, 6'	-577 Jun 25	P	N	
		Rev. III, 6'	-577 Dec 19	P	N	
		Rev. III, 7'–9'	-576 Jun 14	O	N	
		Rev. III, 10'–12'	-576 Dec 8	O	Y	
		Rev. III, 13'–14'	-575 Jun 3	O	Y	
BM 41536	1421	II', 1'	-562 Mar 13	?	N	
		II', 2'–4'	-562 Sep 5	O/P	Y	
		II', 5'–8'	-561 Mar 3	O	Y	
BM 36879		Obv. I, 13'	-528 Nov 17	O	Y	
		Rev. I, 1'	-527 Nov 6	P	N	
		Rev. I, 2'	-526 Apr 4	P	N	
		Rev. I, 3'	-526 Sep 27	P	N	
		Rev. I, 4'	-525 Mar 24	P	Y	
		Rev. I, 5'–11'	-525 Sep 17	O	Y	

Table A.2: Lunar eclipses recorded in the Eclipse Texts

Tablet	LBAT	Lines	Date of Eclipse	O/P	Times	Comments
BM 33066	**1477	Rev. 19–20	-522 Jul 16	?	Y	
		Rev. 21–22	-521 Jan 10	?	Y	
BM 34684 +	1426 +	Rev. I', 1'–3'	-441 Mar 25	O	N	
34787	1427	Rev. I', 6'	-440 Mar 13	P	N	
		Rev. I', 7'	-440 Aug 7	P	N	
		Rev. I', 8'–10'	-439 Feb 2	O	N	
		Rev. I', 11'	-439 Jul 28	P	N	
		Rev. II', 4'–5'	-423 Sep 28	O	Y	
		Rev. II', 6'	-422 Mar 25	P	N	Restoration
		Rev. II', 7'	-422 Aug 19	P	N	
		Rev. II', 8'	-421 Feb 13	P	N	
		Rev. II', 9'	-421 Aug 18	P	N	
		Rev. II', 10'–12'	-420 Feb 2	O	N	
		Obv. 1'	-409 Jun 28	P	Y	
		Obv. 3'–5'	-409 Dec 20	O	Y	
		Obv. 6'–7'	-408 Jun 16	O	N	
		Obv. 8'	-408 Nov 11	P	Y	
		Obv. 9'	-407 May 7	P	N	
		Obv. 10'–12'	-407 Oct 31	O	Y	
		Rev. III', 1–3	-406 Apr 26	P	N	
		Rev. III', 4–6	-406 Oct 21	O	Y	
		Rev. III', 7–8	-405 Apr 15	O	Y	
		Rev. III', 9–10	-405 Oct 10	O	Y	
CBS 11901	**1478	Rev. 4-5	-424 Oct 9	?	Y	From Nippur
BM 37088 +	*1429	I', 1'–2'	-382 Dec 23	P	N	
37652		I', 3'–4'	-381 Jun 18	O	N	
		I', 5'–6'	-381 Dec 12	O	N	
		II', 1'–3'	-363 Jan 2	O	N	
		II', 4'–5'	-363 Jun 29	O	Y	
		II', 6'–8'	-363 Dec 23	O	Y	
		II', 9'–10'	-362 Jun 18	O	Y	
		II', 11'–13'	-362 Dec 12	P	Y	
		II', 14'–15'	-361 Jun 7	P	N	
BM 36612 +		Rev. II, 1	-327 Jan 24	P	N	
37043 +		Rev. II, 2	-327 Jul 20	P	N	Restoration
37107		Rev. II, 3–8	-326 Jan 14	O	Y	
		Rev. II, 9–11	-326 Jul 9	O	N	
		Rev. II, 12–15	-325 Jan 3	O	N	
BM 41800	1452		-283 Mar 17	O	Y	
BM 48090	*1432	1'–3'	-279 Dec 24	P	N	
		4'–5'	-278 Jun 19	P	Y	
		6'–7'	-278 Nov 13	P	Y	
		8'–9'	-277 May 11	P	N	
		10'	-277 Nov 3	P	N	cf. BM 132279 (D)
BM 42053	1439	Obv. 1'–2'	-194 Jun 20	P	Y	cf. BM 41477 (D) cf. BM 34048 (GYT)
		Obv. 3'	-194 Nov 15	?	N	cf. BM 34236 (ET) cf. BM 34048 (GYT)
		Rev. 1–3	-177 Jul 13	O	N	
BM 34236	1436	Obv. II', 1	-194 Nov 15	P	Y	cf. BM 42053 (ET) cf. BM 34048 (GYT)

Table A.2: Lunar eclipses recorded in the Eclipse Texts

Tablet	LBAT	Lines	Date of Eclipse	O/P	Times	Comments
		Obv. II', 2–3	-193 May 11	P	Y	
		Obv. II', 4–6	-193 Nov 5	O	Y	cf. BM 35331 (D)
		Rev. I', 1'–3'	-159 Jan 26	O	Y	cf. BM 46003 (D)
		Rev. I', 4'	-159 Jul 23	P	N	
BM 33643	1437		-189 Feb 28	O	Y	
BM 41129	*1440		-153 Mar 21	O	Y	
BM 33982	1441		-128 Nov 5	O	Y	
BM 45845	1442		-119 Jun 2	O	Y	cf. BM 35759 + ... (D)
BM 34963 +	1334 +	Obv. IV, 6'–8'	-88 Sep 15	O	N	
35198 +	1435 +	Rev. VII, 17'–20'	-86 Mar 1	O	Y	cf Rm 695 + ... (D)
35238	1443					
BM 42145	1444	Obv.	-80 Apr 21	O	N	cf. BM 46227 (D)
						cf. BM 42025 (H)
BM 33562A	**1445		-79 Apr 11	O	Y	
BM 42073	1446		-79 Oct 5	O	Y	
BM 41565	1447		-66 Jan 19	O/P	Y	cf. BM 45628 (ET)
BM 45628	1448		-66 Jan 19	O/P	Y	cf. BM 41565 (ET)
BM 34940	1449		-65 Jan 8	O/P	Y	
BM 32845	*1450		-65 Dec 28	O/P	Y	

Table A.2: Lunar eclipses recorded in the Eclipse Texts

Tablet	LBAT	Lines	Date of Eclipse	O/P	Times	Comments
MM 86.11.369		Obv. 10'	-241 Jun 1	P	D	
BM 41634	1006	Rev. 16'	-217 Mar 9	P	N	
BM 41588	1008	Rev. 14'	-214 Jan 5	P	Y	
BM 45696	1020	Rev. 7'	-199 Mar 20	P	D	
BM 41599	1022	Obv. 3' Rev. 3'	-191 Apr 19 -190 Mar 10	P P	Y D	
BM 34076	1038	Obv. 14	-139 Jun 12	P	Y	
BM 34199	1039	Rev. 5'	-138 Nov 26	P	D	
BM 34056	1043	Obv. 18'	-132 Jul 24	P	N	cf. BM 33867 (A)
BM 33466 + 33448 + 46106 + 47727 + fragments	*1045 + *1046 + 1136	Obv. 8	-128 May 12	P	Y	cf. BM 35551 (A)
BM 34054	1047	Obv. 2'	-127 May 2	P	N	
BM 34078	1051	Rev 17	-122 Feb 7	P	N	
BM 34033	**1055	Obv. 28	-122 Aug 2	P	Y	cf. BM 45947A (D) cf. BM 33044 + ... (D)
		Rev. 53	-122 Dec 29	P	Y	
BM 34758	1057	Obv. 2'	-117 Apr 12	P	N	
BM 34032	**1059	Obv. 12 Rev. 44	-110 May 24 -110 Nov 16	P P	Y Y	cf. BM 35562 (A) cf. BM 35562 (A)

Table A.3: Lunar eclipses recorded in the Normal Star Almanacs

Tablet	LBAT	Lines	Date of Eclipse	O/P	Times	Comments
BM 34232	1122	Rev. 2	-183 Oct 15	P	N	
BM 33873	**1123	Obv. 1-2	-182 Apr 10	P	N	
BM 34345	1129	Obv. 3'	-151 Aug 23	P	N	
BM 35729	1130	Rev. 4'	-149 Dec 28	P	D	
BM 34051	1134	Obv. 12'–13'	-133 Sep 3	P	Y	cf. BM 32843 (D)
BM 33867	1135	Obv. 9	-132 Jul 24	P	N	cf. BM 34056 (NSA)
BM 46106	1136	Rev. 5'	-131 Jan 17	P	Y	
BM 35551	1137	Obv. 4	-128 May 12	P	N	cf. BM 35620 + ... (NS A)
BM 34949	1139	12'	-125 Mar 11	P	N	cf. BM 34112 (D)
BM 45716	*1147	Rev. 7'	-113 Jan 29	P	D	cf. BM 41301 (H)
BM 38212	*1150	Obv. 3'	-111 Jul 2	P	N	cf. BM 46029 + ... (D) cf. BM 46196 + ... (D)
BM 35562	1151	Obv. 5 Rev. 1	-110 May 24 -110 Nov 16	P P	N N	cf. BM 34032 (NSA) cf. BM 34032 (NSA)
BM 34042	1153	Obv. 7'	-102 Jun 23	P	N	cf. BM 34991 + ... (A)
BM 34991 + 35335	1154 + 1155	Obv. 7	-102 Jun 23	P	N	cf. BM 34042 (A)
BM 45698	1174	Obv. 7–8 Rev. 4	-75 Jul 24 -75 Dec 19	P P	Y N	cf. BM 35515 (H)
BM 33798	1182	Obv. 5 Rev. 1	-64 Jun 23 -64 Nov 17	P P	N N	
BM 35570 + 35602	1188 + 1189	Obv. 10	-11 Aug 5	P	N	
BM 35429	1195	Obv. 1 Obv. 14	-6 Apr 14 -6 Oct 8	P P	N N	 cf. BM 34659 (A)
BM 34659	1194	Obv. 2'	-6 Oct 8	P	N	cf. BM 35429 (A)
BM 34614	1193	Rev. 9	-5 Apr 4	P	N	

Table A.4: Lunar eclipses recorded in the Almanacs

Tablet	LBAT	Lines	Date of Eclipse	O/P	Times	Comments
BM 34759	1366	Rev. 3'–8'	-250 Dec 4	O	N	
BM 32154	**1216	Obv. 1–2	-248 Apr 19	P	Y	cf. BM 45723 (D)
		Obv. 5–6	-248 Oct 13	P	Y	
BM 32286	**1218	Rev. 3–5	-239 Nov 3	O	Y	
BM 32222	**1237	Rev. 48–53	-211 Apr 30	O	Y	
		Rev. 56–59	-211 Oct 24	O	Y	
BM 34048	1249	Rev. 8'–9'	-194 Jun 20	P	Y	cf. BM 41477 (D)
						cf. BM 42053 (ET)
		Rev. 10'–12'	-194 Nov 15	P	Y	cf. BM 34236 (ET)
						cf. BM 42053 (ET)
BM 34579	1251	Rev. 14–18	-189 Aug 23	O/P	Y	
		Rev. 23–26	-188 Feb 17	O	Y	
BM 33823	*1256	Rev.	-185 Jun 11	P	Y	
BM 34603	1263	Rev. 8'–10'	-169 Aug 13	P	Y	
		Rev. 11'–12'	-168 Jan 7	P	Y	
BM 34037	1264	Rev. 3'–9'	-162 Mar 31	O	Y	cf. BM 41462 + ... (D)
		Rev. 13'–15'	-162 Sep 23	P	Y	
		Rev. 15'–17'	-161 Feb 18	P	Y	
BM 34615	1266	Rev. 14–17	-161 Aug 14	P	Y	cf. BM 36763 + ... (D)
		Rev. 20–22	-160 Feb 7	P	Y	
BM 35787	1278	Rev. 3'–8'	-142 Feb 17	O	Y	
BM 34034	1285	Rev. 17–23	-135 Apr 1	O	Y	cf. BM 45745 (D)
		Rev. 29–35	-135 Sep 24	O/P	N	
BM 45909	1304	Rev. 3'	-40 Mar 2	O	Y	

Table A.5: Lunar eclipses recorded in the Goal-year Texts

Tablet	LBAT	Lines	Date of Eclipse	O/P	Times	Comments
BM 33382	*1459	Rev. 3–4	-287 Nov 23	?	N	
BM 36620	*1464	Rev. 3–4	-218 Mar 20	?	N	
BM 78089		Rev. 3–4	-125 Sep 5	P	N	
BM 33018	*1468	Rev. 1–3	-124 Aug 24	?	N	
BM 41301		Rev. 8'	-116 Sep 24	?	N	
		Rev 14'	-113 Jan 29	?	N	cf. BM 45716 (A)
BM 34003	1470	10–12	-87 Mar 11	?	N	
BM 42025 + 42164	1472 + 1474	6–7	-80 Apr 21	?	N	cf. BM 46227 (D) cf. BM 42145 (ET)
BM 35515	*1474	Rev. 3–5	-75 Jul 24	?	N	cf. BM 45698 (A)
BM 38104	*1475	Rev. 5–8	-68 Sep 3	O	N	

Table A.6: Lunar eclipses recorded in the Horoscopes

Tablet	LBAT	Lines	Date of Eclipse	O/P	Times	Comments
BM 47725	*168	II, 2'	-381 Jul 3	P	N	
BM 37097 + 37211		Rev. 4'	-368 Apr 11	O	Y	
BM 40122	*218	Rev. 2–3	-366 Mar 20	P	N	
BM 32149 + 32866 + 32252	*183f	III, 28'	-366 Sep 14	?	N	
BM 36913		Rev. 6'	-356 Feb 29	O	N	cf. BM 71537 (ET)
BM 46229	*189	Obv. 28	-345 Jan 29	P	N	
BM 40099	*195	Rev. 10	-332 Oct 27	P	Y	
BM 36761 + 36390	*196	Obv. 12'	-330 Oct 5	P	Y	
BM 45766	*208	Rev. 13'	-322 Oct 7	O	N	
Rm 796 + BM 32240 + 32430 + 32489	*210 + *211	Rev. 32	-321 Apr 2	P	N	
BM 34093 + 35758	212 213	Rev. 21'	-321 Sep 26	O	Y	
BM 34616 + 45901	220	Obv. 21'	-302 Sep 25	P	Y	
BM 32272 + 32288 + 32422 + 32501 + 32624	*224		-291 Aug 25	P	Y	
BM 41660	232	Rev. 9–10	-280 Jan 30	O	Y	
BM 36710 + 92688 + 92689	242f	Rev. 14'	-272 Mar 1	P	N	
BM 45758	247	Obv. 2'	-266 Oct 17	P	Y	
BM 32245 + 32404		Obv. 11'	-261 Dec 21	?	N	
BM 41616 + 41636 + 41645 + 41797 + 42233	255 + 256 + 885 + 895 + 985	Rev. 1	-255 Sep 16	O	Y	
BM 34728 + 35418	596 + 258	Rev. 11-12	-253 Jan 31	O	Y	

Table A.7: Solar eclipses recorded in the Astronomical Diaries

Tablet	LBAT	Lines	Date of Eclipse	O/P	Times	Comments
BM 45723	267	Obv. 8'	-248 May 4	O	N	cf. BM 32154 (GYT)
		Rev. 2'	-248 Oct 28	P	N	
BM 32889 + 32967 + 41614 + 41618	271f	Rev. 3'	-246 Sep 7	P	Y	
BM 132276 + MNB 1884		Rev. 2'	-245 Aug 28	P	N	
Rm 712	276	6'	-241 Jun 15	O	Y	cf. MM 86.11.369 (NSA)
Rm 720 + 732 + BM 41522	*277 + *278 + 883	Obv. 3'	-240 Nov 28	O	Y	
BM 33837	**284	Obv. 22	-232 Nov 30	P	Y	
BM 41647 + 34285 + Rm 728	285 + 550 + 287	Obv. 13	-230 May 15	P	Y	
BM 41871		Rev. 10'	-228 Mar 25	P	Y	
BM 36889		Rev. 12'	-226 Mar 3	P	Y	
BM 41007		Rev. 15'	-225 Feb 20	P	N	
BM 33655		Obv. 11	-225 Jul 17	P	Y	
BM 35110 + 45725 + 45995 + 46145 + 46169 + 46189	292 + 293 + *238	Rev. 2'	-217 Feb 22	P	N	cf. BM 41634 (NSA)
BM 45608 + 45717	299 + *300	Rev. 29	-209 Sep 18	P	Y	
BM 45635	306	Obv. 6'	-202 May 6	O	N	
BM 36591		Obv. 6'	-200 Apr 13	P	Y	
BM 35199 + 35845	317 819	Obv. 2'	-195 Jun 16	P	N	
BM 41477	320	Obv. 2'	-194 Jun 6	O	N	cf. BM 34048 (GYT)
BM 32190	*322	Rev. 5'	-193 May 26	P	Y	
BM 35331	324	Obv. 26'	-193 Nov 19	P	N	
BM 32951		6'	-189 Mar 14	O	N	cf. BM 33812 (ET)
BM 45636 + 45876	332	Obv. 12'	- 186 Dec 31	P	N	
BM 45613 +	340 +	Obv. 5'	-182 Apr 25	O	N	cf. BM 33873 (A)

Table A.7: Solar eclipses recorded in the Astronomical Diaries

Tablet	LBAT	Lines	Date of Eclipse	O/P	Times	Comments
Rm 693 + 734	341					
BM 34773A	348	Rev. 1'–2'	-179 Aug 17	O	N	
BM 34806 + 35610 + 35812 + 55569	361 + 608 + 756 + *358	Rev. 17'	-172 Apr 4	?	N	
BM 32844	*376	Obv. 13–14	-165 May 17	O	Y	
BM 35015 + 35332 + 55531	377 + 645	Obv. 12'	-164 Oct 29	P	Y	
BM 41462 + 41941	380 + 920	Rev. 8'	-162 Mar 15	P	N	
BM 33850 + 47720	*381	Rev. 3	-162 Sep 8	P	Y	cf. BM 34037 (GYT)
BM 36724 + 36792+ 36920	*396	Obv. 15'	-156 May 7	P	N	
BM 34609 + 34788 + 77617 + 78958	403f	Obv. 10'–11'	-144 Sep 19	P	Y	
BM 34050	420	Obv. 28–29	-139 Jan 1	P	Y	
BM 41841 + 41881	430 + 431	Rev. 7	-136 Oct 20	P	Y	
BM 45745	*429	Rev. 13'–15'	-135 Apr 15	O	Y	cf. BM 34034 (GYT)
BM 34669 + 35740	432f 434	Rev. 5'	-133 Feb 24	?	N	
BM 32843	*435	Obv. 3'	-133 Aug 19	O	Y	cf. BM 34051 (A)
BM 34175 + 34039	437 + 436	Obv. 16'–17'	-132 Feb 13	O	Y	cf. BM 34323 (NSA)
BM 34086	441	Rev. 10'	-131 Feb 1	P	Y	
Rm 701 + BM 41478 + 41646	*296 + 446	Obv. 13'	-129 Jun 7	P	N	
BM 45708	447	Rev. 17'	-125 Sep 19	O	Y	
BM 35135	448	Rev. 7'	-124 Mar 15	P	N	
BM 45850	449	Rev. 16	-124 Sep 7	P	Y	cf. BM 34868 (NSA) cf. BM 78089 (H)
BM 45693 +	450 +	Rev. 8'	-123 Feb 3	P	N	

Table A.7: Solar eclipses recorded in the Astronomical Diaries

Tablet	LBAT	Lines	Date of Eclipse	O/P	Times	Comments
45853	451					
BM 33024 + 33045 + 45757	*452	Rev. 3'	-123 Jul 29	P	N	
Rm 756 + BM 42110	972 + *457	6'	-119 May 17	P	N	
BM 41693	459	Obv. 14	-118 May 7	P	Y	
		Rev. 13'–14'	-118 Oct 31	P	Y	
BM 35700	460	Rev. 17'	-116 Mar 17	P	N	
BM 34096	462	Obv. 28	-111 Jun 18	O	Y	
BM 35601	754	Rev. 8'	-107 Sep 29	P	N	
BM 45750 + 45983 + 45984	*472f	Rev. 12'	-106 Mar 27	P	Y	
BM 34852	484	5'	-99 Oct 31	P	N	
BM 34754	401	10'	-95 Aug 19	P	N	
BM 41691 + 42100	507 + 970	Rev. 10	-84 Jan 23	P	Y	
BM 45863	*513	Rev. 6'	-79 Sep 20	P	Y	
BM 43025 + 45689 + 46047	*514	Rev. 26	-77 Aug 30	P	N	
BM 34753	520	Obv. 12	-62 May 18	P	Y	

Table A.7: Solar eclipses recorded in the Astronomical Diaries

Tablet	LBAT	Lines	Date of Eclipse	O/P	Times	Comments
BM 36599 +		Obv. 3	-474 Dec 5	P	N	
36941 + 36737 +		Obv. 5	-473 May 31	P	N	
47912		Obv. 7	-473 Nov 25	P	Y	
		Obv. 9	-472 May 20	P	Y	
		Obv. 11	-472 Nov 13	P	N	
		Obv. 13	-471 May 9	P	Y	
		Obv. 15	-471 Nov 3	P	N	
		Obv. 18	-470 Mar 30	P	N	
		Obv. 20	-470 Sep 23	P	N	
		Obv. 22	-469 Mar 20	P	N	
		Obv. 24	-469 Sep 12	P	N	
		Obv. 26	-468 Mar 8	P	N	
		Obv. 28	-468 Sep 1	P	N	
		Obv. 30	-467 Feb 25	P	N	
		Obv. 32	-467 Aug 21	P	N	
		Obv. 35	-466 Jan 16	P	N	
		Obv. 37	-466 Jul 23	P	N	
		Obv. 39	-465 Jan 5	P	N	
		Obv. 41	-465 Jul 2	P	N	
		Obv. 43	-465 Dec 26	P	N	
		Rev. 1	-464 Jun 20	P	N	
		Rev. 3	-464 Dec 14	P	N	
		Rev. 5	-463 Jun 9	P	N	
		Rev. 8	-463 Nov 4	P	N	
		Rev. 10	-462 Apr 30	P	N	
		Rev. 12	-462 Oct 24	P	N	
		Rev. 14	-461 Apr 20	P	N	
		Rev. 16	-461 Nov 13	P	N	
		Rev. 18	-460 Apr 8	P	N	
		Rev. 20	-460 Oct 2	P	N	
		Rev. 23	-459 Feb 27	P	N	
		Rev. 25	-459 Aug 23	P	N	
		Rev. 27	-458 Feb 16	P	N	
		Rev. 29	-458 Aug 12	P	Y	
		Rev. 31	-457 Feb 5	P	N	
		Rev. 33	-457 Aug 2	P	Y	
		Rev. 35	-456 Jan 26	P	N	
		Rev. 37	-456 Jul 21	P	N	
CBS 11901	**1478	Rev. 7	-424 Oct 23	?	N	From Nippur
BM 71537		Rev. I, 1	-376 Sep 4	?	N	Restoration
		Rev. I, 2–3	-375 Jan 30	P	N	Restoration
		Rev. I, 4–6	-375 Aug 24	P	N	Restoration
		Rev. I, 7	-374 Feb 18	P	N	
		Obv. I, 1–4	-374 Aug 13	P	N	
		Obv. I, 5–7	-373 Feb 7	P	N	
		Obv. I, 8–10	-373 Jul 4	P	N	
		Obv. I, 11–12	-373 Dec 29	P	N	
		Obv. I, 13–14	-372 Jun 23	P	N	Restoration
		Rev. II, 1	-358 Sep 15	P	N	Restoration
		Rev. II, 2–4	-357 Mar 11	P	N	
		Rev. II, 5–6	-357 Sep 4	P	Y	
		Rev. II, 7	-356 Feb 29	?	N	cf. BM 36913 (D)
		Obv. II, 1–4	-356 Aug 23	P	N	
		Obv. II, 5–6	-355 Feb 18	P	Y	
		Obv. II, 8–10	-355 Jul 16	P	N	
		Obv. II, 11-12	-354 Jan 8	P	N	
		Obv. II, 13–14	-354 Jul 4	P	N	

Table A.8: Solar eclipses recorded in the Eclipse Texts

Tablet	LBAT	Lines	Date of Eclipse	O/P	Times	Comments
		Rev. III, 1	-340 Sep 26	P	N	Restoration
		Rev. III, 2–4	-339 Mar 24	P	N	
		Rev. III, 4–6	-339 Sep 15	?	N	
		Rev. III, 7	-338 Mar 11	P	N	
		Obv. III, 1–4	-338 Sep 4	O	N	
		Obv. III, 5–7	-337 Mar 1	P	N	
		Obv. III, 8	-337 Jul 24	P	N	
		Obv. III, 11	-336 Jan 20	P	N	
		Rev. IV, 2–4	-321 Apr 2	P	N	Restoration
		Rev. IV, 4–6	-321 Sep 26	?	N	Restoration
		Rev. IV, 7	-320 Mar 22	P	N	Restoration
		Obv. IV, 1–3	-320 Sep 14	P	N	Restoration
		Obv. IV, 6–7	-319 Mar 11	P	N	
		Obv. IV, 8–10	-319 Aug 4	P	N	
		Obv. IV, 11-12	-318 Jan 30	P	N	
BM 33812	1438		-189 Mar 14	O	Y	cf. BM 32951 (D)
BM 34963 +	1334 +	Rev. V, 1'–7'	-88 Sep 29	O	Y	
35198 +	1435 +	Rev. V, 17'–18'	-87 Feb 25	P	N	Restoration
35238	1443	Rec VI, 20'	-87 Aug 20	P	N	Restoration
		Rev. VII, 13'–14'	-86 Feb 14	P	Y	
BM 35606	1456		-9 Jun 30	O	Y	

Table A.8: Solar eclipses recorded in the Eclipse Texts

Tablet	LBAT	Lines	Date of Eclipse	O/P	Times	Comments
BM 32847 + 41842	*997 + 998	Rev. 12'–13'	-255 Mar 24	P	Y	
MM 86.11.369		Obv. 13'	-241 Jun 15	P	N	cf. Rm 712 (D)
Rm 731	*1003	Obv. 8'	-229 May 5	P	Y	
BM 41880	1005	Rev. 4'	-218 Apr 3	P	Y	cf. BM 36620 (H)
BM 41634	1006	Rev. 13'	-217 Feb 22	P	N	cf. BM 35110 +... (D)
BM 41588	1008	Obv. 6' Rev. 16'	-215 Jul 26 -214 Jan 20	P P	N N	
BM 34080	1010	Rev. 11	-206 Jan 22	P	Y	
BM 40625	*1011	Obv. 10	-206 Jul 17	P	Y	
BM 41545	1012	Rev. 5'	-205 Jan 11	P	Y	
BM 41022 + 41079	*1013 + *1014	Rev. 13	-204 Jan 1	P	N	
BM 41863	1018	Obv. 12'–13'	-204 Jun 25	P	Y	
BM 45696	1020	Obv. 15 Rev. 4'	-200 Oct 8 -199 Mar 4	P P	N Y	
BM 41599	1022	Rev 5'	-190 Mar 24	P	N	
BM 47738	*1024a	Obv. 5'	-183 May 6	P	Y	
BM 35457	1034	Obv. 18'	-154 Oct 10	P	Y	
BM 34076	1038	Obv. 15 Rev. 7'	-139 Jun 27 -139 Dec 21	P P	Y N	
BM 33482 + 33486 + 33487	*1042a + *1042b + *1042c	Rev. 20'	-132 Feb 13	P	N	cf. BM 34175 + ... (D)
BM 34056	1043	Obv. 19'	-132 Aug 7	P	Y	cf. BM 33867 (A)
BM 33466 + 33448+ 46106 47727+ fragments	*1045 + *1046 + 1136 +	Rev. 25'	-127 Apr 16	P	Y	cf. BM 33018 (H)
BM 34868	1049	Obv. 7'	-124 Sep 7	P	D	cf. BM 45850 (D) cf. BM 33018 (H)
BM 34078	1051	Rev. 12'	-122 Feb 7	P	Y	
BM 34033	**1055	Obv. 24 Rev. 55	-122 Jul 19 -121 Jan 12	P P	Y Y	cf. BM 45827 (A)
BM 35637	1056	Rev. 6	-119 Nov 11	P	Y	

Table A.9: Solar eclipses recorded in the Normal Star Almanacs

Tablet	LBAT	Lines	Date of Eclipse	O/P	Times	Comments
BM 34758	1057	Rev. 15'	-116 Mar 16	P	Y	
BM 34032	**1059	Obv. 13	-110 Jun 7	P	Y	cf. BM 35562 (A)
		Rev. 44	-110 Dec 2	P	Y	cf. BM 35562 (A)
BM 41640	1102	Rev. 4	-82 Nov 22	P	N	
BM 32247		Rev. 9'	-76 Jan 24	P	N	cf. BM 35707 + ... (A)

Table A.9: Solar eclipses recorded in the Normal Star Almanacs

Tablet	LBAT	Lines	Date of Eclipse	O/P	Times	Comments
BM 34232	1122	Rev. 3	-183 Oct 29	P	N	
BM 33873	**1123	Obv. 3	-182 Apr 25	P	Y	cf. BM 45613 + ... (D)
		Rev. 3	-181 Mar 15	P	N	
BM 34121	1127	Rev. 9'	-152 Feb 24	P	N	
BM 34051	1134	Obv. 9'–10'	-133 Aug 19	P	N	cf. BM 32843 (D)
BM 33867	1135	Obv. 10	-132 Aug 17	P	N	cf. BM 34056 (NSA)
BM 35551	1137	Rev. 3	-128 Nov 20	P	Y	
BM 45827	1141	Obv. 7'	-122 Jul 19	P	N	cf. BM 34033 (NSA)
BM 45716	*1147	Rev. 9'	-113 Feb 12	P	N	cf. BM 41301 (H)
BM 35562	1151	Obv. 6	-110 Jun 7	P	Y	cf. BM 34032 (NSA)
		Rev. 2	-110 Dec 2	P	N	cf. BM 34032 (NSA)
BM 34042	1153	Obv. 8	-102 Jul 8	P	Y	
BM 35149	1157	Obv. 5'	-91 May 8	P	N	
BM 36016	1159	Rev. 3'	-85 Dec 24	P	N	
BM 35720	1160	Obv. 9'	-78 Sep 9	P	N	
		Rev. 10'	-77 Mar 4	P	Y	
BM 35707 + 36077	1162 + 1163	Rev 4'	-76 Jan 24	P	N	cf. BM 32247 (NSA)
BM 34722	1169	Obv. 5'	-75 Jul 9	P	N	cf. BM 35515 (H) cf. BM 45698 (A)
BM 45698	1174	Obv. 6	-75 Jul 9	P	N	cf. BM 35515 (H) cf. BM 34722 (A)
		Rev. 4	-74 Jan 3	P	N	
BM 33798	1182	Obv. 4	-64 Jun 7	P	N	
		Rev. 2	-64 Dec 2	P	N	
BM 33822	1183	Obv. 7	-63 Jun 25	P	N	
		Rev. 2	-63 Nov 21	P	N	
BM 34159	1185	Rev. 7'	-28 Jan 5	P	N	
BM 41468	1187	Obv. 6'	-14 Sep 22	P	N	
BM 35602	1189	Obv. 8	-11 Jul 21	P	Y	
BM 33797	**1190	Obv. 7	-10 Jul 10	P	N	
BM 35429	1195	Obv. 2	-6 Apr 29	P	Y	
		Obv. 14	-6 Oct 22	P	Y	
BM 34614	1193	Rev. 11	-5 Apr 18	P	N	
DT 143	*1197	Rev. 3	+37 Jan 5	P	Y	

Table A.10: Solar eclipses recorded in the Almanacs

Tablet	LBAT	Lines	Date of Eclipse	O/P	Times	Comments
BM 32154	**1216	Obv. 3–4	-248 May 4	O	Y	cf. Rm 712 (D)
BM 32222	**1237	Rev. 54–55	-211 May 15	P	Y	
BM 34048	1249	Rev. 3'–7'	-194 Jun 6	O	Y	
		Rev. 13'–15'	-194 Nov 30	P	Y	
BM 34579	1251	Rev. 18–20	-189 Sep 7	P	Y	
		Rev. 20–22	-188 Feb 2	P	Y	
BM 33823	*1256	Rev.	-185 May 28	P	N	
		Rev.	-185 Nov 20	P	Y	
BM 34603	1263	Rev. 3'–7'	-169 Jul 28	O	Y	
		Rev. 13'–14'	-168 Jan 22	P	Y	
BM 34037	1264	Rev. 10'–13	-162 Sep 8	P	Y	cf. BM 33850 + ... (D)
		Rev. 17'–19'	-161 Mar 5	P	Y	
BM 34615	1266	Rev. 18–19	-161 Aug 28	P	Y	
BM 35787	1278	Rev. 1'–2'	-143 Sep 8	P	Y	
		Rev. 9'–11'	-142 Mar 5	P	N	
BM 34034	1285	Rev. 24–28	-135 Apr 15	O	Y	cf. BM 45745 (D)
BM 45909	1304	Rev. 1'–2'	-40 Feb 15	P	Y	

Table A.11: Solar eclipses recorded in the Goal-Year Texts

Tablet	LBAT	Lines	Date of Eclipse	O/P	Times	Comments
BM 36620	*1464	Rev. 5–6	-218 Apr 3	P	N	cf. BM 41880 (NSA)
BM 78089		Rev. 4–5	-125 Sep 19	?	N	cf. BM 45708 (D)
BM 33018	*1468	Rev. 4–5	-124 Sep 7	P	N	cf. BM 33466 + ... (NSA) cf. BM 45850 (D)
BM 41301		Rev. 15	-113 Feb 12	?	N	cf. BM 45716 ()
BM 35515	*1474	Rev. 2–3	-75 Jul 9	P	N	cf. BM 34722 (A) cf. BM 45698 (A)
BM 38104	*1475	Rev 2–4	-68 Aug 20	P	N	

Table A.12: Solar eclipses recorded in the Horoscopes

Appendix B

Chinese Sunrise and Sunset Times

The tables on the following pages contain the canonical values of the time of sunrise and sunset at Yang-ch'eng for each of the 24 *Ch'i* transcribed from the standard editions of the dynastic histories. Since the number of *fen* in a *k'o* varies in the different calendars, this is stated in the footnotes to the tables.

Sui-shu, 17

Ch'i	Sunrise	Sunset
Tung-chih	*ch'en* 50 *fen*[1]	*shen* 7 *k'o* 30 *fen*
Hsiao-han	*ch'en* 32 *fen*	*shen* 7 *k'o* 48 *fen*
Ta-han	*mao* 8 *k'o* 49 *fen*	*yu* 1 *fen*
Li-ch'un	*mao* 7 *k'o* 28 *fen*	*yu* 52 *fen*
Yu-shiu	*mao* 6 *k'o* 25 *fen*	*yu* 1 *k'o* 55 *fen*
Ch'ing-che	*mao* 5 *k'o* 30 *fen*	*yu* 3 *k'o* 7 *fen*
Ch'un-fen	*mao* 3 *k'o* 55 *fen*	*yu* 4 *k'o* 25 *fen*
Ch'ing-ming	*mao* 2 *k'o* 37 *fen*	*yu* 5 *k'o* 43 *fen*
Ku-yu	*mao* 1 *k'o* 28 *fen*	*yu* 6 *k'o* 52 *fen*
Li-hsia	*mao* 28 *fen*	*yu* 7 *k'o* 52 *fen*
Hsiao-man	*yin* 8 *k'o* 3 *fen*	*hsu* 17 *fen*
Mang-chung	*yin* 7 *k'o* 36 *fen*	*hsu* 44 *fen*
Hsia-chih	*yin* 7 *k'o* 30 *fen*	*hsu* 50 *fen*
Hsiao-shu	*yin* 7 *k'o* 36 *fen*	*hsu* 44 *fen*
Ta-shu	*yin* 8 *k'o* 3 *fen*	*hsu* 17 *fen*
Li-ch'iu	*mao* 28 *fen*	*yu* 7 *k'o* 52 *fen*
Ch'u-shu	*mao* 1 *k'o* 28 *fen*	*yu* 6 *k'o* 52 *fen*
Po-lu	*mao* 2 *k'o* 37 *fen*	*yu* 5 *k'o* 43 *fen*
Ch'iu-fen	*mao* 3 *k'o* 55 *fen*	*yu* 4 *k'o* 25 *fen*
Han-lu	*mao* 5 *k'o* 30 *fen*	*yu* 3 *k'o* 7 *fen*
Shuang-hsiang	*mao* 6 *k'o* 25 *fen*	*yu* 1 *k'o* 55 *fen*
Li-tung	*mao* 7 *k'o* 28 *fen*	*yu* 52 *fen*
Hsiao-hsueh	*mao* 8 *k'o* 49 *fen*	*yu* 1 *fen*
Ta-hsueh	*ch'en* 32 *fen*	*shen* 7 *k'o* 48 *fen*

1. 60 *fen* = 1 *k'o*.

Chiu T'ang-shu, 32

Ch'i	Sunrise	Sunset
Tung-chih	*ch'en* 23 *fen*[1]	*shen* 7 *k'o* 12 *fen*
Hsiao-han	*ch'en* 13 *fen*	*shen* 7 *k'o* 19 *fen*
Ta-han	*mao* 8 *k'o* 7 *fen*	*yu* 1 *fen*
Li-ch'un	*mao* 7 *k'o* 11 *fen*	*yu* 21 *fen*
Yu-shiu	*mao* 6 *k'o* 10 *fen*	*yu* 1 *k'o* 22 *fen*
Ch'ing-che	*mao* 5 *k'o* 5 *fen*	*yu* 3 *k'o* 3 *fen*
Ch'un-fen	*mao* 3 *k'o* 22 *fen*	*yu* 4 *k'o* 10 *fen*
Ch'ing-ming	*mao* 2 *k'o* 15 *fen*	*yu* 5 *k'o* 17 *fen*
Ku-yu	*mao* 1 *k'o* 11 *fen*	*yu* 6 *k'o* 21 *fen*
Li-hsia	*mao* 12 *fen*	*yu* 7 *k'o* 20 *fen*
Hsiao-man	*yin* 8 *k'o* 1 *fen*	*hsu* 7 *fen*
Mang-chung	*yin* 7 *k'o* 14 *fen*	*hsu* 18 *fen*
Hsia-chih	*yin* 7 *k'o* 12 *fen*	*hsu* 20 *fen*
Hsiao-shu	*yin* 7 *k'o*	*hsu* 18 *fen*
Ta-shu	*yin* 8 *k'o*	*hsu*
Li-ch'iu	*mao* 11 *fen*	*yu* 7 *k'o* 21 *fen*
Ch'u-shu	*mao* 1 *k'o* 11 *fen*	*yu* 6 *k'o* 21 *fen*
Po-lu[2]	*mao* 7 *k'o* 10 *fen*	*yu* 1 *k'o* 22 *fen*
Li-tung	*mao* 7 *k'o* 11 *fen*	*yu* 20 *fen*
Hsiao-hsueh	*mao* 7 *k'o* 7 *fen*	*yu* 1 *fen*
Ta-hsueh	*ch'en* 13 *fen*	*shen* 7 *k'o* 19 *fen*

1. 24 *fen* = 1 *k'o*.
2. Error for *Shuang-hsiang*. *Po-lu*, *Ch'iu-fen*, and *Han-lu* missing.

Sung-shih, 70

Ch'i	Sunrise	Sunset
Tung-chih	*mao* 4 *k'o* 144 *fen*[1]	*shen* 3 *k'o* 51 *fen*
Hsiao-han	*mao* 4 *k'o* 119 *fen*	*shen* 3 *k'o* 76 *fen*
Ta-han	*mao* 4 *k'o* 34 *fen*	*shen* 4 *k'o* 14 *fen*
Li-ch'un	*mao* 3 *k'o* 56 *fen*	*shen* 4 *k'o* 139 *fen*
Yu-shiu	*mao* 2 *k'o* 58 *fen*	*shen* 5 *k'o* 137 *fen*
Ch'ing-che	*mao* 1 *k'o* 40 *fen*	*shen* 7 *k'o* 8 *fen*
Ch'un-fen	*mao* 0 *k'o*	*yu* 0 *k'o*
Ch'ing-ming	*yin* 7 *k'o* 8 *fen*	*yu* 1 *k'o* 40 *fen*
Ku-yu	*yin* 5 *k'o* 127 *fen*	*yu* 2 *k'o* 68 *fen*
Li-hsia	*yin* 4 *k'o* 149 *fen*	*yu* 3 *k'o* 76 *fen*
Hsiao-man	*yin* 3 *k'o* 146 *fen*	*yu* 4 *k'o* 49 *fen*
Mang-chung	*yin* 3 *k'o* 71 *fen*	*yu* 4 *k'o* 124 *fen*
Hsia-chih	*yin* 3 *k'o* 51 *fen*	*yu* 4 *k'o* 144 *fen*
Hsiao-shu	*yin* 3 *k'o* 71 *fen*	*yu* 4 *k'o* 124 *fen*
Ta-shu	*yin* 3 *k'o* 146 *fen*	*yu* 4 *k'o* 49 *fen*
Li-ch'iu	*yin* 4 *k'o* 119 *fen*	*yu* 3 *k'o* 76 *fen*
Ch'u-shu	*yin* 5 *k'o* 127 *fen*	*yu* 2 *k'o* 68 *fen*
Po-lu	*yin* 7 *k'o* 8 *fen*	*yu* 1 *k'o* 40 *fen*
Ch'iu-fen	*mao* 0 *k'o*	*yu* 0 *k'o*
Han-lu	*mao* 1 *k'o* 40 *fen*	*shen* 7 *k'o* 8 *fen*
Shuang-hsiang	*mao* 2 *k'o* 58 *fen*	*shen* 5 *k'o* 137 *fen*
Li-tung	*mao* 3 *k'o* 56 *fen*	*shen* 4 *k'o* 69 *fen*
Hsiao-hsueh	*mao* 4 *k'o* 34 *fen*	*shen* 4 *k'o* 14 *fen*
Ta-hsueh	*mao* 4 *k'o* 119 *fen*	*shen* 3 *k'o* 76 *fen*

1. 150 *fen* = 1 *k'o*.

Sung-shih, 76

Ch'i	Sunrise	Sunset
Tung-chih	*mao* 5 *k'o*	*shen* 3 *k'o* 12 *fen*[1]
Hsiao-han	*mao* 4 *k'o* 50 *fen*	*shen* 3 *k'o* 30 *fen*
Ta-han	*mao* 4 *k'o* 20 *fen*	*shen* 4 *k'o*
Li-ch'un	*mao* 3 *k'o* 22 *fen*	*shen* 4 *k'o* 48 *fen*
Yu-shiu	*mao* 2 *k'o* 30 *fen*	*shen* 5 *k'o* 50 *fen*
Ch'ing-che	*mao* 1 *k'o* 17 *fen*	*shen* 7 *k'o* 3 *fen*
Ch'un-fen	*mao* 0 *k'o*	*yu* 0 *k'o*
Ch'ing-ming	*yin* 7 *k'o* 3 *fen*	*yu* 1 *k'o* 17 *fen*
Ku-yu	*yin* 5 *k'o* 50 *fen*	*yu* 2 *k'o* 20 *fen*
Li-hsia	*yin* 4 *k'o* 48 *fen*	*yu* 3 *k'o* 32 *fen*
Hsiao-man	*yin* 4 *k'o*	*yu* 4 *k'o* 20 *fen*
Mang-chung	*yin* 3 *k'o* 30 *fen*	*yu* 4 *k'o* 50*fen*
Hsia-chih	*yin* 3 *k'o* 20 *fen*	*yu* 5 *k'o*
Hsiao-shu	*yin* 3 *k'o* 30 *fen*	*yu* 4 *k'o* 50 *fen*
Ta-shu	*yin* 4 *k'o*	*yu* 4 *k'o* 20 *fen*
Li-ch'iu	*yin* 4 *k'o* 40 *fen*	*yu* 3 *k'o* 36 *fen*
Ch'u-shu	*yin* 5 *k'o* 50 *fen*	*yu* 2 *k'o* 30 *fen*
Po-lu	*yin* 7 *k'o* 3 *fen*	*yu* 1 *k'o* 17 *fen*
Ch'iu-fen	*mao* 0 *k'o*	*yu* 0 *k'o*
Han-lu	*mao* 1 *k'o* 17 *fen*	*shen* 7 *k'o* 3 *fen*
Shuang-hsiang	*mao* 2 *k'o* 30 *fen*	*shen* 5 *k'o* 50 *fen*
Li-tung	*mao* 3 *k'o* 32 *fen*	*shen* 4 *k'o* 48 *fen*
Hsiao-hsueh	*mao* 4 *k'o* 20 *fen*	*shen* 4 *k'o*
Ta-hsueh	*mao* 4 *k'o* 50 *fen*	*shen* 3 *k'o* 30 *fen*

1. 60 *fen* = 1 *k'o*.

Appendix C

Timed Chinese Eclipse Records

C.1 The *Sung-shu*

- 4 September 434 AD

  "Yuan Chia reign period, 11th year, 7th month, 16th day, night of the full moon. A lunar eclipse was calculated for the hour *mao*. It actually began on the 15th day at the 2nd call of the 4th watch, that is in the initial half of the hour *ch'ou*. It became total at the 4th call, at 15 degrees to the end of *ying-shih*."

  [*Sung-shu*, 12]

- 8 January 437 AD

  "Yuan Chia reign period, 13th year, 12th month, 16th day, night of the full moon. A lunar eclipse was calculated for the hour *yu*. It actually began at the initial half of the hour *hai*. At the 3rd call of the 1st watch it became total. This was at 4 degrees in *kuei*."

  [*Sung-shu*, 12]

- 28 December 437 AD

  "Yuan Chia reign period, 14th year, 11th month, 16th day, night of the full moon. A lunar eclipse was calculated for the hour *hsu*. It actually began at the 4th call of the second watch, that is at the end of the hour *wei*. At the 1st call of the 3rd watch the eclipse became total. This was at 38 degrees in *ching*."

  [*Sung-shu*, 12]

  Note: The standard printed text contains a misprint giving the 12th instead of the 11th month.

- 23 June 438 AD

  "Yuan Chia reign period, 15th year, 5th month, 15th day, night of the full moon. A lunar eclipse was calculated for the hour *hsu*. That day the sun began to reappear and that is all. The light had already started to reappear reaching $\frac{1}{4}$ eclipsed. This was at 16 degrees in *tou*."
  [*Sung-shu*, 12]

- 26 October 440 AD

  "Yuan Chia reign period, 17th year, 9th month, 16th day, night of the full moon. A lunar eclipse was calculated for the start of the hour *tzu*. The eclipse actually began at the end of the 15th day at the 1st call of the 2nd watch. At the 3rd call it was $\frac{12}{15}$ eclipsed. This was at $1\frac{1}{2}$ degrees in *mao*."
  [*Sung-shu*, 12]

C.2 The *Sui-shu*

- 21 January 585 AD

  "K'ai Huang reign period, 4th year, 12th month, 15th day, *kuei-mao*. According to the calendar, the moon was 3 degrees in *kuei*. At the calculated hour *yu*, the moon should have been above the direction of *mao* and have been $\frac{9}{15}$ eclipsed, the loss starting from the north-west side. Now when observed, at the 1st rod of the 1st watch the eclipse started from the north-east side and $\frac{10}{15}$ was covered. At the 4th rod it began to reappear. At the 1st rod of the 2nd watch it was returned to fullness."
  [*Sui-shu*, 17]

- 1 August 585 AD

  "K'ai-Huang reign period, 5th year, 6th month, 30th day. According to the solar eclipse calculations the sun should have been 6 degrees in *ch'i hsing*. At the calculated time of the start of the hour *wu*, the sun should have been $\frac{1}{15}$ eclipsed, the loss beginning from the south-west side. Now when observed, the sun began to be eclipsed after the 6th mark of the hour *wu*. The loss came from the north-east side and the sun was $\frac{6}{15}$ eclipsed. After the 1st mark of the hour *wei* it began to return. At the 5th mark it was returned to fullness."
  [*Sui-shu*, 17]

- 6 July 586 AD

  "K'ai Huang reign period, 6th year, 6th month, 15th day. According to the lunar eclipse calculations at the calculated hour of *yu*, above the direction of *mao* the moon should have been $\frac{9}{15}$ eclipsed, the loss starting from the south-west. At that time thick clouds covered the moon and it was not seen.

Between the hours of *ch'en* and *ssu*, the moon could be seen through the clouds. It was already $\frac{2}{3}$ eclipsed, the loss starting from the north-east. The clouds then returned. Between the hours of *ssu* and *wu*, they parted a little and after the hour of *wu* it was seen within the clouds that the moon was returned to fullness."

[*Sui-shu*, 17]

- 16 December 586 AD

"K'ai Huang reign period, 6th year, 10th month, 30th day, *ting-ch'ou*. According to the solar eclipse calculations the sun should have been at 9 degrees in *tou*. At the calculated time of a little before *ch'en*, the sun should have been $\frac{9}{15}$ eclipsed, the loss starting from the north-east side. Now in the observations it was seen that when the sun had risen 1 *chang* above the mountains, at the 2nd mark of the hour *ch'en*, it began to be eclipsed. The loss started from the west and the sun was $\frac{2}{3}$ eclipsed. After the 2nd mark of the hour *ch'en* it began to recover. It had already set at the 3rd mark in the hour *ssu* before it returned to fullness."

[*Sui-shu*, 17]

- 25 April 590 AD

"K'ai Huang reign period, 10th year, 3rd month, 16th day, *kuei-mao*. According to the calendar the moon should have moved to 7 degrees in *ti*. At the calculated hour of *hsu*, the moon should have been above the direction *ch'en*, and $\frac{7}{15}$ should have been eclipsed, the loss starting from the north-east. Now when observed the moon began to rise from the south already $\frac{1}{2}$ eclipsed. At the start of the direction of *ch'en* it was $\frac{2}{3}$ eclipsed before gradually reappearing. At the end of the direction *ch'en* it was already returned to fullness."

[*Sui-shu*, 17]

- 19 October 590 AD

"K'ai Huang reign period, 10th year ... According to the calendar, in the 9th month, on the 16th day *keng-tzu*, the moon should have moved to 4 degrees in *wei*. At the calculated hour of *ch'ou*, the moon should have been above *wei*. It should have been $\frac{3}{10}$ eclipsed, the loss starting from the east. Now when observed 2 marks after reaching the direction *wu*, the moon was eclipsed, starting from the east, then towards the south. It was above the direction *wei*. The southern part was $\frac{4}{5}$ eclipsed. It gradually recovered and $1\frac{1}{2}$ marks after entering the direction *shen* it was returned to fullness."

[*Sui-shu*, 17]

- 28 August 592 AD

  "K'ai Huang reign period, 12th year, 7th month, 15th day, *chi-wei*. According to the calendar, the moon should have moved to 7 degrees in *shih*. At the calculated hour of *hsu*, the moon should have been above the direction *ch'en*. It should have been $\frac{12}{15}$ eclipsed, the loss starting from the north-west. Now when observed at the 3rd rod of the 1st watch, the loss began from the north-west and it was $\frac{2}{3}$ eclipsed. The calendar comments agree."
  [*Sui-shu*, 17]

- 17 August 593 AD

  "K'ai Huang reign period, 13th year, 7th month, 16th day. According to the calendar the moon should have moved to above the direction *shen*. It should have been less than $\frac{1}{2}$ of a fifteenth eclipsed. The loss should begin from the south-west. On the 15th night from the 4th watch the moon was observed. Before the 1st rod of the 5th watch, the loss began from the north-west. It was more than half eclipsed. It set in clouds and was not seen."
  [*Sui-shu*, 17]

- 23 July 594 AD

  "K'ai-yuang reign period, 14th year, 7th month, 1st day. According to the calendar, at the calculated time of after the hour of *ssu*, more than $12\frac{1}{2}$ fifteenths (of the sun) should have been eclipsed. After the 3rd mark of *wei* the sun was eclipsed, the loss starting on the north-west. It was $\frac{1}{2}$ eclipsed. It set in clouds and was not seen. For an instant it was suddenly seen, at the end when fullness reappeared, but then clouds screened it"
  [*Sui-shu*, 17]

- 22 December 595 AD

  "K'ai Huang reign period, 15th year, 11th month, 16th day *keng-wu*. According to the calendar, the moon should have moved to 17 degrees in *chung*. At the calculated hour of *hai*, the moon should have been above the direction *ssu*. It should have been $\frac{9}{15}$ eclipsed, the loss from the north-west. That night, at the 4th rod of the 1st watch, the moon was above the direction *ch'en*, and it became eclipsed, the loss beginning on the south-east. At the 3rd rod in the 2nd watch, the moon was above the direction *ssu*. It was $\frac{2}{3}$ eclipsed and gradually recovered. At the 1st rod in the 3rd watch, the moon was above the direction *ping* and was returned to fullness."
  [*Sui-shu*, 17]

- 11 December 596 AD

  "K'ai Yuang reign period, 16th year, 11th month, 16th day *i-ch'ou*. According to the calendar, the moon should have moved to 17 degrees in *ching*. At the calculated hour of *ch'ou*, the moon should have been above the direction *wei*. It should have been $\frac{12}{15}$ eclipsed, the loss starting on the south-east. On the 15th day, according to the observations, at the 1st rod of the 3rd watch, the moon was above the direction *ping*. It was seen within the clouds already $\frac{3}{15}$ eclipsed, the loss starting on the east side. Above the direction *ting*, the eclipse became total. Afterwards it reappeared from the south-east side. At the 3rd rod of the 4th watch, the moon was at the end of *wei* and was returned to fullness."

[*Sui-shu*, 17]

C.3 The *Chiu T'ang-shu*

- 5 August 761 AD

"[Shang Yuan reign period], 2nd year, 7th month, *kuei-wei*. On the first day of the month the sun was eclipsed. All of the great stars were visible. Ch'u T'an, the Head of the Astronomy Bureau, proclaimed to the Emperor that on (the day) *kuei-wei*, the sun was dimmed. It began after the 6th mark of *ch'en*. After the 1st mark of *ssu* it was total. It was returned to fullness at the start of the 1st mark of *wu*. This was at 4 degrees in *chang*."

[*Chiu T'ang-shu*, 36]

- 23 March 768 AD

"[Ta Li reign period,] 3rd year, 3rd month, *i-ssu*. On the first day of the month the sun was eclipsed. From the hour of *wu* it began, and at the 1st mark the eclipse had reached $6\frac{1}{2}$ tenths."

[*Chiu T'ang-shu*, 36]

C.4 The *Sung-shih*

- 26 March 1168 AD

"Ch'ien Tao reign period, 4th year, 2nd month, the night of the full moon. At the 5th point of the 2nd watch the moon was eclipsed 9 divisions. It rose above the ground and returned to fullness. I ... said to the prime minister that the moon should have been totally eclipsed when it rose above the ground. The *Chi-yuan-li* also gave the eclipse as total when it rose above the ground. The light should have reappeared at the 2nd mark of the initial half of the hour of *hsu*, and it should have been returned to fullness at the 3rd mark of the central half of the hour of *hsu*. That evening, the moon was concealed by cloud at the time of moonrise. By the time of dusk, it was seen that the moon was already totally eclipsed. By the 3rd mark of the initial half of the hour of *hsu*, the shine had reappeared, and so we may know that the eclipse was total when it rose above the ground. It returned to fullness at the 3rd mark of the central half of the hour *hsu*. This was at the 2nd point of the 2nd watch."

[*Sung-shih*, 82]

Note: The text contains a printing error giving the 5th instead of the 4th year.

- 12 June 1173 AD

"[Ch'ien tao reign period, 9th year,] 5th month. The sun was eclipsed. Officials from the Calendar Making Bureau observed that the sun was eclipsed $4\frac{1}{2}$ divisions. The loss began from the north-west at $5\frac{1}{2}$ marks in the hour *wu*. The eclipse reached its maximum to the north at the 2nd mark of the initial half of *wei*. It was returned to fullness towards the north-east at the 1st mark of the initial half of *shen*."

[*Sung-shih*, 82]

- 18 April 1185 AD

"Shun Hsi reign period, 12th year, 3rd month. At full moon, the moon was eclipsed. This was at the 2nd point of the 3rd watch, but the calendar gave the 2nd point of the 2nd watch. The loss was counted as 4 divisions, but the calendar gave a loss of 5 divisions."

[*Sung-shih*, 82]

- 23 May 1202 AD

"Chia T'ia reign period, 2nd year, 5th month, *chia-ch'en*. On the 1st day of the month the sun was eclipsed. The loss of the sun began at the 1st mark of the initial half of *wu*. It was returned to fullness at the initial mark of *wei*."

[*Sung-shih*, 82]

Note: An additional time of $3\frac{1}{2}$ marks in the initial half of *wei* for the maximum phase is reported by Beijing Observatory (1988: 198).

- 25 July 1245 AD

"Shun Yu reign period, 5th year. Cheng, an astronomical official, calculated that the sun would be eclipsed at the 3rd mark of the initial half of the hour *wei*, but it was observed at the 4th mark of the central half of the hour *wei*. The loss was calculated as 8 divisions, but it was observed to be 6 divisions."

[*Sung-shih*, 82]

C.5 The *Yuan-shih*

C.5.1 Solar Eclipses

- 6 February 547 AD

  "T'ai Ch'ing reign period, 1st year, *ting-mao*, 1st month, *chi-hai*, first day of the month, eclipse at the hour of *shen*. The *Shou-shih-li* gives the eclipse maximum at 1st mark of *shen*. The *Ta-ming-li* gives the eclipse maximum at 3rd mark of *shen*. The *Shou-shih-li* is fairly close. The *Ta-ming-li* is close."
  [*Yuan-shih*, 53]

- 11 July 576 AD

  "Ch'en dynasty, Ta Chien reign period, 8th year, *ping-shen*, 6th month, *wu-shen*, first day of the month, eclipse between the hour of *mao* and the 1st denary hour. The *Shou-shih-li* gives the eclipse maximum at 2nd mark of *mao*. The *Ta-ming-li* gives the eclipse maximum at 4th mark of *mao*. The *Shou-shih-li* is fairly close. The *Ta-ming-li* is far off."
  [*Yuan-shih*, 53]

- 27 November 680 AD

  "T'ang dynasty, Yung Lung reign period, 1st year, *keng-ch'en*, 11th month, *jen-shen*, first day of the month, eclipse maximum at the 4th mark of the hour of *ssu*. The *Shou-shih-li* gives the eclipse maximum at the 7th mark of *ssu*. The *Ta-ming-li* gives the eclipse maximum at the 5th mark of *ssu*. The *Shou-shih-li* is off. The *Ta-ming-li* is close."
  [*Yuan-shih*, 53]

- 16 November 681 AD

  "K'ai Yao reign period, 1st year, *hsin-ssu*, 10th month, *ping-yin*, first day of the month, eclipse maximum at the initial mark of the hour of *ssu*. The *Shou-shih-li* gives the eclipse maximum at the 3rd mark of central half of *ch'en*. The *Ta-ming-li* gives the eclipse maximum at the 1st mark of central half of *ch'en*. The *Shou-shih-li* is close. The *Ta-ming-li* is off."
  [*Yuan-shih*, 53]

- 4 May 691 AD

  "Ssu Sheng reign period, 8th year, *hsin-mao*, 4th month, *jen-yin*, first day of the month, eclipse maximum at the 2nd mark of the hour of *mao*. The *Shou-shih-li* gives the eclipse maximum at the 8th mark of *yin*. The *Ta-ming-li* gives the eclipse maximum at the initial mark of *mao*. Both calendars are fairly close."
  [*Yuan-shih*, 53]

- 23 May 700 AD

  "17th year, *keng-tzu*, 5th month, *chi-yu*, first day of the month, eclipse maximum at the initial half of the hour of *shen*. The *Shou-shih-li* gives the eclipse maximum at the 2nd mark of initial half of *ssu*. The *Ta-ming-li* gives the eclipse maximum at the initial mark of the central half of *ssu*. The *Shou-shih-li* is fairly close. The *Ta-ming-li* is far off."

  [*Yuan-shih*, 53]

- 26 September 702 AD

  "19th year, *jen-yin*, 9th month, *i-ch'ou*, first day of the month, eclipse maximum at the 3rd mark of the hour of *shen*. The *Shou-shih-li* gives the eclipse maximum at the 1st mark of *shen*. The *Ta-ming-li* gives the eclipse maximum at the 4th mark of *shen*. The *Shou-shih-li* is fairly close. The *Ta-ming-li* is close."

  [*Yuan-shih*, 53]

- 4 July 707 AD

  "Ching Lung reign period, 1st year, *ting-wei*, 6th month, *ting-mao*, first day of the month, eclipse maximum at the central half of the hour of *wu*. The *Shou-shih-li* gives the eclipse maximum at the 2nd mark of the central half of *wu*. The *Ta-ming-li* gives the eclipse maximum at the initial mark of the initial half of *wei*. The *Shou-shih-li* is fairly close. The *Ta-ming-li* is far off."

  [*Yuan-shih*, 53]

- 26 September 721 AD

  "K'ai Yuan reign period, 9th year, *hsin-yu*, 9th month, *i-ssu*, first day of the month, eclipse maximum after the 3rd mark of the central half of the hour of *wu*. The *Shou-shih-li* gives the eclipse maximum at the 1st mark of *wu*. The *Ta-ming-li* gives the eclipse maximum at the 2nd mark of *wu*. The *Shou-shih-li* is fairly close. The *Ta-ming-li* is close."

  [*Yuan-shih*, 53]

  Note: The standard printing of the text contains a printing error giving the year as the 1st instead of the 9th.

- 9 April 1046 AD

  "Sung dynasty, Ch'ing Li reign period, 6th year, *ping-hsu*, 3rd month, *hsin-ssu*, first day of the month, eclipse. Return of fullness at the 3rd mark of the central half of the hour of *shen*. The *Shou-shih-li* gives the return of fullness at the 3rd mark of the central half of *shen*. The *Ta-ming-li* gives the return of fullness at the 1st mark of the central half of *shen*. The *Shou-shih-li* is exact. The *Ta-ming-li* is fairly close."

  [*Yuan-shih*, 53]

- 5 February 1049 AD

  "Huang Yu reign period, 1st year, *chi-ch'ou*, 1st month, *chia-wu*, first day of the month, eclipse maximum at the central half of the hour of *wu*. The *Shou-shih-li* gives the eclipse maximum at the 2nd mark of the initial half of *wu*. The *Ta-ming-li* gives the eclipse maximum at the initial mark of the central half of *wu*. The *Shou-shih-li* is close. The *Ta-ming-li* is exact."

  [*Yuan-shih*, 53]

- 13 November 1053 AD

  "5th year, *kuei-ssu*, 10th month, *ping-shen*, first day of the month, eclipse maximum at the 1st mark of the hour of *wei*. The *Shou-shih-li* gives the eclipse maximum at the 3rd mark of *wei*. The *Ta-ming-li* gives the eclipse maximum at the initial mark of *wei*. The *Shou-shih-li* is fairly close. The *Ta-ming-li* is close."

  [*Yuan-shih*, 53]

- 10 May 1054 AD

  "Chih Ho reign period, 1st year, *chia-wu*, 4th month, *chia-wu*, first day of the month, eclipse maximum at the 1st mark of the central half of the hour of *shen*. The *Shou-shih-li* gives the eclipse maximum at the 1st mark of the central half of *shen*. The *Ta-ming-li* gives the eclipse maximum at the 2nd mark of the central half of *shen*. The *Shou-shih-li* is exact. The *Ta-ming-li* is close."

  [*Yuan-shih*, 53]

- 15 February 1059 AD

  "Chia Yu reign period, 4th year, *chia-ssu*, 1st month, *ping-shen*, first day of the month, eclipse. Returned to fullness at the 3rd mark of the hour of *wei*. The *Shou-shih-li* gives the return to fullness at the 2nd mark of the initial half of *wei*. The *Ta-ming-li* gives the return to fullness at the 2nd mark of the initial half of *wu*. Both calendars are close."

  [*Yuan-shih*, 53]

- 20 June 1061 AD

  "6th year, *hsin-ch'ou*, 6th month, *jen-tzu*, first day of the month, eclipse. Beginning of loss at the initial half of the hour of *wei*. The *Shou-shih-li* gives the beginning of loss at the initial mark of *wei*. The *Ta-ming-li* gives the beginning of loss at the 1st mark of *wei*. The *Shou-shih-li* is close. The *Ta-ming-li* is fairly close."

  [*Yuan-shih*, 53]

- 22 September 1066 AD

  "Chia Ping reign period, 3rd year, *ping-wu*, 9th month, *jen-tzu*, first day of the month, eclipse maximum at the 2nd mark of the hour of *wei*. The *Shou-shih-li* gives the eclipse maximum at the 3rd mark of *wu*. The *Ta-ming-li* gives the eclipse maximum at the 4th mark of *wu*. The *Shou-shih-li* is close. The *Ta-ming-li* is fairly close."
  [*Yuan-shih*, 53]

- 21 July 1069 AD

  "Hsi Ning reign period, 2nd year, *chi-yu*, 7th month, *i-ch'ou*, first day of the month, eclipse maximum at the 3rd mark of the hour of *ch'en*. The *Shou-shih-li* gives the eclipse maximum at the 5th mark of *ch'en*. The *Ta-ming-li* gives the eclipse maximum at the 4th mark of *ch'en*. The *Shou-shih-li* is fairly close. The *Ta-ming-li* is close."
  [*Yuan-shih*, 53]

- 14 December 1080 AD

  "Yuan Feng reign period, 3rd year, *keng-shen*, 11th month, *chi-ch'ou*, first day of the month, eclipse maximum at the 6th mark of the hour of *ssu*. The *Shou-shih-li* gives the eclipse maximum at the 5th mark of *wu*. The *Ta-ming-li* gives the eclipse maximum at the 2nd mark of *wu*. The *Shou-shih-li* is close. The *Ta-ming-li* is far off."
  [*Yuan-shih*, 53]

- 19 March 1094 AD

  "Shao Sheng reign period, 1st year, *chia-hsu*, 3rd month, *jen-shen*, first day of the month, eclipse maximum at the 6th mark of the hour of *wei*. The *Shou-shih-li* gives the eclipse maximum at the 5th mark of *wei*. The *Ta-ming-li* gives the eclipse maximum at the 5th mark of *wei*. Both calendars are close."
  [*Yuan-shih*, 53]

- 16 December 1107 AD

  "Ta Kuan reign period, 1st year, *ting-hai*, 11th month, *jen-tzu*, first day of the month, eclipse. Beginning of loss at the 2nd mark of the hour of *wei*, maximum at the 8th mark of *wei*, return to fullness at the 6th mark of *shen*. The *Shou-shih-li* gives the beginning of loss at the 3rd mark of *wei*, maximum at the initial mark of *shen*, and return to fullness at the 6th mark of *shen*. The *Ta-ming-li* gives the beginning of loss at the initial mark of *wei*, maximum at the 7th mark of *wei*, and return to fullness at the 5th mark of *shen*. The *Shou-shih-li* is close for beginning of loss and maximum and exact for return of fullness. The *Ta-ming-li* is fairly close for beginning of loss and close for maximum and return to fullness."
  [*Yuan-shih*, 53]

- 17 January 1162 AD

  "Shao Hsing reign period, 32rd year, *jen-wu*, 1st month, *wu-ch'en*, first day of the month, eclipse. Beginning of loss at the initial half of the hour of *shen*. The *Shou-shih-li* gives the beginning of loss at the 1st mark of *shen*. The *Ta-ming-li* gives the beginning of loss at the 7th mark of *wei*. Both calendars are close."

  [*Yuan-shih*, 53]

- 17 November 1183 AD

  "Shun Hsi reign period, 10th year, *kuei-mao*, 11th month, *jen-hsu*, first day of the month, eclipse maximum at the 2nd mark of the central half of the hour of *ssu*. The *Shou-shih-li* gives the eclipse maximum at the 2nd mark of the central half of *ssu*. The *Ta-ming-li* gives the eclipse maximum at the 1st mark of central half of *ssu*. The *Shou-shih-li* is exact. The *Ta-ming-li* is close."

  [*Yuan-shih*, 53]

- 12 April 1195 AD

  "Ch'ing Yuang reign period, 1st year, *i-mao*, 3rd month, *ping-hsu*, first day of the month, eclipse. Beginning of loss at the 2nd mark of the initial half of the hour of *wu*. The *Shou-shih-li* gives the beginning of loss at the 1st mark of the initial half of *wu*. The *Ta-ming-li* gives the beginning of loss at the 2nd mark of the initial half of *wu*. The *Shou-shih-li* is close. The *Ta-ming-li* is exact."

  [*Yuan-shih*, 53]

- 23 May 1202 AD

  "Chia T'ai reign period, 2nd year, *jen-hsu*, 5th month, *chia-ch'en*, first day of the month, eclipse. Beginning of loss at the 1st mark of the initial half of the hour of *wu*. The *Shou-shih-li* gives the beginning of loss at the 3rd mark of the central half of *ssu*. The *Ta-ming-li* gives the beginning of loss at the 3rd mark of the initial half of *wu*. Both calendars are close."

  [*Yuan-shih*, 53]

- 19 February 1216 AD

  "Chia Ting reign period, 9th year, *ping-tzu*, 2nd month, *chia-shen*, first day of the month, eclipse maximum at the 4th mark of the central half of the hour of *shen*. The *Shou-shih-li* gives the eclipse maximum at the 3rd mark of the central half of *shen*. The *Ta-ming-li* gives the eclipse maximum at the 2nd mark of the central half of *shen*. The *Shou-shih-li* is close. The *Ta-ming-li* is fairly close."

  [*Yuan-shih*, 53]

- 22 March 1243 AD

"Shun Yu reign period, 3rd year, *kuei-mao*, 3rd month, *ting-ch'ou*, first day of the month, eclipse maximum at the 2nd mark of the initial half of the hour of *ssu*. The *Shou-shih-li* gives the eclipse maximum at the 1st mark of the initial half of *ssu*. The *Ta-ming-li* gives the eclipse maximum at the initial mark of the initial half of *ssu*. The *Shou-shih-li* is close. The *Ta-ming-li* is fairly close."

[*Yuan-shih*, 53]

- 12 April 1260 AD

"1st year, *keng-shen*, 3rd month, *wu-ch'en*, first day of the month, eclipse maximum at the 2nd mark of the central half of the hour of *shen*. The *Shou-shih-li* gives the eclipse maximum at the 1st mark of the central half of *shen*. The *Ta-ming-li* gives the eclipse maximum at the 3rd mark of the initial half of *shen*. The *Shou-shih-li* is close. The *Ta-ming-li* is off."

[*Yuan-shih*, 53]

Note: Date given by Sivin (1997) who notes a dating error in the text.

- 28 October 1277 AD

"Chia Yuan reign period, 14th year, *ting-ch'ou*, 10th month, *ping-ch'en*, first day of the month, eclipse. Beginning of loss at the initial mark of the central half of the hour of *wu*. Eclipse maximum at the 1st mark of the initial half of *wei*. Return to fullness at the 2nd mark of the central half of *wei*. The *Shou-shih-li* gives the beginning of loss at the initial of the central half of *wu*. Eclipse maximum at the 1st mark of the initial half of *wei*. Return of fullness at the 1st mark of the central half of *wei*. The *Ta-ming-li* gives the beginning of loss at the 3rd mark of the central half of *wu*. Eclipse maximum at the 1st mark of the central half of *wei*. Return of fullness at the 2nd mark of the initial half of *shen*. The *Shou-shih-li* is exact for beginning of loss and maximum, and close for return of fullness. The *Ta-ming-li* is off for beginning of loss and far off for maximum and return to fullness."

[*Yuan-shih*, 53]

C.5.2 Lunar Eclipses

- 4 September 434 AD

"Sung dynasty, Yuan Chia reign period, 11th year, *chia-hsu*, 7th month, *ping-tzu*, full moon, eclipse. Beginning of loss at the 2nd call of the 4th watch. Eclipse total at the 4th call of the 4th watch. The *Shou-shih-li* gives the beginning of loss at the 3rd point of the 4th watch and totality at the 4th point of the 4th watch. The *Ta-ming-li* gives the loss beginning at the 2nd point of the 4th watch and totality at the 5th point of the 4th watch.

The *Shou-shih-li* is close for beginning of loss and exact for totality. The *Ta-ming-li* is exact for beginning of loss and close for totality."

[*Yuan-shih*, 53]

- 8 January 437 AD

"13th year, *ping-tzu*, 12th month, *kuei-ssu*, full moon, eclipse total at the 3rd call of the 1st watch. The *Shou-shih-li* gives totality at the 3rd point of the 1st watch. The *Ta-ming-li* gives totality at the 4th point of the 1st watch. The *Shou-shih-li* is exact. The *Ta-ming-li* is close."

[*Yuan-shih*, 53]

Note: The text contains a printing error giving the cyclical day as *chi-ssu* instead of *kuei-ssu*.

- 28 December 437 AD

"14th year, *ting-ch'ou*, 11th month, *ting-hai*, full moon, eclipse. Beginning of loss at the 4th call of the 2nd watch. Eclipse total at the 1st call of the 3rd watch. The *Shou-shih-li* gives the beginning of loss at the 5th point of the 2nd watch and totality at the 2nd point of the 3rd watch. The *Ta-ming-li* gives the loss beginning at the 4th point of the 2nd watch and totality at the 2nd point of the 3rd watch. The *Shou-shih-li* is close for beginning of loss and for totality. The *Ta-ming-li* is exact for beginning of loss and close for totality."

[*Yuan-shih*, 53]

- 26 June 530

"Liang dynasty, Chang Ta T'ung reign period, 2nd year, *keng-hsu*, 5th month, *keng-yin*, full moon, eclipse at the hour of *tzu*. The *Shou-shih-li* gives the eclipse maximum at the initial mark of the central half of *tzu*. The *Ta-ming-li* gives the eclipse maximum at the initial mark of the central half of *tzu*. Both calendars are exact."

[*Yuan-shih*, 53]

- 4 May 543 AD

"Ta T'ung reign period, 9th year, *kuei-hai*, 3rd month, *i-ssu*, full moon, eclipse. Beginning of loss at the 3rd call of the 3rd watch. The *Shou-shih-li* gives the beginning of loss at the 1st point of the 3rd watch. The *Ta-ming-li* gives the loss beginning at the 3rd point of the 3rd watch. The *Shou-shih-li* is fairly close. The *Ta-ming-li* is exact."

[*Yuan-shih*, 53]

- 28 August 592 AD

"Sui dynasty, K'ai Huang reign period, 12th year, *jen-tzu*, 7th month, *chi-wei*, full moon, eclipse. Beginning of loss at the 3rd call of the 1st watch. The *Shou-shih-li* gives the beginning of loss at the 4th point of the 1st watch. The *Ta-ming-li* gives the loss beginning at the 5th point of the 1st watch. The *Shou-shih-li* is close. The *Ta-ming-li* is fairly close."

[*Yuan-shih*, 53]

- 22 December 595 AD

"15th year, *i-mao*, 11th month, *keng-wu*, full moon, eclipse. Beginning of loss at the 4th point of the 1st watch. Maximum at the 3rd point of the 2nd watch. Return to fullness at the 1st point of the 3rd watch. The *Shou-shih-li* gives the beginning of loss at the 3rd point of the 1st watch, maximum at the 2nd point of the 2nd watch, and return to fullness at the 5th point of the 2nd watch. The *Ta-ming-li* gives the loss beginning at the 5th point of the 1st watch, maximum at the 3rd point of the 2nd watch, and return to fullness at the 5th point of the 2nd watch. The *Shou-shih-li* is close for beginning of loss, maximum, and return to fullness. The *Ta-ming-li* is close for beginning of loss and return to fullness and exact for maximum."

[*Yuan-shih*, 53]

- 10 December 596 AD

"16th year, *ping-ch'en*, 11th month, *chia-tzu*, full moon, eclipse. Return to fullness at the 3rd rod of the 4th watch. The *Shou-shih-li* gives the return to fullness at the 4th point of the 4th watch. The *Ta-ming-li* gives the return to fullness at the 5th point of the 4th watch. The *Shou-shih-li* is close. The *Ta-ming-li* is fairly close."

[*Yuan-shih*, 53]

- 28 January 948 AD

"Later Han dynasty (of the 5 dynasties), T'ien Fu reign period, 12th year, *ting-wei*, 12th month, *i-wei*, full moon, eclipse. Beginning of loss at the 4th point of the 4th watch. The *Shou-shih-li* gives the beginning of loss at the 5th point of the 4th watch. The *Ta-ming-li* gives the beginning of loss at the 1st point of the 4th watch. The *Shou-shih-li* is close. The *Ta-ming-li* is fairly close."

[*Yuan-shih*, 53]

- 8 December 1052 AD

"Sung dynasty, Huang Yu reign period, 4th year, *jen-ch'en*, 11th month, *ping-ch'en*, full moon, eclipse. Beginning of loss at the 4th mark of *yin*.

The *Shou-shih-li* gives the beginning of loss at the 2nd mark of *yin*. The *Ta-ming-li* gives the beginning of loss at the 1st mark of *yin*. The *Shou-shih-li* is fairly close. The *Ta-ming-li* is off."
[*Yuan-shih*, 53]

- 8 November 1063 AD

"Chia Yu reign period, 8th year, *kuei-mao*, 10th month, *kuei-wei*, full moon, eclipse maximum at the 7th mark of *mao*. The *Shou-shih-li* gives the maximum at the initial mark of *ch'en*. The *Ta-ming-li* gives the maximum at the initial mark of *ch'en*. Both calendars are close."
[*Yuan-shih*, 53]

- 30 December 1069 AD

"Hsi Ning reign period, 2nd year, *chi-yu*, intercalary 11th month, *tung-wei*, full moon, eclipse. Beginning of loss at the 6th mark of *hai*. Maximum at the 5th mark of *tzu*. Return to fullness at the 4th mark of *ch'ou*. The *Shou-shih-li* gives the beginning of loss at the 6th mark of *hai*, maximum at the 5th mark of *tzu*, and return to fullness at the 3rd mark of *ch'ou*. The *Ta-ming-li* gives the loss beginning at the initial mark of *tzu*, maximum at the 6th mark of *tzu*, and return to fullness at the 4th mark of *ch'ou*. The *Shou-shih-li* is exact for beginning of loss and maximum and close for the return to fullness. The *Ta-ming-li* is fairly close for beginning of loss, close for maximum, and exact for return to fullness."
[*Yuan-shih*, 53]

- 9 December 1071 AD

"4th year, *hsin-hai*, 11th month, *ping-shen*, full moon, eclipse. Beginning of loss at the 2nd mark of *mao*. Maximum at the 6th mark of *mao*. The *Shou-shih-li* gives the beginning of loss at the initial mark of *mao*, and maximum at the 5th mark of *mao*. The *Ta-ming-li* gives the loss beginning at the 4th mark of *mao*, and maximum at the 7th mark of *mao*. Both calendars are fairly close for beginning of loss and close for maximum."
[*Yuan-shih*, 53]

- 24 April 1073 AD

"6th year, *kuei-jen*, 3rd month, *wu-wu*, full moon, eclipse. Beginning of loss at the 1st mark of *hai*. Maximum at the 6th mark of *hai*. Return to fullness at the 4th mark of *tzu*. The *Shou-shih-li* gives the beginning of loss at the 7th mark of *hsu*, maximum at the 5th mark of *hai*, and return to fullness at the 3rd mark of *tzu*. The *Ta-ming-li* gives the loss beginning at the 2nd mark of *hai*, maximum at the 7th mark of *hai*, and return to fullness at the 4th mark of *tzu*. The *Shou-shih-li* is fairly close for beginning of loss and close for maximum and return to fullness. The *Ta-ming-li* is close for beginning of loss and maximum, and exact for return to fullness."
[*Yuan-shih*, 53]

- 7 October 1074 AD

  "7th year, *chia-yin*, 9th month, *chi-yu*, full moon, eclipse. Beginning of loss at the 5th point of the 4th watch. Eclipse total at the 3rd point of the 5th watch. The *Shou-shih-li* gives the beginning of loss at the 5th point of the 4th watch and totality at the 3rd point of the 5th watch. The *Ta-ming-li* gives the loss beginning at the 3rd point of the 4th watch and totality at the 2nd point of the 5th watch. The *Shou-shih-li* is exact for beginning of loss and totality. The *Ta-ming-li* is fairly close for beginning of loss and close for totality."

  [*Yuan-shih*, 53]

- 21 January 1106 AD

  "Ch'ung Ning reign period, 4th year, *i-yu*, 12 month, *wu-yin*, full moon, eclipse. Maximum at the 3rd mark of *yu*. Return to fullness at the initial mark of *hsu*. The *Shou-shih-li* gives the maximum at the 1st mark of *yu*, and return to fullness at the 7th mark of *yu*. The *Ta-ming-li* gives the maximum at the 3rd mark of *yu*, and return to fullness at the 2nd mark of *hsu*. The *Shou-shih-li* is fairly close for maximum and return to fullness. The *Ta-ming-li* is exact for maximum and fairly close for return to fullness."

  [*Yuan-shih*, 53]

- 7 April 1270 AD

  "Present (Yuan) dynasty, Chih Yuan reign period, 7th year, *keng-wu*, 3rd month, *i-mao*, full moon, eclipse. Beginning of loss at the 3rd mark of *ch'ou*. Maximum at the initial mark of *yin*. Return to fullness at the 6th mark of *yin*. The *Shou-shih-li* gives the beginning of loss at the 2nd mark of *ch'ou*, maximum at the initial mark of *yin*, and return to fullness at the 6th mark of *yin*. The *Ta-ming-li* gives the loss beginning at the 4th mark of *ch'ou*, maximum at the 1st mark of *yin*, and return to fullness at the 7th mark of *yin*. The *Shou-shih-li* is exact for beginning of loss, maximum, and return to fullness. The *Ta-ming-li* is close for beginning of loss, maximum, and return to fullness."

  [*Yuan-shih*, 53]

- 10 August 1272 AD

  "9th year, *jen-shen*, 7th month, *hsin-wei*, full moon, eclipse. Beginning of loss at the initial mark of *ch'ou*. Maximum at the 6th mark of *ch'ou*. Return to fullness at the 3rd mark of *yin*. The *Shou-shih-li* gives the beginning of loss at the 7th mark of *tzu*, maximum at the 4th mark of *ch'ou*, and return to fullness at the 1st mark of *yin*. The *Ta-ming-li* gives the loss beginning at the 2nd mark of *ch'ou*, maximum at the 6th mark of *ch'ou*, and return to fullness at the 2nd mark of *yin*. The *Shou-shih-li* is close for beginning of loss and fairly close for maximum and return to fullness. The *Ta-ming-li* is fairly close for beginning of loss, exact for maximum, and close for return to fullness."

[*Yuan-shih*, 53]

- 18 May 1277 AD

"14th year, *ting-ch'ou*, 4th month, *kuei-yu*, full moon, eclipse. Beginning of loss at the 6th mark of *tzu*. Totality at the 3rd mark of *ch'ou*. Reappearance of light at the 5th mark of *ch'ou*. Return to fullness at the 4th mark of *yin*. The *Shou-shih-li* gives the beginning of loss at the 6th mark of *tzu*, totality at the 4th mark of *ch'ou*, maximum at the 5th mark of *ch'ou*, reappearance of light at the 6th mark of *ch'ou*, and return to fullness at the 4th mark of *yin*. The *Ta-ming-li* gives the loss beginning at the initial mark of *ch'ou*, totality at the 7th mark of *ch'ou*, maximum at the 7th mark of *ch'ou*, reappearance of light at the 8th mark of *ch'ou*, and return to fullness at the 6th mark of *yin*. The *Shou-shih-li* is exact for beginning of loss, maximum, and return to fullness, and close for totality and reappearance of light. The *Ta-ming-li* is fairly close for beginning of loss, maximum, and return to fullness, far off for totality and close for reappearance of light."

[*Yuan-shih*, 53]

- 29 March 1279 AD

"16th year, *chi-mao*, 2nd month, *kuei-ssu*, full moon, eclipse. Beginning of loss at the 5th mark of *tzu*. Maximum at the 2nd mark of *ch'ou*. Return to fullness at the 7th mark of *ch'ou*. The *Shou-shih-li* gives the beginning of loss at the 5th mark of *tzu*, maximum at the 2nd mark of *ch'ou*, and return to fullness at the 7th mark of *ch'ou*. The *Ta-ming-li* gives the loss beginning at the 7th mark of *tzu*, maximum at the 3rd mark of *ch'ou*, and return to fullness at the 7th mark of *ch'ou*. The *Shou-shih-li* is exact for beginning of loss, maximum, and return to fullness. The *Ta-ming-li* is fairly close for beginning of loss, exact for maximum and return to fullness."

[*Yuan-shih*, 53]

Note: The text contains a printing error giving the cyclical day as *kuei-yu* instead of *kuei-ssu*.

- 21 October 1279 AD

"8th month, *chi-ch'ou*, full moon, eclipse. Beginning of loss at the 5th mark of *ch'ou*. Maximum at the initial mark of *yin*. Return to fullness at the 4th mark of *yin*. The *Shou-shih-li* gives the beginning of loss at the 3rd mark of *ch'ou*, maximum at the initial mark of *yin*, and return to fullness at the 4th mark of *yin*. The *Ta-ming-li* gives the loss beginning at the 7th mark of *ch'ou*, maximum at the 2nd mark of *yin*, and return to fullness at the 4th mark of *yin*. The *Shou-shih-li* is fairly close for beginning of loss and exact for maximum and return to fullness. The *Ta-ming-li* is fairly close for beginning of loss and maximum, and exact for return to fullness."

[*Yuan-shih*, 53]

- 10 October 1280 AD

  "17th year, *keng-ch'en*, 8th month, *chia-shen*, full moon, eclipse in daylight. Return to fullness at the 1st mark of *hsu*. The *Shou-shih-li* gives the return to fullness at the 1st mark of *hsu*. The *Ta-ming-li* gives the return to fullness at the 4th mark of *hsu*. The *Shou-shih-li* is exact. The *Ta-ming-li* is close."

  [*Yuan-shih*, 53]

C.6 The *Wen-hsien T'ung-k'ao*

C.6.1 Solar Eclipses

- 15 February 1040 AD

  "Pao Yuan reign period, 3rd year, 1st month, *ping-ch'en*. On the 1st day of the month (the sun) was eclipsed 6 divisions. At the 1st mark of *shen* it was returned (to fullness)."

  [*Wen-hsien T'ung-k'ao*, 283]

  Note: The standard printing of the text contains a misprint giving the 1st instead of the 3rd year.

- 9 April 1046 AD

  "[Ch'ing Li reign period,] 6th year, 3rd month, *hsin-ssu*. On the 1st day of the month (the sun) was eclipsed $4\frac{1}{2}$ divisions. It was returned (to fullness) at the 3rd mark of *shen*."

  [*Wen-hsien T'ung-k'ao*, 283]

- 24 November 1052 AD

  "[Huang Yu reign period,] 4th year, 11th month, *jen-yin*. On the 1st day of the month (the sun) was eclipsed by a little more than 2 divisions. It was returned (to fullness) at the 1st mark of *wei*."

  [*Wen-hsien T'ung-k'ao*, 283]

- 13 November 1053 AD

  "[Huang Yu reign period,] 5th year, 10th month, *ping-shen*. On the 1st day of the month, at the 1st mark of the central half of *wei*, (the sun) was eclipsed by $4\frac{1}{2}$ divisions."

  [*Wen-hsien T'ung-k'ao*, 283]

- 10 May 1054 AD

  "Chih Ho reign period, 1st year, 4th month, *chia-wu*. On the 1st day of the month, (the sun) was eclipsed. It waned by more than 9 divisions from the south-west. At the 1st mark of the central half of *shen*, it reached its maximum. That day, it rained..."

  [*Wen-hsien T'ung-k'ao*, 283]

- 15 February 1059 AD

  "[Chia Yu reign period,] 4th year, 1st month, *ping-shen*. On the 1st day of the month (the sun) was eclipsed by more than 3 divisions. It was returned to fulness at the 3rd mark of the initial half of *wei*."

  [*Wen-hsien T'ung-k'ao*, 283]

- 20 June 1061 AD

  "[Chia Yu reign period,] 6th year, 6th month, *jen-tzu*. On the 1st day of the month, at the 1st mark in *wei*, (the sun) was eclipsed 4 divisions. It then set and was not seen."

  [*Wen-hsien T'ung-k'ao*, 283]

- 19 March 1094 AD

  "[Yuan Yu reign period,] 9th year, 3rd month, *jen-shen*. On the 1st day of the month, according to the Astronomer Royal, the sun should have been eclipsed, but on account of thick clouds it was not (fully) seen. The loss began (to be seen) at the 3rd mark of *wei*. It was seen through the clouds that the sun was eclipsed on the south-western side in excess of 1 division. At the 6th mark it reached a maximum of 7 divisions. On account of the clouds, its recovery was not seen."

  [*Wen-hsien T'ung-k'ao*, 283]

C.6.2 Lunar Eclipses

- 8 November 1063 AD

  "Chia Yu reign period, 8th year, 10th month, *kuei-wei*. (The moon) was totally eclipsed. The eclipse reached its maximum at the 7th mark of *mao*. It then set and was not seen."

  [*Wen-hsien T'ung-k'ao*, 285]

- 3 March 1067 AD

  "Reign of Shen Tsung, Chih Ping reign period, 4th year, 2nd month, *ch-ia-wu*. The moon was eclipsed. At the 4th mark of *ch'ou*, the loss was seen from the west. This was at the 5th degree of *chi*. By the 6th mark, the

eclipse had reached its maximum of more than 8 divisions. It then set in the direction *yu* and was not seen."
[*Wen-hsien T'ung-k'ao*, 285]

- 15 August 1068 AD

"Hsi Ning reign period, 1st year, 7th month, *i-yu*. The moon was eclipsed at the 5th mark in *ch'ou*. This was at 10 degrees in *wei*. The loss was seen from the north-eastern side. It was eclipsed $2\frac{1}{2}$ divisions and then set and was not seen."
[*Wen-hsien T'ung-k'ao*, 285]

- 30 December 1069 AD

"[Hsu Ning reign period], 2nd year, intercalary 11th month, *ting-wei*. The moon was eclipsed. At the 1st mark of *hai*, the loss was seen on the north-eastern side. At the initial mark of *tzu*, the eclipse reached its maximum of 8 divisions. This was within *ching*. At the 3rd mark it was returned (to fullness)."
[*Wen-hsien T'ung-k'ao*, 285]

- 9 December 1071 AD

"[Hsu Ning reign period, 4th year,] 11th month, *ping-shen*. The moon was eclipsed at the 2nd mark of *mao*. The loss passed from the south-east side to the west side. At the 6th mark, the eclipse reached its maximum of $4\frac{1}{2}$ divisions. This was at 1 degree in *tung-ching*. As dawn broke, the moon set, and the end of the eclipse could not be discerned."
[*Wen-hsien T'ung-k'ao*, 285]

- 24 April 1073 AD

"[Hsu Ning reign period,] 6th year, 3rd month, *wu-wu*. The moon was eclipsed at the 1st mark on *hai*. The loss was seen from the south-east. At the 6th mark, the eclipse reached its maximum of 7 divisions. At the 4th mark in *tzu*, it was returned (to fullness)."
[*Wen-hsien T'ung-k'ao*, 285]

- 18 October 1073 AD

"[Hsu Ning reign period, 6th year,] 9th month, *i-mao*. The moon was eclipsed at the 4th mark of *ch'ou*. The loss was seen on the north-eastern side. At the 1st mark of *yin*, the eclipse reached its maximum of 6 divisions. It had recovered to 3 divisions short of fullness when it set and its return (to fullness) was not seen."
[*Wen-hsien T'ung-k'ao*, 285]

- 7 October 1074 AD

  "[Hsu Ning reign period,] 7th year, 9th month, *chi-yu*. The moon was eclipsed. At the 1st mark of *ch'ou* the loss began on the eastern side. At the 6th mark, the eclipse reached its maximum. This was at 2 degrees in *lou*. Dawn broke and its return (to fullness) was not seen."
  [*Wen-hsien T'ung-k'ao*, 285]

- 10 February 1077 AD

  "[Hsu Ning reign period,] 10th year, 1st month, *ping-yin*. The moon was eclipsed. At the 3rd mark in *tzu*, the loss began on the south-eastern side. At the 7th mark, the eclipse reached its maximum of 7 divisions. This was in *chang*. At the 3rd mark of *ch'ou* it was returned (to fullness)."
  [*Wen-hsien T'ung-k'ao*, 285]

- 30 January 1078 AD

  "Yuan Feng reign period, 1st year, 1st month, *keng-shen*. The moon was eclipsed at the 5th mark in *ch'ou*. It was seen on the south-eastern side."
  [*Wen-hsien T'ung-k'ao*, 285]

- 27 July 1078 AD

  "[Yuan Feng reign period, 1st year,] 6th month, *wu-wu*. The moon was eclipsed. At the 1st mark of *hsu* it was seen through the clouds in the east to have reached a magnitude of $7\frac{1}{2}$ divisions. At the 2nd mark the eclipse became total. This was in *hsu*. At $3\frac{1}{2}$ marks in *hai*, it was returned (to fullness)."
  [*Wen-hsien T'ung-k'ao*, 285]
  Note: The standard printing of the text contains a misprint giving the cyclical day as *hsu-wu* instead of *wu-wu*.

- 25 May 1081 AD

  "[Yuan Feng reign period,] 4th year, 4th month, *hsin-wei*. The moon was totally eclipsed. At the 2nd mark of *hsu* it rose from out of the gloom with 1 division on the east side remaining bright. This left 9 divisions on the west side covered. This was within *wei*. At the 6th mark it was restored."
  [*Wen-hsien T'ung-k'ao*, 285]

- 8 November 1082 AD

  "[Yuan Feng reign period,] 5th year, 10th month, *kuei-hai*. The moon was eclipsed. From the 2nd mark of *yu*, the loss was seen from the north. At the 7th mark, the eclipse reached a maximum of 3 divisions. This was at a little more than 7 degrees in *mao*. At the 3rd mark of *hsu* it was returned (to fullness)."
  [*Wen-hsien T'ung-k'ao*, 285]

- 6 September 1085 AD

  "[Yuan Feng reign period,] 8th year, 8th month, *ping-tzu*. The moon was eclipsed. At the 3rd mark of *hsu* the loss began. At the 7th mark the eclipse was total. At the 1st mark of *tzu* it was returned (to fullness)."
  [*Wen-hsien T'ung-k'ao*, 285]

- 6 July 1088 AD

  "[Reign of Che Tsung, Yuan Yu reign period], 1st year, 12th month, *wu-hsu*. The moon was eclipsed. The loss began at the 5th mark of *hai*. At the 6th mark of *tzu* the eclipse was total. At the 4th mark of *ch'ou* it was returned (to fullness). This was within *tou*."
  [*Wen-hsien T'ung-k'ao*, 285]

- 25 June 1089 AD

  "[Yuan Yu reign period], 4th year, 5th month, *chia-shen*. The moon was eclipsed, but the beginning was not seen on account of thick clouds. At the initial mark of ch'ou, the eclipse had reached 9 divisions. This was within *tou*. At the 6th mark, it was returned (to fullness)."
  [*Wen-hsien T'ung-k'ao*, 285]

- 24 April 1092 AD

  "[Yuan Yu reign period], 7th year, 3rd month, *wu-hsu*. The moon was eclipsed. The loss began at the 1st mark of the initial half of *hai*. At the 7th mark, the eclipse was total. At the 7th mark of *tzu*, it was restored to fullness. The eclipse reached its maximum within *ti*."
  [*Wen-hsien T'ung-k'ao*, 285]

- 30 January 1097 AD

  "Shao Sheng reign period, 4th year, 1st month, *keng-tzu*. The moon was eclipsed but on account of clouds the beginning of loss could not be seen. By the 7th mark of *yu*, the eclipse had reached 3 divisions. At the 1st mark of *hsu*, the eclipse had reached 4 divisions. During the return (to fullness), the clouds returned and veiled (the moon)."
  [*Wen-hsien T'ung-k'ao*, 285]

- 5 June 1099 AD

  "Yuan Fu reign period, 2nd year, 5th month, *ping-chen*. The moon was eclipsed at the 3rd mark of *tzu*. At the 2nd mark of *ch'ou* the eclipse was total. At the 2nd mark of *yin*, it was returned (to fullness). This was within *chi*."
  [*Wen-hsien T'ung-k'ao*, 285]

- 30 November 1099 AD

  "[Yuan Fu reign period, 2nd year,] 10th month, *chia-yin*. The moon was eclipsed. The loss began at the 4th mark of *hai*. At the 4th mark of *tzu*, the eclipse was total. At the 4th mark of *ch'ou*, it was returned (to fullness). This was in *shen*."

  [*Wen-hsien T'ung-k'ao*, 285]

C.7 Other sources

- 6 February 1068 AD

  "The sun was eclipsed. According to the Astronomers, that day at the 8th mark of the hour of *ssu*, the sun waned, the loss beginning from the south-western side. After the 5th mark in the hour of *wu*, the eclipse had reached (a maximum) of slightly less than 6 divisions. At the 3rd mark in the hour of *wei*, it was returned to fullness."

  [*Sung-hui-yao Chi-k'ao*]

- 11 May 1100 AD

  "The Astronomer Royal proclaimed that the sun was eclipsed 4 divisions from the north-west in the initial half of the hour of *ch'en*. At the 3rd mark of the central half of *ssu*, it was returned to fullness."

  [*Sung-hui-yao Chi-k'ao*]

- 10 July 1572 AD

  "... the sun was eclipsed from the 3rd mark of the central half of *mao* to the 3rd mark of the initial half of *ssu*. It was not wholly divided into a half. The orbit brought it within *ching*."

  [*Ming-shen-tsung Shih-lu*, 2]

- 2 April 1577 AD

  "... At the 3rd mark in the central half of *ch'ou*, the moon was eclipsed. It was total."

  [*Ming-shih-lu*, 60]

- 23 March 1578 AD

  "... That night, at the 3rd mark in the central half of *hsu*, the moon was eclipsed."

  [*Ming-shih-lu*, 72]

- 24 March 1587 AD

  "On the night of the full moon, the moon was eclipsed $9\frac{1}{2}$ divisions ... At the 3rd mark of the central half of *hai*, it returned (to fullness.)"
  [*Ming-shih-lu*, 183]

- 22 October 1596 AD

  "... At the 2nd mark in the central half of *ssu*, the sun began to be eclipsed. At the 4th mark of the central half of *wu*, it was eclipsed $9\frac{1}{2}$ divisions."
  [*Ming-shen-tsung Shih-lu*, 301]

- 15 December 1610 AD

  "The sun was eclipsed $7\frac{1}{2}$ divisions. It began at the 1st mark of the central half of *wei*. The eclipse reached its maximum at the 2nd mark in the initial half of *shen*. At the 2nd mark in the initial half of *yu*, it had returned."
  [*Kuo-tsio*]

- 20 February 1617 AD

  "The moon was eclipsed at the 2nd mark of the initial half of *hsu*. It was total. The eclipse reached its maximum at the 3rd mark of *hsu*."
  [*Han-yeh-lu*]

References

Aaboe A., 1969, A Computed List of New Moons for 391 BC to 316 BC from Babylon: BM 40094, *Det Kongelige Danske Videnskabernes Selskab Matematisk-fysiske Meddelelser*, 37/3, 1–25

Aaboe A., 1972, Remarks on the Theoretical Treatment of Eclipses in Antiquity, *Journal for the History of Astronomy*, 3, 105–118

Aaboe A., 1974, Scientific Astronomy in Antiquity, *Philosophical Transactions of the Royal Society of London, Series A*, 276, 21–42

Aaboe A., 1980, Observation and Theory in Babylonian Astronomy, *Centaurus*, 24, 14–35

Aaboe A., Britton J. P., Henderson J. A., Neugebauer O., Sachs A. J., 1991, *Saros Cycle Dates and Related Babylonian Astronomical Texts*. American Philosophical Society, Philadelphia

Aaboe A., Henderson J. A., 1975, The Babylonian Theory of Lunar Latitude and Eclipses According to System A, *Archives Internationales d'Histoire des Sciences*, 25, 181–222

Al-Hassan A. Y., Hill D. R., 1986, *Islamic Technology*. Cambridge University Press, Cambridge

Al-Rawi F. N. H., George A. R., 1992, Enūma Anu Enlil XIV and Other Early Astronomical Tables, *Archiv für Orientforschung*, 39, 52–73

Ali J., 1967, *The Determination of the Coordinates of Cities*. The American University of Beirut, Beirut

Aston W. G., 1972, *Nihongi: Chronicles of Japan from the Earliest Times to* AD *697*. Charles E. Tuttle, Rutland, Vermont

Aveni A. F., 1980, *Skywatchers of Ancient Mexico*. University of Texas Press, Austin, Texas

Beaujouan G., 1974, Observations et Calculs Astronomiques de Jean de Murs, in *Actes du XIVᵉ Congrès International d'Histoire des Sciences.* Tokyo and Kyoto, 27–30

Beaulieu P.-A., Britton J. P., 1994, Rituals for an Eclipse Possibility in the 8th year of Cyrus, *Journal of Cuneiform Studies*, 46, 73

Beck B. J. M., 1990, *The Treatises of Later Han.* E. J. Brill, Leiden

Beer A., Ho Ping-yu, Lu Gwei-djen, Needham J., Pulleyblank E. G., Thompson G. I., 1961, An 8th Century Gnomon Line: I-Hsing's Chain of Gnomons and the Pre-history of the Metric System, *Vistas in Astronomy*, 4, 3–28

Beijing Observatory, 1988, *Zhongguo Gudai Tianxian Jilu Zongji.* Kexue Jishi Chuanshe, Kiangxu

Bernard H., 1973, *Matteo Ricci's Scientific Contribution to China.* Hyperion Press, Westport, Connecticut

Bielenstein H., 1950, An Interpretation of the Portents in the *Ts'ien-han-shu*, *Bulletin of the Museum of Far Eastern Antiquities*, 22, 127–143

Bielenstein H., 1984, Han Portents and Prognostications, *Bulletin of the Museum of Far Eastern Antiquities*, 56, 97–112

Bielenstein H., Sivin N., 1977, Further Comments on the Use of Statistics in the Study of Han Dynasty Portents, *Journal of the American Oriental Society*, 97, 185–187

Bo Shuren, 1997, An Outline of Guo Shoujing's Astronomical Work, in Nha Il-Seong, Stephenson F. R., (ed), *Oriental Astronomy from Guo Shoujing to King Sejong.* Yonsei University Press, Seoul, 15–24

Brack-Bernsen L., 1997, *Zur Entstehung der Babylonischen Mondtheorie.* Franz Steiner Verlag, Stuttgart

Brack-Bernsen L., 1999, Goal-Year Tables: Lunar Data and Predictions, in Swerdlow N. M., (ed), *Ancient Astronomy and Celestial Divination.* The MIT Press, Cambridge, Massachusetts, in press

Brack-Bernsen L., Schmidt O., 1994, On the Foundations of the Babylonian Column Φ: Astronomical Significance of Partial Sums of the Lunar Four, *Centaurus*, 37, 183–209

Bretagnon P., Simon J. L., Laskar J., 1985, Presentation of New Solar and Planetary Tables of Interest for Historical Calculations, *Journal for the History of Astronomy*, 16, 39–50

Bricker H. M., Bricker V. R., 1983, Classic Maya Prediction of Solar Eclipses, *Current Anthropology*, 24, 1–23

Brinkman J. A., 1968, *A Political History of Post-Kassite Babylonia*. Pontificium Institutum Biblicum, Rome

Brinkman J. A., 1990, The Babylonian Chronicle Revisited, in Abusch T., Huehnergard J., Steinkeller P., (ed), *Lingering Over Words: Studies in Ancient Near Eastern Literature in Honor of William L. Moran*. Scholars Press, Atlanta, Georgia, 73–104

Britton J. P., 1989, An Early Function for Eclipse Magnitudes in Babylonian Astronomy, *Centaurus*, 32, 1–52

Britton J. P., 1992, *Models and Precision: The Quality of Ptolemy's Observations and Parameters*. Garland Publishing Inc., New York

Britton J. P., 1993, Scientific Astronomy in Pre-Seleucid Babylon, in Galter H. D., (ed), *Die Rolle der Astronomie in den Kulturen Mesopotamiens*. Grazer Morgenländische Studien, Graz, 61–76

Britton J. P., 1999, Babylonian Theories of Lunar Anomaly, in Swerdlow N. M., (ed), *Ancient Astronomy and Celestial Divination*. The MIT Press, Cambridge, Massachusetts, in press

Brown D., 1999a, The Cuneiform Conception of Celestial Space and Time, *Cambridge Archaeological Journal*, in press

Brown D., 1999b, *Neo-Assyrian and Neo-Babylonian Planetary Astronomy-Astrology (747–612 BC)*. Styx, Groningen (forthcoming)

Brown D., Fermor J., Walker C. B. F., 1999, The Water Clock in Mesopotamia, *Archiv für Orientforschung*, in press

Brown D., Linsenn M., 1999, BM 134761 = 1965-10-14, 1 and the Hellenistic Period Eclipse Ritual from Uruk, *Revue d'Assyriologie et d'Archéologie Orientale*, in press

Budge E. A. W., 1925, *The Rise and Progress of Assyriology*. Hopkinson and Co., London

Caussin C., 1804, *Le livre de la Grande Table Hakemite par ebn Iuonus*. Paris

Chapman A., 1983, A Study of the Accuracy of Scale Graduations on a Group of European Astrolabes, *Annals of Science*, 40, 473–488

Clark D. H., Stephenson F. R., 1977, *The Historical Supernovae*. Pergamon Press, Oxford

Closs M. P., 1989, Cognitive Aspects of Ancient Maya Eclipse Theory, in Aveni A. F., (ed), *World Archaeoastronomy*. Cambridge University Press, Cambridge, 389–415

Cohen A. P., Newton R. R., 1983, Solar Eclipses Recorded in China During the Tarng Dynasty, *Monumenta Serica*, 35, 347–430

Cullen C., 1993, Motivations for Scientific Change in Ancient China: Emperor Wu and the Grand Inception Astronomical Reforms of 104 BC, *Journal for the History of Astronomy*, 24, 185–203

de Kuyper J., 1993, Mesopotamian Astronomy and Astrology as seen by Greek Literature: The Chaldaeans, in Galter H. D., (ed), *Die Rolle der Astronomie in den Kulturen Mesopotamiens*. Grazer Morgenländische Studien, Graz, 135–137

de Meis S., Hunger H., 1998, *Astronomical Dating of Assyrian and Babylonian Reports*. Istituto Italiano per l'Africa e l'Oriente, Rome

de Sélincourt A., 1979, *Herodotus: The Histories*. Penguin, London

deB. Beaver D., 1970, Bernard Walther: Inovator in Astronomical Observation, *Journal for the History of Astronomy*, 1, 39–43

Debarnot M.-T., 1987, The *Zīj* of Ḥabash al-Ḥāsib: A Survey of MS Istanbul Yeni Cami 784/2, in King D. A., Saliba G., (ed), *From Deferent to Equant: A Volume of Studies in the History of Science in the Ancient and Medieval Near East in Honor of E. S. Kennedy*. The New York Academy of Sciences, New York, 35–69

d'Elia P. M., 1960, *Galileo in China*. Harvard University Press, Cambridge, Massachusetts

Depuydt L., 1995, More Valuable Than All Gold: Ptolemy's Royal Canon and Babylonian Chronology, *Journal of Cuneiform Studies*, 47, 97–117

Dickey J. O. et al., 1994, Lunar Laser Ranging: A Continuing Legacy of the Apollo Program, *Science*, 265, 482–490

Dobrzycki J., Kremer R. L., 1996, Peurbach and Marāgha Astronomy? The Ephemerides of Johannes Angelus and their Implications, *Journal for the History of Astronomy*, 27, 187–237

Dold-Samplonius Y., 1974, al-Māhānī, in Gillispie C. C., (ed), *Dictionary of Scientific Biography, Volume 9*. Charles Scribner's Sons, New York, 21–22

Dreyer J. L. E., 1890, *Tycho Brahe: A Picture of Scientific Life and Work in the Sixteenth Century*. Adam & Charles Black, Edinburgh

Dreyer J. L. E., 1920, On the Original Form of the Alphonsine Tables, *Monthly Notices of the Royal Astronomical Society*, 80, 243–267

Dreyer J. L. E., 1923, *Tychonis Brahe Dani Opera Omnia, Tomus X*. Copenhagen Libraria Gyldendaliana, Copenhagen

Dreyer J. L. E., 1924, *Tychonis Brahe Dani Opera Omnia, Tomus XI.* Copenhagen Libraria Gyldendaliana, Copenhagen

Dreyer J. L. E., 1925, *Tychonis Brahe Dani Opera Omnia, Tomus XII.* Copenhagen Libraria Gyldendaliana, Copenhagen

Dreyer J. L. E., 1926, *Tychonis Brahe Dani Opera Omnia, Tomus XIII.* Copenhagen Libraria Gyldendaliana, Copenhagen

Dubs H. H., 1938, *The History of the Former Han Dynasty, Volume 1.* Waverly Press, Baltimore

Dubs H. H., 1944, *The History of the Former Han Dynasty, Volume 2.* Waverly Press, Baltimore

Dubs H. H., 1955, *The History of the Former Han Dynasty, Volume 3.* Waverly Press, Baltimore

Duncan A. M., 1976, *Copernicus: On the Revolutions of the Heavenly Spheres.* David & Charles, Newton Abbot

Eberhard W., 1957, The Political Function of Astronomy and Astronomers in Han China, in Fairbank J. K., (ed), *Chinese Thought and Institutions.* The University of Chicago Press, Chicago, 33–70

Eberhard W., Müller R., 1936, Contributions to the Astronomy of the Han Period III, *Harvard Journal of Asiatic Studies*, 1, 194–241

Eirich R., 1987, Bernard Walther (1430–1504) und sein Familie, *Mitteilurgen des Vereins für die Deschichte der Stadt Nurnberg*, 74, 77–128

Epping J., 1889, *Astronomisches aus Babylon.* Herder'sche Verlagshandlung, Freiburg im Breisgau

Epping J., 1890, Sachliche Erklärung des Tablets No. 400 der Cambyses-Inscriften, *Zeitschrift für Assyriologie*, 5, 281–288

Fatoohi L. J., Stephenson F. R., 1996, Accuracy of Lunar Eclipse Observations made by Jesuit Astronomers in China, *Journal for the History of Astronomy*, 27, 61–67

Fermor J., Steele J. M., 1999, The Design of Babylonian Waterclocks: Astronomical and Experimental Evidence, in preparation

Foley N., 1989, *A Statistical Study of the Solar Eclipses Recorded in Chinese and Korean History During the pre-Telescopic Era.* MSc Thesis, University of Durham

Forke A., 1907, *Lun Heng, Philosophical Essays of Wang Ch'ung. Volume 1.* Kelly & Walsh, Shanghai

Forte A., 1988, *Mingtang and Buddhist Utopias in the History of the Astronomical Clock.* Istituto Italiano per il Medio ed Estremo Oriente, Rome

Fotheringham J. K., 1920a, Note on the Secular Accelerations of the Sun and Moon as Determined from the Ancient Lunar and Solar Eclipses, Occultations, and Equinox Observations, *Monthly Notices of the Royal Astronomical Society*, 80, 578–581

Fotheringham J. K., 1920b, A Solution of the Ancient Eclipses of the Sun, *Monthly Notices of the Royal Astronomical Society*, 81, 104–126

Fotheringham J. K., 1932, *The Observatory*, 55, 338–340

Freeman-Grenville G. S. P., 1977, *The Muslim and Christian Calendars.* Oxford University Press, Oxford

Gadd C. J., 1967, Omens Expressed in Numbers, *Journal of Cuneiform Studies*, 21, 52–63

Gasche H., Armstrong J. A., Cole S. W., Gurzadyan V. G., 1998, *Dating the Fall of Babylon.* University of Ghent and the Oriental Institute of the University of Chicago, Ghent and Chicago

Ginzel F. K., 1899, *Spezieller Kanon der Sonnen- und Mondfinsternisse.* Mayer & Müller, Berlin

Goldstein B. R., 1963, A Medieval Table for Reckoning Time from Solar Altitude, *Scripta Mathematica*, 27, 61–66

Goldstein B. R., 1972, Theory and Observation in Medieval Astronomy, *Isis*, 63, 39–47

Goldstein B. R., 1974, *The Astronomical Tables of Levi ben Gerson.* The Connecticut Academy of Arts and Sciences, New Haven, Connecticut

Goldstein B. R., 1979, Medieval Observations of Solar and Lunar Eclipses, *Archives Internationales d'Histoire des Sciences*, 29, 101–156

Goldstein B. R., Bowen A. C., 1995, Pliny and Hipparchus's 600-Year Cycle, *Journal for the History of Astronomy*, 26, 155–158

Grayson A. K., 1975, *Assyrian and Babylonian Chronicles.* J. J. Augustin, Locust Valley, New York

Gushee L., 1969, New Sources for the Biography of Johannes de Muris, *Journal of the American Musicological Society*, 22, 3–36

Hallo W. W., 1988, The Nabonassar Era and Other Epochs in Mesopotamian Chronology and Chronography, in Leichty E., Ellis M. D., Geraddi P., (ed), *A Scientific Humanist: Studies in Memory of Abraham Sachs*. Occasional Publications of the Samuel Noah Kramer Fund, Philadelphia, 175–190

Hamilton H. C., Falconer W., 1903, *The Geography of Strabo, Volume 1*. George Bell and Sons, London

Hamilton H. C., Falconer W., 1906, *The Geography of Strabo, Volume 2*. George Bell and Sons, London

Hamilton N. T., Swerdlow N. M., 1981, Judgement on Ptolemy, *Journal for the History of Astronomy*, 12, 59–63

Han Yu-shan, 1955, *Elements of Chinese Historiography*. W. M. Hawley, Hollywood, California

Hartner W., 1970, al-Battānī, in Gillispie C. C., (ed), *Dictionary of Scientific Biography, Volume 1*. Charles Scribner's Sons, New York, 507–516

Hellman C. D., 1970, Brahe, in Gillispie C. C., (ed), *Dictionary of Scientific Biography, Volume 2*. Charles Scribner's Sons, New York, 401–416

Hellman C. D., Swerdlow N. M., 1978, Peurbach, in Gillispie C. C., (ed), *Dictionary of Scientific Biography, Volume 15*. Charles Scribner's Sons, New York, 473–479

Hirshfeld A., Sinnott R. W., 1982, *Sky Catalogue 2000.0, Volume 1: Stars to Magnitude 8.0*. Cambridge University Press, Cambridge

Ho Peng Yoke, 1966, *The Astronomical Chapters of the Chin Shu*. Mouton & Co, Paris

Ho Peng Yoke, 1970, The Astronomical Bureau in Ming China, *Journal of Asian History*, 3, 137–157

Horowitz W., 1994, Two New Ziqpu-Star Texts and Stellar Circles, *Journal of Cuneiform Studies*, 46, 89–98

Horowitz W., 1998, *Mesopotamian Cosmic Geography*. Eisenbrauns, Winona Lake, Indiana

Høyrup J., 1998, A Note on Water-Clocks and on the Authority of Texts, *Archiv für Orientforschung*, 44/45, 192–194

Hsueh Chung-san, Ou-yang I, 1956, *A Sino-Western Calendar for Two Thousand Years*. San-lien Shu-tien, Beijing

Hua Tongxu, 1997, The Water Clock of the Ming Dynasty, in Nha Il-Seong, Stephenson F. R., (ed), *Oriental Astronomy from Guo Shoujing to King Sejong*. Yonsei University Press, Seoul, 325–328

Huber P. J., 1973, *Babylonian Eclipse Observations: 750* BC *to 0.* Unpublished manuscript

Huber P. J., 1987, Dating by Lunar Eclipse Omina with Speculations on the Birth of Omen Astrology, in Bergren J. L., Goldstein B. R., (ed), *From Ancient Omens to Statistical Mechanics: Essays on the Exact Sciences Presented to Asger Aaboe.* University Library, Copenhagen, 3-13

Huber P. J., Sachs A. J., Stol M., Whiting R. M., Leichty E., Walker C. B. F., van Driel G., 1982, Astronomical Dating of Babylon I and Ur III, *Occasional Papers on the Near East*, 1/4, 108–199

Hunger H., 1969, Kryptographische Astrologische Omina, in Dietrich M., Röllig W., (ed), *lišān mitḫurti: Festschrift Wolfram Freiherr von Soden.* Verlag Butzon & Bercker Kevelaer, Neukirchen-Vluyn, 133–145

Hunger H., 1992, *Astrological Reports to Assyrian Kings.* Helsinki University Press, Helsinki

Hunger H., 1999, Non-Mathematical Astronomical Texts and Their Relationships, in Swerdlow N. M., (ed), *Ancient Astronomy and Celestial Divination.* The MIT Press, Cambridge, Massachusetts, in press

Hunger H., Pingree D., 1989, *MUL.APIN: An Astronomical Compendium in Cuneiform.* Verlag Ferdinand Berger & Sohne, Horn

IAU, 1968, The Epoch of Ephemeris Time, *Transactions of the International Astronomical Union*, 13B, 48

Jones A., 1990, *Ptolemy's First Commentator.* American Philosophical Society, Philadelphia

Jones A., 1991, The Adaption of Babylonian Methods in Greek Numerical Astronomy, *Isis*, 82, 441–453

Jones A., 1993, Evidence for Babylonian Arithmetical Schemes in Greek Astronomy, in Galter H. D., (ed), *Die Rolle der Astronomie in den Kulturen Mesopotamiens.* Grazer Morgenländische Studien, Graz, 77–94

Jones A., 1994, The Place of Astronomy in Roman Egypt, in Barnes T. D., (ed), *The Sciences in Greco-Roman Society.* Academic Printing and Publishing, Edmonton, 25–51

Jones A., 1996, Later Greek and Byzantine Astronomy, in Walker C. B. F., (ed), *Astronomy Before the Telescope.* British Museum Publications, London, 98–109

Jones A., 1997, Studies in the Astronomy of the Roman Period I: The Standard Lunar Scheme, *Centaurus*, 39, 1–36

Jones A., 1999, *Astronomical Papyri from Oxyrhynchus*. American Philosophical Society, Philadelphia

Jones A., 2000, Calendrica I: New Callippic Dates, *Zeitscrift für Papyrologie und Epigraphik*, in press

Kanda Shigeru, 1935, *Nihon Tenmon Shiryō*. Koseisha, Tokyo

Kennedy E. S., 1956, A Survey of Islamic Astronomical Tables, *Transactions of the American Philosophical Society*, 46, 123–177

Kennedy E. S., 1970, al-Bīrūnī, in Gillispie C. C., (ed), *Dictionary of Scientific Biography, Volume 2*. Charles Scribner's Sons, New York, 147–158

Kennedy E. S., Destombes M., 1966, Introduction, in Kennedy E. S., Destombes M., (ed), *Al-Ṣūfi's Kitāb al-ʿAmal biʾl-Asṭurlāb*. Osmania Oriental Publications, Hyderabad, 1–44

Kilmer A. D., 1978, A Note on the Babylonian Mythological Explanation of the Lunar Eclipse, *Journal of the Americal Oriental Society*, 98, 372–374

King D. A., 1973, Ibn Yūnus' Very Useful Tables for Recknoning Time by the Sun, *Archive for History of Exact Science*, 10, 342–394

King D. A., 1976, Ibn Yūnus, in Gillispie C. C., (ed), *Dictionary of Scientific Biography, Volume 14*. Charles Scribner's Sons, New York, 574–580

King D. A., 1996, Islamic Astronomy, in Walker C. B. F., (ed), *Astronomy Before the Telescope*. British Museum Publications, London, 143–174

Knobel E. B., 1905, On the Astronomical Observations Recorded in the Nihongi, the Ancient Chronicle of Japan, *Monthly Notices of the Royal Astronomical Society*, 66, 67–74

Koch J., 1989, *Neue Untersuchungen zur Topographie des Babylonischen Fixsternhimmels*. Otto Harrassowitz, Wiesbaden

Koch-Westenholz U., 1995, *Mesopotamian Astrology*. Museum Tusculanum Press, Copenhagen

Kremer R. L., 1980, Bernard Walther's Astronomical Observations, *Journal for the History of Astronomy*, 11, 174–191

Kremer R. L., 1981, The Use of Bernard Walther's Astronomical Observations: Theory and Observation in Early Modern Astronomy, *Journal for the History of Astronomy*, 12, 124–132

Kremer R. L., 1983, Walther's Solar Observations: A Reply to R. R. Newton, *Quarterly Journal of the Royal Astronomical Society*, 24, 36–47

Kremer R. L., Dobrzycki J., 1998, Alfonsine Meridians: Tradition Versus Experience in Astronomical Practice c. 1500, *Journal for the History of Astronomy*, 29, 187–199

Kugler F. X., 1900a, *Die Babylonische Mondrechnung*. Herder'sche Verlagschandlung, Freiburg im Breisgau

Kugler F. X., 1900b, Zur Erklärung der Babylonischen Mondtafeln I: Mond- und Sonnenfinsternisse, *Zeitschrift für Assyriologie*, 15, 178–209

Kugler F. X., 1903, Eine Rätselvolle Astronomische Keilinschrift (Strm. Kambys. 400), *Zeitschrift für Assyriologie*, 17, 203–238

Kugler F. X., 1907, *Sternkund und Sterndienst in Babel, Buch I*. Aschendorffsche Verlagsbuchhandlung, Münster in Westfalen

Kugler F. X., 1909, *Sternkund und Sterndienst in Babel, Buch II, Teil I*. Aschendorffsche Verlagsbuchhandlung, Münster in Westfalen

Kugler F. X., 1912, *Sternkund und Sterndienst in Babel, Buch II, Teil II, Heft I*. Aschendorffsche Verlagsbuchhandlung, Münster in Westfalen

Kugler F. X., 1913, *Sternkund und Sterndienst in Babel, Ergänzungen zum Ersten und Zweiten Buch, Teil I*. Aschendorffsche Verlagsbuchhandlung, Münster in Westfalen

Kugler F. X., 1914, *Sternkund und Sterndienst in Babel, Ergänzungen zum Ersten und Zweiten Buch, Teil II*. Aschendorffsche Verlagsbuchhandlung, Münster in Westfalen

Kugler F. X., 1924, *Sternkund und Sterndienst in Babel, Buch II, Teil II, Heft II*. Aschendorffsche Verlagsbuchhandlung, Münster in Westfalen

Kugler F. X., Schaumberger J., 1935, *Sternkund und Sterndienst in Babel, 3. Ergänzungsheft zum Ersten und Zweiten Buch*. Aschendorffsche Verlagsbuchhandlung, Münster in Westfalen

Kuniji Saito, 1979, Glimpses of Ancient Japanese Astronomy, *Sky and Telescope*, 58, 108–109

Lambert W. G., 1975, The Cosmology of Sumer and Babylon, in Blacker C., Loewe M., (ed), *Ancient Cosmologies*. George Allen & Unwin, London, 42–65

Lanfranchi G. L., Parpola S., 1990, *The Correspondance of Sargon II, Part 2: Letters from the Northern and Northeastern Provinces*. Helsinki University Press, Helsinki

Langdon S., 1935, *Babylonian Menologies and the Semitic Calendars*. Oxford University Press, London

Langdon S., Fotheringham J. K., 1928, *The Venus Tablets of Ammizaduga.* Oxford University Press, London

Larsen M. T., 1996, *The Conquest of Assyria.* Routledge, London

Lee Eun-Hee, 1997, The Ch'iljŏngsan Naepiŏn, An Adopted Version of the Shoushi-li and Datong-li, in Nha Il-Seong, Stephenson F. R., (ed), *Oriental Astronomy from Guo Shoujing to King Sejong.* Yonsei University Press, Seoul, 339–348

Liu Bao-lin, Stephenson F. R., 1998, The Chinese Calendar and its Operational Rules, *Orion*, 286, 16–19

Livingstone A., 1986, *Mystical and Mythological Explanatory Works of Assyrian and Babylonian Scholars.* Oxford University Press, Oxford

Lounsbury F. G., 1976, Maya Numeration, Computation, and Calendrical Astronomy, in Gillispie C. C., (ed), *Dictionary of Scientific Biography, Volume 15.* Charles Scribner's Sons, New York, 759–818

Maspero H., 1939, Les Instruments Astronomiques des Chinois au Temps des Han, *Mélanges Chinois et Bouddhiques*, 6, 183–370

Mayer L. A., 1956, *Islamic Astrolabists and their Works.* Albert Kundig, Genève

McCluskey S. C., 1998, *Astronomies and Cultures in Early Medieval Europe.* Cambridge University Press, Cambridge

McEwan G. J. P., 1981, *Priest and Temple in Hellenistic Babylonia.* Franz Steiner Verlag, Wiesbaden

Morrison L. V., Lukac M. R., Stephenson F. R., 1981, Catalogue of Observations of Occultations of Stars by the Moon for the Years 1623 to 1942 and Solar Eclipses for the Years 1621 to 1806, *Royal Greenwich Observatory Bulletin*, 186

Morrison L. V., Ward C. G., 1975, An Analysis of the Transits of Mercury, *Monthly Notices of the Royal Astronomical Society*, 173, 183–206

Muller P. M., 1975, *An Analysis of the Ancient Astronomical Observations with the Implications for Geophysics and Cosmology.* PhD Thesis, University of Newcastle

Nakayama Shigeru, 1966, Characteristics of Chinese Astrology, *Isis*, 57, 442–454

Nakayama Shigeru, 1969, *A History of Japanese Astronomy.* Harvard University Press, Cambridge, Massachusetts

Needham J., 1954, *Science and Civilisation in China, Volume 1.* Cambridge University Press, Cambridge

Needham J., 1959, *Science and Civilisation in China, Volume 3.* Cambridge University Press, Cambridge

Needham J., 1965, *Science and Civilisation in China, Volume 4, Part II*. Cambridge University Press, Cambridge

Needham J., 1974, Astronomy in Ancient and Medieval China, *Philosophical Transactions of the Royal Society of London, Series A*, 276, 67–82

Needham J., Lu Gwei-Djen, Combridge J. H., Major J. S., 1986, *The Hall of Heavenly Records*. Cambridge University Press, Cambridge

Needham J., Wang Ling, de Solla Price D. J., 1986, *Heavenly Clockwork*. Cambridge University Press, Cambridge

Neugebauer O., 1941, Some Fundamental Concepts in Ancient Astronomy, in *Studies in the History of Science*. University of Pennsylvania Press, Philadelphia, 13–29

Neugebauer O., 1945, Studies in Ancient Astronomy VII: Magnitudes of Lunar Eclipses in Babylonian Mathematical Astronomy, *Isis*, 36, 10–15

Neugebauer O., 1947a, Studies in Ancient Astronomy VIII: The Water Clock in Babylonian Astronomy, *Isis*, 37, 37–43

Neugebauer O., 1947b, A Table of Solstices from Uruk, *Journal of Cuneiform Studies*, 1, 143–148

Neugebauer O., 1948, Solstices and Equinoxes in Babylonian Astronomy During the Seleucid Period, *Journal of Cuneiform Studies*, 2, 209–222

Neugebauer O., 1955, *Astronomical Cuneiform Texts*. Lund Humphries, London

Neugebauer O., 1957, *The Exact Sciences in Antiquity*. Brown University Press, Providence

Neugebauer O., 1963, The Survival of Babylonian Methods in the Exact Sciences of Antiquity and the Middle Ages, *Proceedings of the American Philosophical Society*, 107, 528–535

Neugebauer O., 1975, *A History of Ancient Mathematical Astronomy*. Springer-Verlag, Berlin

Neugebauer O., 1989, From Assyriology to Renaissance Art, *Proceedings of the American Philosophical Society*, 133, 391–403

Neugebauer O., Parker R., 1969, *Egyptian Astronomical Texts, Volume 3: Decans, Planets, Constellations and Zodiacs*. Brown University Press, Providence, Rhode Island

Neugebauer O., Parker R. A., Zauzich K.-T., 1981, A Demotic Lunar Eclipse Text of the First Century BC, *Proceedings of the American Philosophical Society*, 125, 312–327

Neugebauer O., Sachs A. J., 1967, Some Atypical Astronomical Cuneiform Texts I, *Journal of Cuneiform Studies*, 21, 183–218

Neugebauer O., van Hoesen H. B., 1959, *Greek Horoscopes*. The American Philosophical Society, Philadelphia

Newcomb S., 1878, Researches on the Motion of the Moon, Part 1, *Washington Observations for 1875*, Appendix 2

Newcomb S., 1895, Tables of the Motion of the Earth on its Axis and Around the Sun, *Astronomical Papers Prepared for the use of the American Ephemeris and Nautical Almanac*, 6

Newton R. R., 1970, *Ancient Astronomical Observations and the Accelerations of the Sun and Moon*. Johns Hopkins University Press, Baltimore

Newton R. R., 1972, *Medieval Chronicles and the Rotation of the Earth*. Johns Hopkins University Press, Baltimore

Newton R. R., 1977, *The Crime of Claudius Ptolemy*. Johns Hopkins University Press, Baltimore

Newton R. R., 1982, An Analysis of the Solar Observations of Regiomontanus and Walther, *Quarterly Journal of the Royal Astronomical Society*, 23, 67–93

North J. D., 1974, The Astrolabe, *Scientific American*, 230, 96–106

Oates J., 1986, *Babylon*. Thames and Hudson, London

Oldfather C. H., 1933, *Diodorus of Sicily, Books I–II*. William Heinemann, London

Oppenheim A. L., 1969, Divination and Celestial Observation in the Last Assyrian Empire, *Centaurus*, 14, 97–135

Oppenheim A. L., 1974, A Babylonian Diviner's Manual, *Journal of Near Eastern Studies*, 33, 197–220

Oppert J., 1891, Un Texte Babylonien Astronomique et sa Traduction Greque d'Après Claude Ptolémée, *Zeitschrift für Assyriologie*, 6, 103–123

Parker R. A., 1959, *A Vienna Demotic Papyrus on Eclipse- and Lunar-Omina*. Brown University Press, Providence

Parker R. A., Dubberstein W. H., 1956, *Babylonian Chronology 626* BC – AD *75*. Brown University Press, Providence

Parker R. A., Zauzich K.-T., 1981, The Seasons in the First Century BC, in *Studies Presented to Hans Jakob Polotsky*. Pirtle and Polsen, East Gloucester, 472–479

Parpola S., 1970, *Letters from Assyrian Scholars to the Kings Esarhaddon and Assurbanipal, Part 1: Texts*. Verlag Butzon & Bercker Kevelaer, Neukirchen-Vluyn

Parpola S., 1983, *Letters from Assyrian Scholars to the Kings Esarhaddon and Assurbanipal, Part 2: Commentary and Appendices*. Verlag Butzon & Bercker Kevelaer, Neukirchen-Vluyn

Parpola S., 1993, *Letters from Assyrian and Babylonian Scholars*. Helsinki University Press, Helsinki

Pedersen O., 1974, *A Survey of the Almagest*. Odense University Press, Odense

Pinches T. G., 1888, An Astronomical or Astrological Tablet from Babylon, *The Babylonian and Oriental Record*, 9, 202–207

Pingré A.-G., 1901, *Annales Célestes du Dix-Septième Siècle*. Gauthier-Villars, Paris

Pingree D., 1973, The Mesopotamian Origin of Early Indian Mathematical Astronomy, *Journal for the History of Astronomy*, 4, 1–12

Pingree D., 1987, Babylonian Planetary Theory in Sanskrit Omen Texts, in Bergren J. L., Goldstein B. R., (ed), *From Ancient Omens to Statistical Mechanics: Essays on the Exact Sciences Presented to Asger Aaboe*. University Library, Copenhagen, 91–99

Pingree D., 1996, Astronomy in India, in Walker C. B. F., (ed), *Astronomy Before the Telescope*. British Museum Publications, London, 123–142

Pingree D., 1998, Legacies in Astronomy and Celestial Omens, in Dalley S., (ed), *The Legacy of Mesopotamia*. Cambridge University Press, Cambridge, 125–137

Pingree D., Reiner E., 1977, A Neo-Babylonian Report on Seasonal Hours, *Archiv für Orientforschung*, 25, 50–55

Poulle E., 1973, John of Murs, in Gillispie C. C., (ed), *Dictionary of Scientific Biography, Volume 8*. Charles Scribner's Sons, New York, 128–133

Poulle E., 1984, *Les Tables Alphonsine avec le Canons de Jean de Saxe*. Centre National de la Resherche Scientifique, Paris

Poulle E., 1988, The Alfonsine Tables and Alfonso X of Castille, *Journal for the History of Astronomy*, 19, 97–113

Procter E. S., 1945, The Scientific Works of the Court of Alfonso X of Castille: The King and his Collaborators, *Modern Language Review*, 40, 12–29

Rackham H., 1937, *Pliny: Natural History, Books I–II*. William Heinemann, London

Rackham H., 1942, *Pliny: Natural History, Books III–VII*. William Heinemann, London

Raeder H., Strömgren E., Strömgren B., 1946, *Tycho Brahe's Description of his Instruments and Scientific Work*. Kobenhavn, Munksgaard

Ravn O. E., 1942, *Herodotus' Description of Babylon*. Arnold Busck, Copenhagen

Reade J. E., 1986, Rassam's Babylonian Collection: The Excavations and the Archives, in Leichty E., (ed), *Catalogue of the Babylonian Tablets in the British Museum, Volume VI, Tablets from Sippar I*. British Museum Publications, London, xii–xxxvi

Reiner E., 1995, *Astral Magic in Babylonia*. American Philosophical Society, Philadelphia

Reiner E., Pingree D., 1975, *Babylonian Planetary Omens I: Enūma Anu Enlil Tablet 63: The Venus Tablet of Ammiṣaduqa*. Undena, Malibu, California

Reiner E., Pingree D., 1981, *Babylonian Planetary Omens II: Enūma Anu Enlil Tablets 50–51*. Undena, Malibu, California

Reiner E., Pingree D., 1998, *Babylonian Planetary Omens III*. Styx, Groningen

Robbins F. E., 1940, *Ptolemy: Tetrabiblos*. William Heinmann, London

Rochberg-Halton F., 1984, New Evidence for the History of Astrology, *Journal of Near Eastern Studies*, 43, 115–140

Rochberg-Halton F., 1987, The Assumed 29th *AḪÛ* Tablet of *ENŪMA ANU ENLIL*, in Rochberg-Halton F., (ed), *Language, Literature, and History: Philological and Historical Studies Presented to Erica Reiner*. American Oriental Society, New Haven, Connecticut, 327–350

Rochberg-Halton F., 1988, *Aspects of Babylonian Celestial Divination: The Lunar Eclipse Tablets of Enūma Anu Enlil*. Verlag Ferdinand Berger & Sohne, Horn

Rochberg-Halton F., 1989a, Babylonian Horoscopes and their Sources, *Orientalia*, 58, 102–123

Rochberg-Halton F., 1989b, Babylonian Seasonal Hours, *Centaurus*, 32, 146–170

Rochberg F., 1993, The Cultural Locus of Astronomy in Late Babylonia, in Galter H. D., (ed), *Die Rolle der Astronomie in den Kulturen Mesopotamiens*. Grazer Morgenländische Studien, Graz, 31–45

Rochberg F., 1998, *Babylonian Horoscopes*. American Philosophical Society, Philadelphia

Rochberg F., 1999a, Empiricism in Babylonian Omen Texts and the Classification of Mesopotamian Divination as Science, *Journal of the American Oriental Society*, 119, in press

Rochberg F., 1999b, Scribes and Scholars: The *ṭupšar Enūma Anu Enlil*, in Neumann H., (ed), *Festscrift for J. Oelsner*, Berlin, in press

Rosen E., 1959, *Three Copernican Treatises*. Dover, New York

Rosen E., 1975, Regiomontanus, in Gillispie C. C., (ed), *Dictionary of Scientific Biography, Volume 11*. Charles Scribner's Sons, New York, 348–352

Sachs A. J., 1948, A Classification of the Babylonian Astronomical Tablets of the Seleucid Period, *Journal of Cuneiform Studies*, 2, 271–290

Sachs A. J., 1952a, Babylonian Horoscopes, *Journal of Cuneiform Studies*, 6, 49–75

Sachs A. J., 1952b, Sirius Dates in Babylonian Astronomical Texts of the Seleucid Period, *Journal of Cuneiform Studies*, 6, 105–114

Sachs A. J., 1955, *Late Babylonian Astronomical and Related Texts Copied by T. G. Pinches and J. N. Strassmaier*. Brown University Press, Providence

Sachs A. J., 1974, Babylonian Observational Astronomy, *Philosophical Transactions of the Royal Society of London, Series A*, 276, 43–50

Sachs A. J., 1976, The Latest Datable Cuneiform Tablets, in Eichler B., (ed), *Kramer Anniversary Volume: Cuneiform Studies in Honour of Samuel Noah Kramer*. Verlag Butzon & Bercker Kevelaer, Neukirchen-Vluyn, 379–398

Sachs A. J., Hunger H., 1988, *Astronomical Diaries and Related Texts from Babylonia, Volume 1*. Österreichische Akademie der Wissenschaften, Wien

Sachs A. J., Hunger H., 1989, *Astronomical Diaries and Related Texts from Babylonia, Volume 2*. Österreichische Akademie der Wissenschaften, Wien

Sachs A. J., Hunger H., 1996, *Astronomical Diaries and Related Texts from Babylonia, Volume 3*. Österreichische Akademie der Wissenschaften, Wien

Sachs A. J., Hunger H., 1999, *Astronomical Diaries and Related Texts from Babylonia, Volume 5*. Österreichische Akademie der Wissenschaften, Wien (forthcoming)

Sachs A. J., Walker C. B. F., 1984, Kepler's View of the Star of Bethlehem and the Babylonian Almanac for 7/6 BC, *Iraq*, 46, 43–56

Said S. S., Stephenson F. R., 1996, Solar and Lunar Eclipse Measurements by Medieval Muslim Astronomers, I: Background, *Journal for the History of Astronomy*, 27, 259–273

Said S. S., Stephenson F. R., 1997, Solar and Lunar Eclipse Measurements by Medieval Muslim Astronomers, II: Observations, *Journal for the History of Astronomy*, 28, 29–48

Said S. S., Stephenson F. R., Rada W., 1989, Records of Solar Eclipses in Arabic Chronicles, *Bulletin of the School of Oriental and African Studies*, 52, 38–64

Saliba G., 1986, The Determination of New Planetary Parameters at the Maragha Observatory, *Centaurus*, 29, 247–271

Saliba G., 1987, Theory and Observation in Islamic Astronomy: The Work of Ibn al-Shāṭir of Damascus, *Journal for the History of Astronomy*, 18, 35–43

Samsó J., 1973, Levi ben Gerson, in Gillispie C. C., (ed), *Dictionary of Scientific Biography, Volume 7.* Charles Scribner's Sons, New York, 279–282

Samsó J., 1991, Andalusian Astronomy: Its Main Characteristics and Influence in the Latin West, in *Islamic Civilization and the Sciences.* Consiglio Regionale della Toscana, Florence, 1–23

Sayili A., 1960, *The Observatory in Islam.* Turk Tarih Kurumu Basimevi, Ankara

Schaefer B. E., 1993, Astronomy and the Limits of Vision, *Vistas in Astronomy*, 36, 311–361

Schaumberger J., 1952, Die *Zigpu*-Gestirne nach neuen Keilschrifttexten, *Zeitschrift für Assyriologie*, 50, 214–229

Schaumberger J., 1955, Anaphora und Aufgangskalender in neuen Ziqpu-Texten, *Zeitschrift für Assyriologie*, 55, 237–251

Schnabel P., 1930, Die Entstehungsgeschichte des Kartographischen Erdbildes des Klaidios Ptolemaios, *Sitzungsberichte der Preussischen Akademie der Wissenschaften, phil.-hist. Klasse*, 14, 214–250

Schoener J., 1544, *Scripta Clarissimi Mathematici M. Ioannis Regiomontanus.* Nuremburg (Reprinted in facsimilie by Minerva, Frankfurt, 1976)

Sellers J. B., 1992, *The Death of Gods in Ancient Egypt.* Penguin, London

Sivin N., 1966, Chinese Conceptions of Time, *The Earlham Review*, 1, 82–92

Sivin N., 1969, Cosmos and Computation in Early Chinese Mathematical Astronomy, *T'oung Pao*, 55, 1–73

Sivin N., 1986, On the Limits of Empirical Knowledge in the Traditional Chinese Sciences, in Fraser J. T., Lawrence N., Haber F. C., (ed), *Time, Science, and Society in China and the West.* The University of Massachusetts Press, Amherst, 151–169

Sivin N., 1997, *Yuan History Astronomical Treatise.* Unpublished manuscript

Smith S., 1969, Babylonian Time Reckoning, *Iraq*, 31, 74–81

Steele J. M., 1997, Solar Eclipse Times Predicted by the Babylonians, *Journal for the History of Astronomy*, 28, 133–139

Steele J. M., 1998a, On the Use of the Chinese *Hsuan-ming* Calendar to Predict the Times of Eclipses in Japan, *Bulletin of the School of Oriental and African Studies*, 61, 527–533

Steele J. M., 1998b, Predictions of Eclipse Times Recorded in Chinese History, *Journal for the History of Astronomy*, 29, 275–285

Steele J. M., 1999a, Eclipse Prediction in Mesopotamia, *Archive for History of Exact Science*, in press

Steele J. M., 1999b, A Re-analysis of the Eclipse Observations in Ptolemy's *Almagest*, *Centaurus*, in press

Steele J. M., Stephenson F. R., 1997, Lunar Eclipse Times Predicted by the Babylonians, *Journal for the History of Astronomy*, 28, 119–131

Steele J. M., Stephenson F. R., 1998a, Astronomical Evidence for the Accuracy of Clocks in pre-Jesuit China, *Journal for the History of Astronomy*, 29, 35–48

Steele J. M., Stephenson F. R., 1998b, Eclipse Observations made by Regiomontanus and Walther, *Journal for the History of Astronomy*, 29, 331–344

Steele J. M., Stephenson F. R., Morrison L. V., 1997, The Accuracy of Eclipse Times Measured by the Babylonians, *Journal for the History of Astronomy*, 28, 337–345

Stephenson F. R., 1968, Early Japanese Astronomical Observations, *Monthly Notices of the Royal Astronomical Society*, 141, 69–75

Stephenson F. R., 1974, Late Babylonian Observations of 'Lunar Sixes', *Philosophical Transactions of the Royal Society of London, Series A*, 276, 118

Stephenson F. R., 1994, Chinese and Korean Star Maps and Catalogues, in Harley J. B., Woodward D., (ed), *The History of Cartography, Volume 2, Book 2: Cartography in the Traditional East and Southeast Asian Societies*. The University of Chicago Press, Chicago, 511–578

Stephenson F. R., 1997a, Accuracy of Medieval Chinese Measurements of Lunar and Solar Eclipse Times, in Nha Il-Seong, Stephenson F. R., (ed), *Oriental Astronomy from Guo Shoujing to King Sejong*. Yonsei University Press, Seoul, 159–187

Stephenson F. R., 1997b, *Historical Eclipses and Earth's Rotation*. Cambridge University Press, Cambridge

Stephenson F. R., Fatoohi L. J., 1993, Lunar Eclipse Times Recorded in Babylonian History, *Journal for the History of Astronomy*, 24, 255–267

Stephenson F. R., Fatoohi L. J., 1994, The Babylonian Unit of Time, *Journal for the History of Astronomy*, 25, 99–110

Stephenson F. R., Fatoohi L. J., 1995, Accuracy of Solar Eclipse Observations made by Jesuit Astronomers in China, *Journal for the History of Astronomy*, 26, 227–236

Stephenson F. R., Morrison L. V., 1984, Long Term Changes in the Rotation of the Earth: 700 BC – AD 1980, *Philosophical Transactions of the Royal Society of London, Series A*, 313, 47–70

Stephenson F. R., Morrison L. V., 1995, Long Term Fluctuations in the Earth's Rotation: 700 BC to AD 1990, *Philosophical Transactions of the Royal Society of London, Series A*, 351, 165–202

Stephenson F. R., Said S. S., 1991, Precision of Medieval Islamic Eclipse Measurements, *Journal for the History of Astronomy*, 22, 195–207

Stephenson F. R., Said S. S., 1997, Records of Lunar Eclipses in Medieval Arabic Chronicles, *Bulletin of the School of Oriental and African Studies*, 60, 1–34

Stephenson F. R., Walker C. B. F., 1985, *Halley's Comet in History*. British Museum Publications, London

Stephenson F. R., Yau K. K. C., 1992, Astronomical Records in the *Ch'un-ch'iu* Chronicle, *Journal for the History of Astronomy*, 23, 31–51

Strassmaier J. N., 1890, *Inschriften von Cambyses, König von Babylon*. Eduard Pfeiffer, Leipzig

Subbarayappa B. V., Sarma K. V., 1985, *Indian Astronomy: A Source Book*. Nehru Centre, Bombay

Sugimoto Masayoshi, Swain D. L., 1978, *Science and Culture in Traditional Japan*. The MIT Press, Cambridge, Massachusetts

Swerdlow N. M., 1979, Ptolemy on Trial, *American Scholar*, 48, 523–531

Swerdlow N. M., 1990, Regiomontanus on the Critical Problems of Astronomy, in Levere T. H., Shea W. R., (ed), *Nature, Experiment, and the Sciences*. Kluwer Academic Publishers, Dordrecht, 165–195

Swerdlow N. M., 1996, Astronomy in the Renaissance, in Walker C. B. F., (ed), *Astronomy Before the Telescope*. British Museum Press, London, 187–230

Swerdlow N. M., 1998, *The Babylonian Theory of the Planets*. Princeton University Press, Princeton, New Jersey

Swerdlow N. M., Neugebauer O., 1984, *Mathematical Astronomy in Copernicus's De Revolutionibus*. Springer-Verlag, Berlin

Tekeli S., 1972, Ḥabash al-Ḥāsib, in Gillispie C. C., (ed), *Dictionary of Scientific Biography, Volume 5*. Charles Scribner's Sons, New York, 612–620

Thomas P. D., 1970, Alfonso el Sabio, in Gillispie C. C., (ed), *Dictionary of Scientific Biography, Volume 1*. Charles Scribner's Sons, New York, 122

Thompson R. C., 1900, *The Reports of the Magicians and Astrologers of Nineveh and Babylon in the British Museum*. Luzac and Co., London

Thoren V. E., 1973, New Light on Tycho's Instruments, *Journal for the History of Astronomy*, 4, 25–45

Thoren V. E., 1990, *The Lord of Uraniborg*. Cambridge University Press, Cambridge

Thoren V. E., Grant E., 1974, Extracts from the Alphonsine Tables and Rules for Their Use, in Grant E., (ed), *A Source Book in Medieval Science*. Harvard University Press, Cambridge, Massachusetts, 465–487

Thorndike L., 1951, Predictions of Eclipses in the Fourteenth Century, *Isis*, 42, 301–302

Thorndike L., 1952, A Record of Eclipses for the Years 1478 to 1506, *Isis*, 43, 252–256

Thorndike L., 1957, Eclipses in the Fourteenth and Fifteenth Centuries, *Isis*, 48, 51–57

Thureau-Dangin F., 1937, La Clepsydre Babylonienne, *Revue d'Assyriologie et d'Archéologie Orientale*, 34, 144

Toomer G. J., 1968, A Survey of the Toledan Tables, *Osiris*, 15, 5–174

Toomer G. J., 1977, Theon of Alexandria, in Gillispie C. C., (ed), *Dictionary of Scientific Biography, Volume 13*. Charles Scribner's Sons, New York, 321–325

Toomer G. J., 1984, *Ptolemy's Almagest*. Duckworth, London

Toomer G. J., 1988, Hipparchus and Babylonian Astronomy, in Leichty E., Ellis M. D., Geraddi P., (ed), *A Scientific Humanist: Studies in Memory of Abraham Sachs*. Occasional Publications of the Samuel Noah Kramer Fund, Philadelphia, 353–362

Tuge Hideomi, 1969, *Historical Development of Science and Technology in Japan*. Kokusai Bunka Shinkokai, Tokyo

van der Spek R. J., 1985, The Babylonian Temple During the Macedonian and Partian Domination, *Bibliotheca Orientalis*, 42, 541–562

van der Waerden B. L., 1951, Babylonian Astronomy III: The Earliest Astronomical Computations, *Journal of Near Eastern Studies*, 10, 20–34

van der Waerden B. L., 1974, *Science Awakening II: The Birth of Astronomy*. Noordhoff, Leyden

van Soldt W. H., 1995, *Solar Omens of Enuma Anu Enlil*. Nederlands Historisch-Archaeologisch Instituut, Istanbul

von Oppolzer T., 1887, *Canon der Finsternisse*. Denkschriften der Akademie der Wissenschaften, Wien

Walker C. B. F., 1997, Achaemenid Chronology and the Babylonian Sources, in Curtis J., (ed), *Mesopotamia and Iran in the Persian Period: Conquest and Imperialism 539–331* BC. British Museum Press, London, 17–25

Walker C. B. F., Roughton N. A., 1999, Astronomical Texts, in Lambert W. G., (ed), *Cuneiform Texts in the Metropolitan Museum of Art, Volume 3* (forthcomming)

Wang Lianhe, 1988, The Ancient Japanese Lunisolar Ofice and the Ancient Calendar, *Vistas in Astronomy*, 31, 811–817

Weidner E., 1944a, Die Astrologische Serie Enûma Anu Enlil, *Archiv für Orientforschung*, 14, 175–195

Weidner E., 1944b, Die Astrologische Serie Enûma Anu Enlil, *Archiv für Orientforschung*, 14, 308–318

Weidner E., 1956, Die Astrologische Serie Enûma Anu Enlil, *Archiv für Orientforschung*, 17, 71–89

Weidner E., 1969, Die Astrologische Serie Enûma Anu Enlil, *Archiv für Orientforschung*, 22, 65–75

Weir J. D., 1972, *The Venus Tablets of Ammizaduga*. Nederlands Historisch-Archaeologisch Instituut, Istanbul

Wesley W. G., 1978, The Accuracy of Tycho Brahe's Instruments, *Journal for the History of Astronomy*, 9, 42–53

Wong T., 1902, *Chronological Tables of the Chinese Dynasties*. Shanghai Printing Co., Shanghai

Xu Zhen-tao, Stephenson F. R., Jiang Yao-tiao, 1995, Astronomy on Oracle Bone Inscriptions, *Quarterly Journal of the Royal Astronomical Society*, 36, 397–406

Xu Zhen-tao, Yau K. K. C., Stephenson F. R., 1989, Astronomical Records on the Shang Dynasty Oracle Bones, *Archaeoastronomy*, 14, 61–72

Yabuuti Kiyoshi, 1963a, Astronomical Tables in China from the Han to the T'ang Dynasties, in Yabuuti Kiyoshi, (ed), *Chūgoku Chūsei Kagaku Gijutsushi No Kenkyū*, Tokyo, 445–492

Yabuuti Kiyoshi, 1963b, Astronomical Tables in China from the Wutai to the Ch'ing Dynasties, *Japanese Studies in the History of Science*, 2, 94–100

Yabuuti Kiyoshi, 1973, Chinese Astronomy: Development and Limiting Factors, in Nakayama Shigeru, Sivin N., (ed), *Chinese Science: Explorations of an Ancient Tradition.* The MIT Press, Cambridge, Massachusetts, 91–103

Yabuuti Kiyoshi, 1974, The Calendar Reforms in the Han Dynasties and Ideas in their Background, *Archives Internationales d'Histoire des Sciences*, 24, 51–65

Yabuuti Kiyoshi, 1979, Researches on the *Chiu-chih li* — Indian Astronomy under the T'ang Dynasty, *Acta Asiatica*, 36, 7–48

Yabuuti Kiyoshi, van Dalen B., 1997, Islamic Astronomy in China during the Yuan and Ming Dynasties, *Historia Scientarum*, 7, 11–43

Zinner E., 1990, *Regiomontanus: His Life and Work.* North-Holland, Amsterdam

Subject Index

Index of Texts

Archimedes

NEW STUDIES IN THE HISTORY AND PHILOSOPHY OF SCIENCE AND TECHNOLOGY

---

1. J.Z. Buchwald (ed.): *Scientific Credibility and Technical Standards. In 19th and Early 20th Century Germany and Britain.* 1996 ISBN 0-7923-4241-0
2. K. Gavroglu (ed.): *The Sciences in the European Periphery During the Enlightenment.* 1999 ISBN 0-7923-5548-2
3. P. Galison and A. Roland (eds.): *Atmospheric Flight in the Twentieth Century*, 2000 ISBN 0-7923-6037-0
4. J.M. Steele: *Observations and Predictions of Eclipse Times by Early Astronomers.* 2000 ISBN 0-7923-6298-5

KLUWER ACADEMIC PUBLISHERS – DORDRECHT/BOSTON/LONDON